D0088323

Membrane Formation
and Modification

ACS SYMPOSIUM SERIES **744**

Membrane Formation and Modification

I. Pinnau, EDITOR
Membrane Technology and Research, Inc.

B. D. Freeman, EDITOR
North Carolina State University

American Chemical Society, Washington, DC

Library of Congress Cataloging-in-Publication Data

Membrane formation and modification / editors, I. Pinnau, B. D. Freeman.

p.   cm.—(ACS symposium series ; 744)

Includes bibliographical references and index.

ISBN 0–8412–3604–6

1. Membranes (Technology) Congress.

I. Pinnau, I. (Ingo) II. Freeman, B. D. III. Series.

TP159.M4M443   1999
660′.2842—dc21   99-39049
                                                                    CIP

The paper used in this publication meets the minimum requirements of American National Standard for Information Sciences—Permanence of Paper for Printed Library Materials, ANSI Z39.48–1984.

Foreword

The ACS Symposium Series was first published in 1974 to provide a mechanism for publishing symposia quickly in book form. The purpose of the series is to publish timely, comprehensive books developed from ACS sponsored symposia based on current scientific research. Occasionally, books are developed from symposia sponsored by other organizations when the topic is of keen interest to the chemistry audience.

Before agreeing to publish a book, the proposed table of contents is reviewed for appropriate and comprehensive coverage and for interest to the audience. Some papers may be excluded in order to better focus the book; others may be added to provide comprehensiveness. When appropriate, overview or introductory chapters are added. Drafts of chapters are peer-reviewed prior to final acceptance or rejection, and manuscripts are prepared in camera-ready format.

As a rule, only original research papers and original review papers are included in the volumes. Verbatim reproductions of previously published papers are not accepted.

ACS Books Department

Contents

Preface xi

1. **Formation and Modification of Polymeric Membranes: Overview** 1
   I. Pinnau and B. D. Freeman

2. **Formation of Anisotropic and Asymmetric Membranes via Thermally-Induced Phase Separation** 23
   H. Matsuyama, S. Berghmans, M. T. Batarseh, and D. R. Lloyd

3. **Thermally-Induced Phase Separation Mechanism Study for Membrane Formation** 42
   Sung Soo Kim, Min-Oh Yeom, and In-Sok Cho

4. **A New Type of Asymmetric Polyimide Gas Separation Membrane Having Ultrathin Skin Layer** 65
   Hisao Hachisuka, Tomomi Ohara, and Kenichi Ikeda

5. **Influence of Surface Skin Layer of Asymmetric Polyimide Membrane on Gas Permselectivity** 79
   Hiroyoshi Kawakami and Shoji Nagaoka

6. **Effect of Surfactant as an Additive on the Formation of Asymmetric Polysulfone Membranes for Gas Separation** 87
   A. Yamasaki, R. K. Tyagi, A. E. Fouda, K. Jonnason, and T. Matsuura

7. **Preparation of Poly(ether sulfone) and Poly(ether imide) Hollow Fiber Membranes for Gas Separation: Effect of Internal Coagulant** 96
   D. Wang, K. Li, and W. K. Teo

8. **Effect of Solvent Exchange on the Morphology of Asymmetric Membranes** 110
   H. C. Park, Y. S. Moon, H. W. Rhee, J. Won, Y. S. Kang, and U. Y. Kim

9. **Thin-Film Composite Membranes Prepared from Sulfonated Poly(phenylene oxide): Preparation, Characterization, and Performance** 125
   G. Chowdhury, S. Singh, C. Tsang, and T. Matsuura

10. **Functionalized Polysulfones: Methods for Chemical Modification and Membrane Applications** 137
   Michael D. Guiver, Gilles P. Robertson, Masakazu Yoshikawa, and Chung M. Tam

11. **Polyphosphazene-Based Cation-Exchange Membranes: Polymer Manipulation and Membrane Fabrication** 162
   Qunhui Guo, Hao Tang, Peter N. Pintauro, and Sally O'Connor

12. **Development of Radiation-Grafted Membranes for Fuel Cell Applications Based on Poly(ethylene-*alt*-tetrafluoroethylene)** 174
   H. P. Brack, F. N. Büchi, J. Huslage, M. Rota, and G. G. Scherer

13. **Surface Modification of Microporous Polypropylene Membranes by Polyelectrolyte Multilayers** 189
   T. Rieser, K. Lunkwitz, S. Berwald, J. Meier-Haack, M. Müller, F. Cassel, Z. Dioszeghy, and F. Simon

14. **Postsynthesis Method for Development of Membranes Using Ion Beam Irradiation of Polyimide Thin Films** 205
   X. L. Xu, M. R. Coleman, U. Myler, and P. J. Simpson

15. **Preparation and Properties of Surface-Modified Polyacrylonitrile Hollow Fibers** 228
   A. Higuchi, P. Wang, T.-M. Tak, T. Nohmi, and T. Hashimoto

16. **Facilitated Transport Membranes Incorporating Metal Affinity for Recovery of Amino Acids** 238
   C. Oxford, D. Crookston, R. R. Beitle, and M. R. Coleman

17. **Membranes from Modified Poly(1-trimethylsilyl-1-propyne)** 252
   C. Merano, M. Andreotti, A. Turturro, F. Vigo, and G. Costa

18. **Pervaporation Properties of Surface-Modified Poly[(1-trimethylsilyl)-1-propyne] Membranes** 263
   Tadashi Uragami, T. Doi, and T. Miyata

19. **Simple Surface Modifications of Poly(dimethylsiloxane) Membranes by Polymer Additives and Their Permselectivity for Aqueous Ethanol Solutions** 280
   Takashi Miyata, Yuichi Nakanishi, and Tadashi Uragami

20. **Carbon Molecular Sieve Membranes: Preparation, Characterization, and Gas Permeation Properties** 295
    H. Suda and K. Haraya

21. **Preparation of Carbon Molecular Sieve Membranes and Their Gas Separation Properties** 314
    Ken-ichi Okamoto, Makoto Yoshino, Kenji Noborio,
    Hiroshi Maeda, Kazuhiro Tanaka, and Hidetoshi Kita

22. **Preparation and Pervaporation Properties of X- and Y-Type Zeolite Membranes** 330
    Hidetoshi Kita, Hidetoshi Asamura, Kazuhiro Tanaka,
    and Ken-ichi Okamoto

Author Index 343

Subject Index 344

Preface

Membrane processes have become competitive with conventional separation methods, such as distillation, absorption, or adsorption, in a wide variety of large-scale industrial applications during the past 20 years. The economic feasibility of membrane-based separation processes was made possible by the development of (1) high-performance membrane materials, (2) microporous membranes with well-defined microstructure (pore size and pore size distribution), (3) dense, ultrathin (<0.2 μm) membranes, (4) highly efficient membrane modules, and (5) innovative membrane system designs.

This book provides an overview of recent developments related to novel membrane materials, membrane fabrication methods, and membrane modification techniques. The 22 chapters of this book were presented at the Symposium on Chemistry and Materials Science of Synthetic Membranes held at the Fall National American Chemical Society (ACS) Meeting in Las Vegas, Nevada, September 8–11, 1997.

We thank Air Products and Chemicals, Inc., Daicel Chemical Industries, Ltd., the ACS Petroleum Research Fund, and the ACS Division of Polymeric Materials Science and Engineering, for their generous financial support for the symposium. Furthermore, we thank Anne Wilson and Kelly Dennis of the ACS Books Department for their guidance during the preparation of this book. Most of all, we express our sincere gratitude to the authors of the chapters.

We dedicate this book to Bill Koros and Klaus-Viktor Peinemann. Their commitment to excellence in materials research continues to inspire us and many others in the fascinating field of membrane science.

I. PINNAU
Membrane Technology and Research, Inc.
1360 Willow Road, Suite 103
Menlo Park, CA 94025

B. D. FREEMAN
Chemical Engineering Department
North Carolina State University
113 Riddick Hall, Campus Box 7905
Raleigh, NC 27695-7905

Chapter 1

Formation and Modification of Polymeric Membranes: Overview

I. Pinnau[1] and B. D. Freeman[2]

[1]Membrane Technology and Research, Inc., 1360 Willow Road, Suite 103, Menlo Park, CA 94025
[2]Department of Chemical Engineering, North Carolina State University, Raleigh, NC 27695–7905

> Since the early 1960s, synthetic membranes have been used successfully in a wide variety of large-scale industrial applications. The rapid adoption of membranes in industry resulted from breakthrough developments in (i) membrane materials, (ii) membrane structures, and (iii) large-scale membrane production methods. This overview provides a brief introduction to some of the most common membrane formation and modification methods used for production of commercial membranes.

A membrane is a thin barrier that permits selective mass transport (*1*). Membranes can be fabricated from a wide variety of organic (e.g. polymers, liquids) or inorganic (e.g. carbons, zeolites etc.) materials. Currently, the vast majority of commercial membranes are made from polymers. Some of the most commonly used polymer membranes are listed in Table I. The properties of a membrane are controlled by the membrane material and the membrane structure. To be useful in an industrial separation process, a membrane must exhibit at least the following characteristics (*2*):

- high flux.
- high selectivity (rejection).
- mechanical stability.
- tolerance to all feed stream components (fouling resistance).
- tolerance to temperature variations.
- manufacturing reproducibility.
- low manufacturing cost.
- ability to be packaged into high surface area modules.

Of the above requirements, flux and selectivity (rejection) determine the selective mass transport properties of a membrane. The higher the flux of a membrane at a given driving force, the lower is the membrane area required for a given feed flow rate, and, therefore, the lower are the capital costs of a membrane system. The selectivity determines the extent of separation. Membranes with higher selectivity are desirable because a higher product purity can be achieved in a separation process.

Typically, porous membranes are used in dialysis, ultrafiltration, and microfiltration applications. Optimum porous membranes have high porosity and a narrow pore size distribution. Membranes having a dense, selective layer are applied in reverse osmosis, pervaporation, and gas separation processes. Permeation through dense membranes occurs by a solution/diffusion mechanism (1). Ideal dense membranes should have a very thin selective layer, because flux is inversely proportional to the membrane thickness. In addition, the thin separating layer should be molecularly dense, because even a very small area fraction of defects in the membrane can cause a significant decrease in selectivity.

Table I. Common polymers used for production of commercial membranes.

Membrane material	Membrane process
Cellulose regenerated	D, UF, MF
Cellulose nitrate	MF
Cellulose acetate	GS, RO, D, UF, MF,
Polyamide	RO, NF, D, UF, MF
Polysulfone	GS, UF, MF
Poly(ether sulfone)	UF, MF
Polycarbonate	GS, D, UF, MF
Poly(ether imide)	UF, MF
Poly(2,6-dimethyl-1,4-phenylene oxide)	GS
Polyimide	GS
Poly(vinylidene fluoride)	UF, MF
Polytetrafluoroethylene	MF
Polypropylene	MF
Polyacrylonitrile	D, UF, MF
Poly(methyl methacrylate)	D, UF
Poly(vinyl alcohol)	PV
Polydimethylsiloxane	PV, GS

MF = microfiltration; UF = ultrafiltration; NF = nanofiltration; D = dialysis; PV = pervaporation; GS = gas separation.

During the past twenty years, significant progress has been made in the development of tailor-made polymers having higher selectivity and permeability than currently available commercial membrane materials (3,4). However, only very few of these novel materials have been made into membranes on a commercial scale and used in industrial membrane processes. The reason for the very slow adoption of new membrane materials is that all other performance requirements listed above are at least as important for commercialization of new membranes in an industrial process. For example, the mechanical strength, thermal stability, and chemical resistance must be taken into consideration for the practical use of a membrane material. Membranes must exhibit stable long-term separation properties (flux and selectivity/rejection) under industrial operating conditions. Many of the currently available commercial polymer membranes can be operated over a relatively wide temperature range (-20-150°C) and can withstand pressure differences of up to 2000 psig. However, many membranes show a decline in separation performance (flux and/or selectivity) over time and must, therefore, be replaced on a regular basis. In particular, fouling, swelling or even degradation of the membrane limit the long-term use of many current membrane types.

Membrane Types and Membrane Formation Methods

Membranes can be categorized according to their (i) geometry, (ii) bulk structure, (iii) production method, (iv) separation regime, and (v) application. This basic classification scheme is shown in Figure 1. Membranes can be produced in flat-sheet or tubular (hollow-fiber) geometry. Flat-sheet membranes are packaged either in plate-and-frame or spiral-wound modules, whereas tubular membranes are packaged in hollow-fiber modules. Although hollow-fiber modules have the highest membrane packing density per module volume, spiral-wound and plate-and-frame modules are also commonly used in large-scale separation processes (*4*).

Membranes either have a symmetric (isotropic) or an asymmetric (anisotropic) structure. Symmetric membranes have a uniform structure throughout the entire membrane thickness, whereas asymmetric membranes have a gradient in structure. The separation properties of symmetric membranes are determined by their entire structure. On the other hand, the separation properties of asymmetric membrane are determined primarily by the densest region in the membrane. The most common symmetric and asymmetric membrane types are shown in Figure 2.

Symmetric Membranes

Porous Symmetric Membranes. Typically, symmetric porous membranes have cylindrical, sponge-, web- or slit-like structures, and can be can be made by a variety of techniques. The most important methods for production of symmetric porous membranes are: (i) irradiation, (ii) stretching of a melt-processed semi-crystalline polymer film, iii) vapor-induced phase separation, and (iv) temperature-induced phase separation (*6-10*).

Symmetric membranes with a cylindrical porous structure can be produced by an irradiation-etching process (*11*). Membranes made by this process are often referred to as nucleation track membranes. In the first step of this process, a dense polymer film, such as polycarbonate, is irradiated with charged particles, which induce polymer chain scission along nucleation tracks throughout the entire film thickness. In a second step, the film is passed through an etching medium, typically a sodium hydroxide solution. During this process, pores are formed by etching the partially degraded polymer along the nucleation tracks. Membranes made by this method have a very uniform pore size. The porosity and pore size of nucleation track membranes can be controlled by the irradiation and etching time, respectively (*4*).

Geometry	Bulk Structure	Production Method	Separation Regime	Application

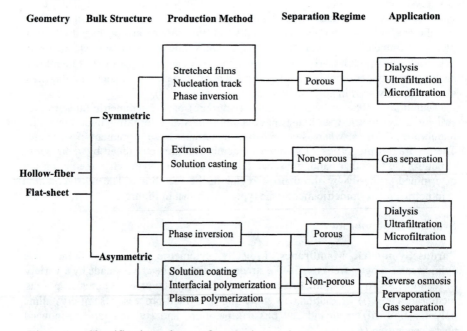

Figure 1. Classification scheme of synthetic membranes based on their geometry, bulk structure, production method, separation regime, and application.

5

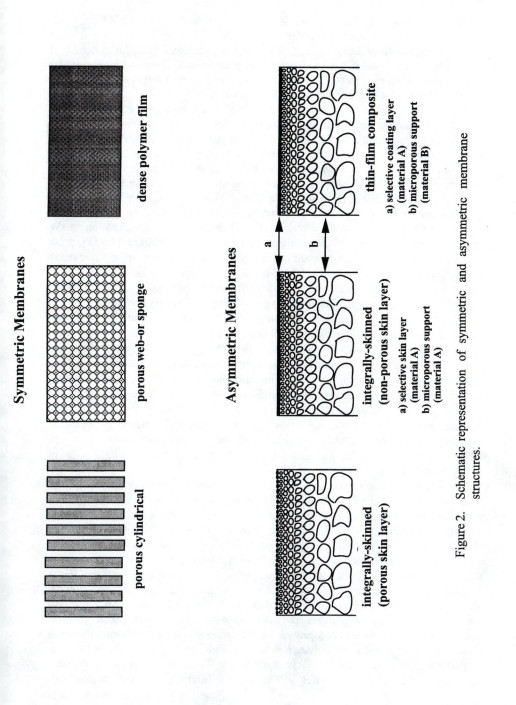

Figure 2. Schematic representation of symmetric and asymmetric membrane structures.

Membranes with a symmetric slit-like porous structure can be made from semi-crystalline polymers, such as polyethylene and polypropylene, using a melt extrusion/stretching process (*12,13*). First, a row nucleated lamellar structure is formed by melt-extrusion of a semi-crystalline polymer and re-crystallization under high stress. In the second step, slit-like pores are formed between the stacked lamellae by stretching the membrane in the machine direction (*13*).

Symmetric membranes with sponge- or web-like pore structures can be made by a vapor-precipitation/evaporation process (*14-18*). In this process, a solution containing polymer, solvents and non-solvents is cast onto a suitable substrate and is then exposed to a water-vapor-saturated air stream. The water vapor induces phase separation in the initially stable polymer solution. After phase separation, the solvent and non-solvent components are evaporated by blowing a hot air stream across the membrane. The cross-section of a typical membrane made by the vapor-precipitation/evaporation process is shown in Figure 3. The porosity and pore size of this membrane type can be controlled by (i) the polymer concentration in the casting solution and (ii) the composition of the vapor atmosphere (*17-18*). Low polymer concentration, high humidity, and addition of solvent-vapor to the casting atmosphere lead to membranes with high porosity and large pore size. Formation of membranes by phase separation of an initially stable polymer solution is often referred to as phase inversion process (*5*) and is discussed in detail below.

In the phase inversion process, the membrane structure is formed by bringing a thermodynamically stable polymer solution to an unstable state. A change in temperature, composition or pressure that leads to a decrease in the Gibbs free energy of mixing of the solution will cause phase separation of the initially stable solution into two or more phases with different compositions. The resulting membrane structure depends primarily on the concentration gradient of the polymer in the cast solution at the onset of phase separation (*6-8*). Because membranes made by the vapor-precipitation/evaporation process have an essentially constant polymer concentration profile throughout the entire membrane thickness at the onset of phase separation, the resulting membranes are porous and have a fairly symmetric structure.

Symmetric porous membranes can also be made by a thermally-induced phase inversion process (*19-27*). A schematic phase diagram for a binary solution containing an amorphous polymer and a solvent displaying an upper critical solution temperature (UCST) behavior, is shown in Figure 4. The figure shows the stable, meta-stable, and unstable regions of the solution as a function of the polymer volume fraction. The binodal line is the boundary between the stable and the meta-stable region of the polymer solution, whereas the spinodal line sets the limits between the meta-stable and the unstable region in the phase diagram. The binodal and spinodal lines can be calculated using the Flory-Huggins theory (*28-29*). Phase separation of an initially stable polymer solution can be the result of two mechanisms: (i) nucleation and growth or (ii) spinodal decomposition. Figure 4 shows the quench paths of polymer solutions with compositions A, B, and C, respectively, from temperature T_1 to temperature T_2. If the temperature is kept constant at T_2, solution A and B will reside within the meta-stable region and,

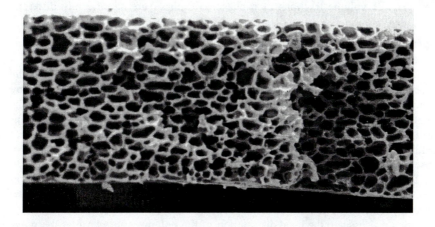

Figure 3. Cross-section of a symmetric porous polysulfone membrane made by the vapor-precipitation/evaporation process.

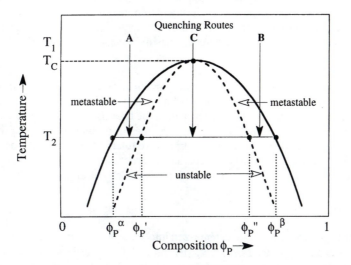

Figure 4. Schematic diagram of a binary polymer-solvent system with an upper critical solution temperature (UCST). Reproduced with permission from reference 40.

therefore, phase separation will occur by nucleation and growth. Solution A forms nuclei with composition ϕ_p^β, whereas solution B will form nuclei with composition ϕ_p^α. The resulting equilibrium phases are composed of ϕ_p^α and ϕ_p^β in both cases. However, the volume fractions of α-phase to β-phase are significantly different. Solution A consists of a small volume fraction of the polymer-rich phase, ϕ_p^β, dispersed in a polymer-poor phase with composition ϕ_p^α, whereas the opposite applies for solution B.

A thermal quench of solution C passes directly into the unstable region of the phase diagram. In this case, the resulting phase separation process is determined by spinodal decomposition. Phase separation proceeds instantaneously and results initially in a regular, highly interconnected structure, which tends to coarsen during the later stages of spinodal decomposition (*30*). Typical morphologies of membranes made from solutions A, B and C, respectively, at various stages of phase separation are schematically shown in Figure 5 (*31*).

The above discussion provides a description for liquid-liquid phase separation phenomena and their relationship to the evolution of membrane structures. However, the final membrane structure depends also on the kinetics of the phase separation process and the local distribution of the polymer-rich phase at the point of vitrification. Arnauts *et al.* (*32*) and Hikmet *et al.* (*33*) demonstrated that solidification of a binary polymer/solvent system occurs when the binodal line intersects the curve defining the glass transition, as shown in Figure 6. The location of the vitrification point, VP, is very important for the formation of membranes. If liquid-liquid phase separation occurs during the course of cooling an initially stable solution, continuing phase separation and/or coarsening of the resulting phases will be arrested at the temperature where the tie-line intersects the vitrification point (*33*).

Dense Symmetric Membranes. Dense symmetric membranes with thicknesses larger than 10 μm can be made by solution casting and subsequent solvent evaporation or by melt extrusion. This membrane type is used for characterization of material-specific transport properties (permeability and selectivity) of new membrane materials. Because these membranes are relatively thick, they have found only limited application in gas separation processes (*34*). However, dense symmetric membranes are commonly used in electrodialysis applications (*1*).

Asymmetric Membranes

Asymmetric membranes can be categorized into three basic structures: (i) integral-asymmetric with a porous skin layer, (ii) integral-asymmetric with a dense skin layer, and (iii) thin-film composite membranes. Porous integral-asymmetric membranes are made by the phase inversion process (*5-7*) and are applied in dialysis, ultrafiltration, and microfiltration applications, whereas integral-asymmetric membranes with a dense skin layer are applied in reverse osmosis and gas separation applications.

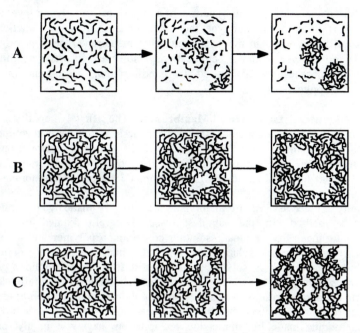

Figure 5.   Structures of membranes resulting from (a) nucleation and growth of a polymer-rich phase, (b) nucleation and growth of a polymer-poor phase, and (c) spinodal decomposition. Reproduced with permission from reference 31.

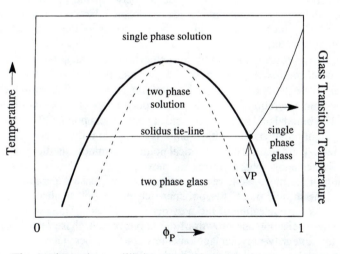

Figure 6.   Thermodynamic equilibrium and glass transition temperature of a binary polymer-solvent system as a function of solution composition. VP= vitrification point. Reproduced with permission from reference 40.

Thin-film composite membranes consist of a thin, selective polymer layer atop a porous support. In this membrane type, the separation and mechanical functions are assigned to different layers in the membrane. This membrane type was originally developed for reverse osmosis applications; however, nowadays thin-film composite membranes are also used in nanofiltration, gas separation, and pervaporation applications.

Integrally-Skinned Asymmetric Membranes. The first integrally-skinned asymmetric cellulose acetate membranes for water desalination were developed by Loeb and Sourirajan in the early 1960s at UCLA (*42*). The desalination properties of these membranes were significantly better than those of any other existing membrane at that time and led to the rapid commercialization of membrane-based desalination.

Integrally-skinned asymmetric membranes are typically made by an immersion precipitation method. In the simplest case, integrally-skinned asymmetric membranes are made from a binary solution containing a polymer and a solvent. Upon immersion of the cast solution into a liquid, which is a non-solvent for the polymer but miscible with the solvent, an asymmetric structure with either a porous or non-porous skin layer is formed. The structural gradient in integrally-skinned asymmetric membranes results from a very steep polymer concentration gradient in the nascent membrane at the onset of phase separation (*4-6*). The structure of a typical membrane made by immersion precipitation having a highly porous substructure and a thin skin layer is shown in Figures 7a and 7b.

In the immersion precipitation process, phase separation can be induced by (i) solvent evaporation and/or (ii) solvent/non-solvent exchange during the quench step. A typical isothermal ternary phase diagram is illustrated schematically in Figure 8. The diagram shows (i) the stable region, located between the polymer/solvent axis and the binodal, (ii) the meta-stable region, located between the binodal and spinodal, and (iii) the unstable region, located between the spinodal and the polymer/non-solvent axis.

For a given average composition located within the meta-stable or unstable region, the equilibrium composition of the polymer-poor phase and that of the polymer-rich phase is determined by the intercept of the tie-line with the binodal curve. The volumetric ratio of polymer-rich to polymer-poor phase is given by the lever rule. The binodal and spinodal coincide at the critical point, CP. For most high molecular weight polymers, the critical point is located at polymer concentrations of 5 vol.% or less. The location of the critical point determines whether the polymer-rich or polymer-poor phase forms a new phase. If the average polymer concentration in the meta-stable region at the onset of phase separation is smaller than that of the critical point, the polymer-rich phase will nucleate, whereas the opposite occurs for nucleation of the polymer-poor phase. Thus, at low polymer concentrations, nucleation and growth of the polymer-rich phase leads to polymer powder or low integrity polymer agglomerates. On the other hand, nucleation and growth of the polymer-poor phase at high polymer concentration in the upper meta-stable region results in a sponge-like cell structure (*35*).

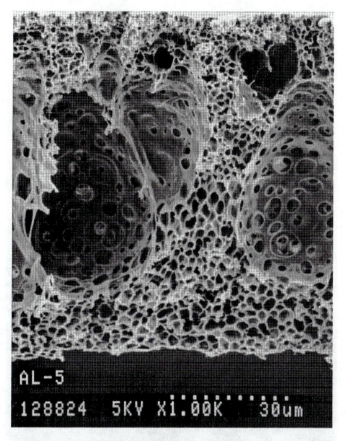

a

Figure 7. (a) Porous bulk structure and (b) skin layer of an integrally-skinned asymmetric polysulfone membrane made by the immersion precipitation process.

Continued on next page.

b

Figure 7. *Continued.*

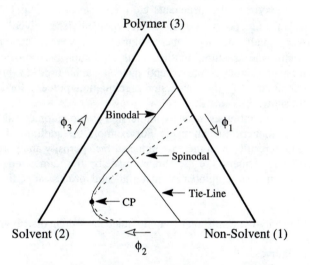

Figure 8. Schematic phase diagram of a ternary polymer-solvent-non-solvent system. CP = critical point. Reproduced with permission from reference 40.

Typically, the formation of membranes made by the immersion precipitation method occurs over a very short time scale, typically less than a few seconds. Different membrane structures can be obtained by careful control of the thermodynamic and kinetic variables involved in the immersion precipitation process. Some progress has been made in modeling the extremely complex thermodynamic and kinetic processes involved in the immersion precipitation process (*36-39*). However, the predictive capabilities of these models are limited, and, therefore, optimization of membranes made by immersion precipitation is still primarily based on empirically developed protocols.

Most commercial membranes made by the immersion precipitation method are made from multi-component solutions containing polymer, solvent(s), and non-solvent(s) or additives. In many cases, the porosity, pore size, and skin layer thickness can be modified by the addition of non-solvents to the casting solution (e.g. alcohols, carboxylic acids, surfactants etc.), inorganic salts (e.g. $LiNO_3$ or LiCl etc.) or polymers (e.g. polyvinylpyrrolidone, polyethylene glycol etc.) (*5-10, 40,41*). Even very small amounts of these solution additives can have a significant effect on the membrane structure, and hence, its separation performance. Examples of multi-component casting solutions and quench media used to prepare high-performance membranes by the immersion precipitation process for a variety of applications are shown in Table II.

The structure of membranes made by immersion precipitation can also be altered by using multi-component quench media. For example, the addition of a solvent to the quench medium results in an increase in the surface porosity and pore size of the membrane (*10*). The formation of membranes made by the immersion precipitation process depends on a large number of material- and process-specific parameters including:

- choice of the polymer (molecular weight, molecular weight distribution).
- choice of the solvent(s).
- choice of additives.
- composition of the casting solution.
- temperature of the casting solution.
- choice of the quench medium.
- temperature of the quench medium.
- composition of the casting atmosphere.
- temperature of the casting atmosphere.
- evaporation conditions.
- casting thickness.
- casting or spinning speed.
- membrane support material (type of woven or non-woven)
- drying conditions.

Thin-Film Composite Membranes. Composite membranes consist of at least two structural elements made from different materials. The two basic types of thin-film composite membranes are shown schematically in Figure 9. A single-layer composite membrane, shown in Figure 9a, consists of a porous support and a thin,

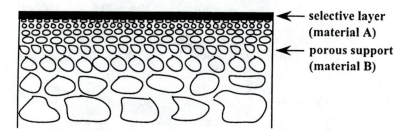

← selective layer
   (material A)
← porous support
   (material B)

a) single-layer composite

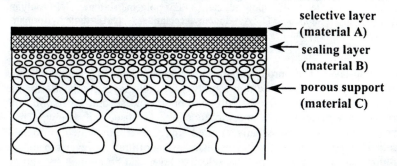

selective layer
← (material A)
← sealing layer
   (material B)
← porous support
   (material C)

b) multi-layer composite

Figure 9.   Schematic diagram of (a) single-layer and (b) multi-layer thin-film composite membranes.

Table II. Multi-component casting or spinning solutions for production of membranes by the immersion precipitation process.

Polymer	Solvent	Non-solvent or additive	Quench medium	Application
22.2 wt% CA	66.7 wt% acetone	10.0 wt% water + 1.1 wt% MgClO$_4$	water	RO (*42*)
16.2 wt% PSF	79 wt% DMAc	4.8 wt% PVP	70.5 wt% IPA + 29.5 wt% water	UF (*43*)
10.46 wt% PES	69.72 wt% DMF	19.82 wt% *t*-amyl alcohol	water	MF (44)
37 wt% PSF	36 wt% NMP	27 wt% propionic acid	water	GS (*45*)
18 wt% PI	82 wt% *p*-chlorophenol	-	35 wt% water + 65 wt% ethanol	GS (*46*)

CA=cellulose acetate; PSF=polysulfone; PES=polyethersulfone; PI=polyimide; PVP=polyvinylpyrrolidone; DMAc=dimethylacetamide; DMF=dimethylformamide; NMP=N-methylpyrrolidone;

selective top-layer. The porous support provides mechanical strength, whereas the separation is performed by the thin top-layer. A multi-layer composite membrane, shown in Figure 9b, consist of a porous support and several layers of different materials, each performing a specific function, as discussed below. Thin-film composite membranes are usually applied in processes in which permeation is controlled by the solution-diffusion mechanism (nanofiltration, reverse osmosis, gas separation, and pervaporation). The selective layer can be applied by lamination (*47,48*), solution coating (*49-51*), interfacial polymerization (*52-55*), or plasma polymerization (*56-57*) methods. Compared to integrally-skinned asymmetric membranes, composite membranes offer several significant advantages:

- Independent selection of materials from which the separating layer and the porous support are formed.
- Independent preparation of the separating layer and the porous support membrane, thereby making it possible to optimize each structural element.
- The ability to control the thickness of the separating layer.
- Very expensive membrane materials (>1,000 $/lb) can be used because only a very small amount of polymer is required for the formation of the thin separation layer (~ 1 g polymer/m^2 of membrane for a 1-μm-thick selective layer).

In most cases, porous, ultrafiltration-type membranes made by the immersion precipitation process are used as mechanical support for thin-film composite membranes. Optimum porous supports for thin-film composite membranes should have the following properties:

- The porous support must be chemically resistant against the solvent or solvent mixture from which the thin separating layer is formed.
- The porous support should have a high surface porosity and small pore size. High surface porosity is important because the support should not provide any significant resistance to mass transport in a composite membrane. A small pore size is required for the deposition of ultrathin, defect-free coatings.

The two most important methods for the commercial production of thin-film composite are based on interfacial polymerization and solution coating methods. The first interfacially polymerized thin-film composite membranes were developed by Cadotte at the North Star Research Institute and represented a breakthrough in membrane performance for reverse osmosis applications (52,53,55). The original interfacial polymerization process involved soaking a microporous polysulfone support in an aqueous solution of a polymeric amine and then immersing the amine-impregnated membrane into a solution of a di-isocyanate in hexane. The membrane was then cross-linked by heat-treatment at 110°C. The resulting polyurea membrane had better salt rejection than that of an integrally-skinned asymmetric cellulose acetate membrane and high water flux. Modifications in the chemistry of the original interfacial polymerization reaction scheme resulted in further significant improvement in performance of thin-film composite membranes for reverse osmosis applications (55).

The solution coating method involves deposition of a dilute polymer solution onto the surface of a porous membrane and subsequent drying of the thin liquid film (49-51). The simplicity of this process is very attractive for the production of membranes on a commercial scale. However, it is generally very difficult to produce defect-free thin-film composite membranes with thickness of less than 1μm by the solution coating process. These defects are caused by incomplete coverage of surface pores in the support membrane after complete evaporation of the solvent. The difficulty in completely covering surface pores results from penetration of the coating solution into the porous support membrane structure. Because capillary forces in the porous membrane tend to pull the thin liquid polymer solution into the bulk support membrane, the coating layer can be disrupted easily. Several methods have been proposed to overcome problems with the formation of the thin, selective layer by the solution-coating process. One method is to use ultrahigh molecular weight polymers for the formation of the selective layer. Rezac and Koros suggested that defect-free composite membranes can be made if the polymer chain dimensions of the coating material are larger than the largest pores in the surface of the support membrane (58). In this case, the formation of the thin, selective layer occurs via a sieving mechanism in which the surface pores are sealed with the coating material. Typically polymers have coil diameters between 200 and 500 Å (58). Therefore, the pore size of optimum support membranes should be smaller than about 200 Å. However, many polymeric supports used for composite membranes have a fairly broad pore size distribution and contain even a few larger defects on the order of 0.1 μm or more. In this case, even ultrahigh molecular weight polymers can not completely seal the surface pores by a simple sieving mechanism.

An alternative approach for eliminating defects in the thin selective coating layer is to fabricate multi-layer composite membranes (*59-61*). These membranes, shown schematically in Figure 9b, consist of (i) a porous support, (ii) a sealing layer, and (iii) an ultrathin, selective coating. The function of the sealing layer is to plug the pores in the support membrane and to provide a smooth surface onto which the thin coating layer can be applied. In addition, the sealing layer helps in channeling the permeating components to the surface pores, thereby rendering the entire surface area available for mass transport (*61*). The sealing layer should not provide a significant mass transport resistance in a multi-layer composite membrane. Hence, the sealing layer material should be significantly more permeable than the thin, selective top-layer.

Membrane Modification Methods

The development of high-performance membranes involves the selection of a suitable membrane material and the formation of this material into a desired membrane structure. However, it is often necessary to modify the membrane material or the structure to enhance the overall performance of the membrane. Generally, the objectives for modification of pre-formed membranes are (i) increasing flux and/or selectivity and (ii) increasing chemical resistance (solvent resistance, swelling, or fouling resistance). Some of the most commonly practiced membrane modification methods are listed in Table III.

The first reported membrane modification method involved annealing of porous membranes by heat-treatment. Zsigmondy and Bachmann demonstrated that the pore size of a pre-formed nitrocellulose membrane could be decreased with a hot water or steam treatment (*14*). Loeb and Sourirajan used the same method to improve the salt rejection of integrally-skinned asymmetric cellulose acetate reverse osmosis membranes (*42*). The properties of gas separation membranes can also be improved by annealing with a heat-treatment, as shown by Hoehn and Kusuki *et al.* (*46,62*) Heat-treatment of gas separation membranes typically leads to an increase in selectivity, because of elimination of micro-defects in the thin separating layer. An alternative method of reducing the number of defects in the separating layer in a membrane is based on a solvent-swelling technique (*63,64*). In this method, micro-defects in a membrane can be eliminated by swelling the thin separating layer with a vapor or a liquid. As a result of swelling, the modulus of the polymer decreases sharply. It has been suggested that capillary forces can pull the swollen polymer layer together, and, thereby, eliminating membrane defects (*64*).

During the development of integrally-skinned asymmetric cellulose acetate gas separation membranes it was found that water-wet membranes collapse and form an essentially dense film upon drying. This collapse occurs because of the strong capillary forces within the finely porous structure during the drying process. This phenomenon can be described by the well-known Young-Laplace relationship ($\Delta p = 2\gamma/r$ in the case of perfect wetting of the liquid in the pores). Hence, the capillary pressure is directly proportional to the surface tension of a liquid, but inversely proportional to the pore radius. If the modulus of the membrane material (in the swollen state) is lower than the capillary force of the liquid in the pore space, the pores will collapse and form a dense polymer film. Because water has a very high surface tension, it is often difficult to dry water-wet membranes without collapsing the membrane structure. An exchange of water with liquids having lower surface tension, such as alcohols or aliphatic hydrocarbons, results in maintaining the original membrane structure upon drying. Typical solvent-exchange methods involve replacing water first with *iso*-propanol and then with *n*-hexane (*65-67*). Other methods of eliminating the collapse of finely porous membrane structures include freeze-drying and addition of surfactants to the water prior to drying of the wet membranes (*68-70*).

Table III. Membrane modification methods.

Modification method	Goal	Application
Annealing		RO, GS, UF
• heat-treatment	elimination of membrane defects	
• solvent treatment	control of pore size	
Solvent-exchange	elimination of membrane defects	GS, UF
Surface coating	elimination of membrane defects	GS
	improvement of fouling resistance	RO, NF, UF
Chemical treatment		
• fluorination	improvement of flux and selectivity	GS
• cross-linking	improvement of chemical resistance	UF
• pyrolysis	improvement of flux and selectivity	RO, GS, PV

In the 1970s, commercialization of gas separation membranes was severely limited by the very poor reproducibility of making ultrathin, defect-free membranes on a large scale. Methods for production of thin-film composite membranes as well as integrally-skinned asymmetric membranes with separating layer thicknesses of less than 0.2 μm were known. However, production of these membranes without defects was not possible. Defects as small as 20 Å over an area fraction of less than 10^{-4}% can severely reduce the selectivity of gas separation membranes. However, a thin coating of a highly permeable polymer, such as polydimethylsiloxane, can render defective membranes suitable for gas separations. Modification methods developed by Browall for thin-film composite membranes (*71*) and, in particular, Henis and Tripodi for integrally-skinned asymmetric membranes (*72,73*) resulted in rapid commercialization of gas separation membranes. Surface coatings are also applicable in improving the fouling resistance of membranes for ultrafiltration or nanofiltration applications (*74-76*).

Chemical surface modification methods of gas separation membranes include treatment with fluorine, chlorine, bromine, or ozone (77-81). Typically, these treatments result in an increase in membrane selectivity coupled with a decrease in flux. Cross-linking of polymers is often applied to improve the chemical stability and selectivity of membranes for reverse osmosis (54), pervaporation (82), and gas separation applications (50).

Pyrolysis of polymeric precursors is a relatively new modification method that can lead to significantly improved separation performance of synthetic membranes. Specifically, molecular sieve membranes made from pyrolized polyacrylonitrile and polyimide (83-85) as well as selective surface flow membranes made from polyvinylidene chloride-acrylate terpolymer (86) have significantly better separation performance than polymeric membranes in gas separation applications. However, the use of these membranes in industrial applications is currently limited by their poor mechanical stability, susceptibility to fouling, and very high production costs. If these problems can be solved in future work, inorganic membranes made from polymeric precursors will present a new generation of high-performance membranes for the next millennium.

Literature Cited

1. Cabasso, I. In *Encyclopedia of Polymer Science and Engineering Vol. 9*, John Wiley & Sons, New York, 1987, pp 509-579.
2. Pinnau. I. *Polym. Adv. Tech.* **1994**, *5*, 733.
3. Koros, W.J.; Fleming, G.K.; Jordan, S.M.; Kim, T.H.; Hoehn, H.H. *Prog. Polym. Sci.* **1988**, *13*, 339.
4. Baker, R.W.; Cussler, E.L.; Eykamp, W.; Koros, W.J.; Riley, R.L.; Strathmann, H. *Membrane Separation Systems-Recent Developments and Future Directions*, Noyes Data Corporation, Park Ridge, NJ, 1991.
5. Kesting, R.E. *Synthetic Polymeric Membranes*, McGraw-Hill Book Company, New York, 1971.
6. Strathmann, H. *Trennung von molekularen Mischungen mit Hilfe synthetischer Membranen*, Dr. Dietrich Steinkopff Verlag, Darmstadt, 1979.
7. Strathmann, H. In *Handbook of Industrial Membrane Technology*, Porter M.C., Ed.; Noyes Publications, Park Ridge, NJ, 1985, pp 1-60.
8. Strathmann, H. In *Materials Science of Synthetic Membranes*, Lloyd, D.R, Ed.; ACS Symp. Ser. 269: ACS, Washington DC, 1985, pp 165-195.
9. Kesting, R.E. In *Materials Science of Synthetic Membranes*, Lloyd, D.R, Ed.; ACS Symp. Ser. 269: ACS, Washington DC, 1985, pp 131-164.
10. Mulder, M. *Basic Principles of Membrane Technology*, 2nd Ed., Kluwer Academic Publishers, Boston, MA 1996.
11. Fleischer, R.L.; Price, P.B.; Walker, R.M. *Ann. Rev. Nucl. Sci.* **1965**, *15*, 1.
12. Bierebaum, H.S., Isaacson, R.B.; Druin, M.L.; Plovan, S.G. *Ind. Eng. Chem. Prod. Res. Dev.* **1974**, *13*, 2.
13. Chen, R.T.; Saw, C.K.; Jamieson, M.G.; Aversa, T.R.; Callahan, R.W. *J. Appl. Polym. Sci.* **1994**, *533*, 471.

14. Zsigmondy, R.; Bachmann, W. *U.S. Patent* 1,421,341 (1922).

15. Goetz, A. *U.S. Patent* 2,926,104 (1960).

16. Maier, K.; Scheuermann, E. *Kolloid-Z. u. Z. Polymere* **1960**, *171*, 122.

17. Park, H.C.; Kim, Y.P.; Kim, H.Y.; Kang, Y.S. *J. Membrane Sci.* **1999**, *156*, 169.

18. Bjerrum, N.; Manegold, E. *Kolloid-Z. u. Z. Polymere* **1927**, *75*, 795.

19. Castro, A.J. *U.S. Patent* 4,247,498 (1981).

20. Vitzhum, G.H.; Davis, M.A. *U.S. Patent* 4,490,431 (1984).

21. Hiatt, W.C.; Vitzthum, G.H.; Wagener, K.B.; Gerlach, K.; Josefiak, C. In *Materials Science of Synthetic Membranes*, Lloyd, D.R, Ed.; ACS Symp. Ser. 269: ACS, Washington DC, 1985, pp 229-244.

22. Lloyd, D.R.; Kinzer, K.E.; Tseng, H.S. *J. Membrane Sci.* **1990**, *52*, 239.

23. Lloyd, D.R.; Kim, S.S.; Kinzer, K.E. *J. Membrane Sci.* **1991**, *64*, 1.

24. Caneba, G.T.M.; Soong, D.S. *U.S. Patent* 4,659,470 (1987).

25. Kinzer, K.E. *U.S. Patent* 4,867,881 (1989).

26. Damrow, P.A.; Mahoney, R.D.; Beck, N.; Sonnenschein, M.F. *U.S. Patent* 5,205,968 (1993).

27. Berghmans, S.; Berghmans, H.; Meijer, H.E.H. *J. Membrane Sci.* **1996**, *116*, 171.

28. Tompa, H. *Polymer Solutions*, Butterworths., London, 1956.

29. Yilmaz, L.; McHugh, A.J. *J.Appl. Polym. Sci.* **1986**, *31*, 997.

30. Song, S.-W.; Torkelson, J. M. *J. Membrane Sci.* **1995**, *98*, 209.

31. Pinnau, I. *Ph.D. dissertation*, The University of Texas, 1991.

32. Arnauts, J.; Berghmans, H. *Polym. Com.* **1987**, *28*, 66.

33. Hikmet, R.M.; Callister, S.; Keller, A. *Polymer* **1988**, *29*, 1378.

34. Skiens, W.E.; Lipps, B.J.; McLain, E.A.; Dubocq, D.E. *U.S. Patent* 3,798,185 (1974).

35. Kimmerle, K.; Strathmann, H. *Desalination* **1990**, *79*, 283.

36. Reuvers, A.J.; van den Berg, J.W.A.; Smolders, C.A. *J. Membrane Sci.* **1987**, *34*, 45.

37. Tsay, C.S.; McHugh, A.J. *J. Polym. Sci. Polym. Phys. Ed.* **1991**, *29*, 1261.

38. Radovanovic, P.; Thiel, S.W.; Hwang, S.-T. *J. Membrane Sci.* **1992**, *65*, 213.

39. Termonia, Y. *J. Membrane Sci.* **1995**, *104*, 173.

40. Koros, W.J.; Pinnau, I. In *Polymeric Gas Separation Membranes*, Paul, D.R.; Yampol'skii, Eds. CRC Press, Boca Raton, 1994, pp 209-271.

41. Kesting, R.E.; Fritzsche, A.K. *Polymeric Gas Separation Membranes*, John Wiley & Sons, Inc., New York, 1993.

42. Loeb, S.; Sourirajan, S. *Adv. Chem. Ser.* **1962**, *38*, 117.

43. Wenthold, R.M.; Hall, R.T.; Brinda, P.D.; Cosentino, L.C.; Reggin, R.F.; Pigott, D.T. *U.S. Patent* 5,762,798 (1998).

44. Wang, I.-F. *U.S. Patent* 5,886,059 (1999).

45. Kesting, R.E.; Fritzsche, A.K.; Murphy, M.K.; Handerman, A.C.; Cruse, C.A.; Malon, R.F. *U.S. Patent* 4,871,494 (1989).

46. Kusuki, Y.; Yoshinaga, T.; Shimazaki, H. *U.S. Patent* 5,141,642 (1992).

47. Browall, W.R. *U.S. Patent* 3,874,986 (1975).

48. Salemme, R.M.; Browall, W.R. *U.S. Patent* 4,155,793 (1979).

22

49. Riley, R.L. *U.S. Patent* 3,648,845 (1972).
50. Riley, R.L.; Grabowski, R.L. *U.S. Patent* 4,243,701 (1981).
51. Kazuse, Y.; Chikura, S. *U.S. Patent* 4,590,098 (1986).
52. Cadotte, J.E. In *Materials Science of Synthetic Membranes*, Lloyd, D.R, Ed.; ACS Symp. Ser. 269: ACS, Washington DC, 1985, pp 273-294.
53. Petersen, R.J.; Cadotte, J.E. In *Handbook of Industrial Membrane Technology*, Porter M.C., Ed.; Noyes Publications, Park Ridge, NJ, 1985, pp 307-348.
54. Sasaki, T.; Fujimaki, H.; Uemura, T.; Kurihara, M. *U.S. Patent* 4,758,343 (1988).
55. Petersen, R.J. *J. Membrane Sci.* **1993**, *83*, 81.
56. Yasuda, H.; Marsh, H.C.; Tsai, J. *J. Appl. Polm. Sci.* **1975**, *19*, 2157.
57. van der Scheer, A. *U.S. Patent* 4,581,043 (1986).
58. Rezac, M.; Koros, W.J. *J. Appl. Polym. Sci.* **1992**, *46*, 1927.
59. Salemme, R.M.; Browall, W.R. *U.S. Patent* 3,874,986 (1975).
60. Cabasso, I.; Lundy, K.A *U.S. Patent* 4,602,922 (1986).
61. Lundy, K.A.; Cabasso, I. *Ind. Eng. Chem. Res.* **1989**, *28*, 742.
62. Hoehn, H. *U.S. Patent* 3,822,202 (1974).
63. Pinnau, I.; Wind, J. *U.S. Patent* 5,007,944 (1991).
64. Rezac, M.E.; Le Roux, J.D.; Chen, H.; Paul, D.R.; Koros, W.J. *J.Membrane Sci.* **1994**, *90*, 213.
65. Rowley, M.E.; Slowig, W.D. *U.S. Patent* 3,592,672 (1971).
66. Manos, P. *U.S. Patent* 4,120,098 (1978).
67. Admassu, W. *U.S. Patent* 4,843,733 (1989).
68. Gantzel, P.K.; Merten, U. Ind. Eng. *Chem. Proc. Des. Dev.* **1970**, *9*, 331.
69. Hayes, R.A. *U.S. Patent* 5,032,149 (1991).
70. Jensvold, J.A. Cheng, T.; Schmidt, D.L. *U.S. Patent* 5,141,530 (1992).
71. Browall, W.R. *U.S. Patent* 3,980,456 (1976).
72. Henis, J.M.S.; Tripodi, M.K. *U.S. Patent* 4,230,463 (1980).
73. Henis, J.M.S.; Tripodi, M.K. *J. Membrane Sci.* **1981**, *8*, 233.
74. Li, R.H.; Barbari, T.A. *J. Membrane Sci.* **1995**, *105*, 71.
75. Nunes, S.P.; Sforça, M.L.; Peinemann, K.-V. *J. Membrane Sci.* **1995**, *106*, 49.
76. Ulbricht, M.; Belfort, G. *J. Membrane Sci.* **1996**, *111*, 193.
77. Langsam, M. *U.S. Patent* 4,657,564 (1987).
78. Mohr, J.M.; Paul, D.R.; Mlsna, T.E.; Lagow, R.J. *J. Membrane Sci.* **1991**, *55*, 131.
79. Leväsalmi, J.-M.; McCarthy, T.J. *Macromolecules* **1995**, *28*, 1733.
80. Barbari, T.A.; Datwani, S.S. *J. Membrane Sci.* **1995**, *107*, 263.
81. Kramer, P.W.; Murphy, M.K.; Stookey, D.S.; Henis, J.M.S.; Stedonsky, E.R. *U.S. Patent* 5,215,554 (1993).
82. Brüschke, H. *U.S. Patent* 4,755,299 (1988).
83. Koresh, J.E.; Soffer, A. *Sep. Sci. Technol.* **1987**, *22*, 973.
84. Jones, C.W.; Koros, W.J. *Carbon* **1994**, *324*, 1419.
85. Kusuki, Y.; Shimazaki, H.; Tanihara, N.; Nakanishi, S.; Yoshinaga, T. *J. Membrane Sci.* **1997**, *134*, 245.
86. Rao, M.B.; Sircar, S.; Golden, T.C. *U.S. Patent* 5,507,860 (1996).

Chapter 2

Formation of Anisotropic and Asymmetric Membranes via Thermally-Induced Phase Separation

H. Matsuyama[1], S. Berghmans[2], M. T. Batarseh[2], and D. R. Lloyd[2]

[1]Department of Environmental Chemistry and Materials, Okayama University, Okayama 700, Japan
[2]Department of Chemical Engineering, University of Texas at Austin, Austin, TX 78712

Anisotropic (gradient in pore size) and asymmetric (integrally-skinned) membranes were formed by the thermally-induced phase separation (TIPS) process. To form such structures, evaporation of diluent was allowed from one side of an isotactic polypropylene (iPP)-diphenyl ether melt-blend, thereby creating a polymer concentration gradient in the sample before cooling and phase separation. The quench temperature was used to influence the cell size. The combined use of temperature gradient and polymer concentration gradient produced pronounced asymmetric structures with a skin layer at the top surface. The evaporation process was analyzed by solving appropriate mass transfer and heat transfer equations. The membrane structures are discussed in detail based on the calculated polymer volume fraction profiles in the membranes.

The thermally-induced phase separation (TIPS) process is a valuable way of making microporous materials, such as membranes and foams (*1-37*). In this process, a polymer is dissolved in a diluent at high temperature. Upon, removal of the thermal energy by cooling or quenching, phase separation occurs in the solution.

While these membranes are useful in many applications, it is often desirable to have an anisotropic structure (a gradient in cell size from small cells at the feed-side to large cells at the permeate-side of the membrane) or asymmetric structure (a relatively dense skin on an isotropic or anisotropic microporous support) (*38*). These structures reduce the blocking of internal pores by suspended particulate or colloidal materials and facilitate membrane cleaning via cross flow. Furthermore, anisotropic membranes show significant improvement in permeability and in throughput compared to isotropic membranes with similar retention. However, there are few reports of using the TIPS process to produce membranes with either an anisotropic structure *(13-16)* or an integral skin (*17-22*).

Prior and current work in our laboratories *(23-30)* and work conducted elsewhere *(13,32-37)* has demonstrated that the kinetics of droplet growth in liquid-liquid TIPS is (i) dependent on the polymer-diluent interfacial tension, (ii) increases with increasing volume fraction of the droplet phase, and (iii) decreases with increasing viscosity of the polymer-rich matrix phase. Smaller cells are obtained when the polymer concentration is higher (that is, lower diluent concentration), whereas larger cells result when the concentration is lower. Thus, inducing a concentration gradient across the thickness of the membrane prior to phase separation will result in a gradient in structure development kinetics and thus a gradient in cell size across the thickness of the membrane.

In this work, production of anisotropic and asymmetric TIPS membranes was achieved by inducing a polymer concentration gradient in the membrane solution prior to cooling or quenching the solution. The concentration gradient was established by controlled evaporation of diluent from one side of the polymer-diluent solution at a temperature above the binodal. Because of the evaporation, the polymer concentration increases at the top surface of the membrane. Subsequent cooling of the membrane produces an anisotropic or asymmetric structure with smaller pores or skin on the top of the membrane.

This research studied the effect of quench temperature on the morphology of membranes produced from nascent membranes containing a polymer concentration gradient. In actual membrane production, a polymer-diluent melt-blend is extruded through a slit die to form a flat sheet or a spinneret to form a hollow-fiber or tube. The melt-blend passes through an air gap of controlled distance and is then cooled either in a quench bath or on a chill roll. The rate at which the melt-blend cools in the air and in the quench bath can be altered by changing the temperature, the flow rate, or the flow configuration of these cooling media. A higher temperature will decrease the cooling rate, whereas a lower temperature will increase it. The effect of different media temperatures combined with the effect of a concentration gradient was studied in this research.

This paper also reports on the effects of imposing a temperature gradient on the melt-blend. A temperature gradient has been previously used the production of anisotropic structures *(13,14,20)*. In the research reported here, a flat melt-blend sample with polymer concentration gradient was taken out of an oven and the top side of the sample was immersed in water to induce rapid cooling, while the bottom side of the sample, which was in contact with a glass support, cooled more slowly. The effect of combined temperature and concentration gradients is analyzed. The evaporation process was mathematically simulated. The calculated polymer volume fraction profiles in the solution prior to phase separation were used to understand the anisotropic and asymmetric membrane structures.

Principle of Formation of Anisotropic and Asymmetric Structures Based on a Polymer Concentration Gradient. Figure 1 shows the phase diagram for the

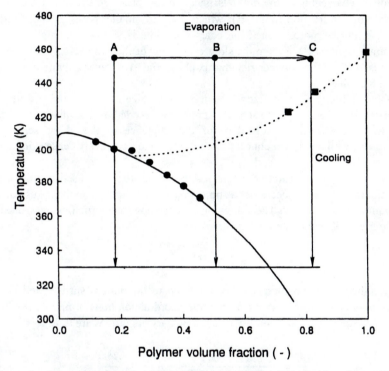

Figure 1. Phase diagram for iPP/DPE system (*23*). (●) cloud points; (■) experimental equilibrium melting points; (—) theoretical coexistence curve; (---) theoretical melting point depression curve.

isotactic polypropylene (iPP)-diphenyl ether (DPE) system (23). Point A shows the initial solution conditions. When evaporation occurs only from the top surface in the melt-blended solution, the polymer volume fraction at the top surface increases and moves to point B, whereas the volume fraction at the bottom surface remains at point A. Then the sample is cooled to the desired temperature to introduce the phase separation, as shown by the arrows in Figure 1. If the cooling rate is the same at both surfaces, then the amount of time spent under the binodal curve is shorter in the case of point B than in the case of point A. Therefore, there is less time for coarsening of the droplets generated by phase separation at the higher polymer volume fraction (point B), which leads to smaller pores in the finished membrane. Furthermore, even if the coarsening time is similar, the high polymer concentration (point B) leads to the slow coarsening of the droplets (37) due to the higher viscosity of the polymer-rich matrix phase and also lower volume fraction of the droplets phase.

If the polymer volume fraction at the top surface moves to point C due to diluent evaporation, L-L phase separation may be preceded by S-L phase separation (polymer crystallization). Regardless of the sequence of the phase separation mechanisms, a high polymer concentration may lead to a relatively dense skin.

Analysis of the Evaporation Process. A binary system was used in this work. The membrane thickness decreases as time progresses in this system because of the evaporation of the diluent. The following coordination transformation was used in order to immobilize the interface position (39).

$$\xi = z / L(t) \tag{1}$$

Here, z is the vertical position coordinate relative to the bottom surface, and $L(t)$ is the membrane thickness at time t. Using this coordination transformation, the mass transfer equation, the initial condition, and boundary conditions are (39):

$$\frac{\partial \phi_2}{\partial t} = \frac{\zeta}{L(t)} \frac{dL(t)}{dt} \frac{\partial \phi_2}{\partial \zeta} + \frac{1}{L(t)^2} \frac{\partial}{\partial \zeta} \left\{ D_{23} \frac{\partial \phi_2}{\partial \zeta} \right\} \tag{2}$$

$$t = 0; \; \phi_2 = \phi_{20} \tag{3}$$

$$\zeta = 0; \; \partial \phi_2 / \partial \zeta = 0 \tag{4}$$

$$\zeta = 1; \quad \phi_{20} L_0 - L(t) \int_0^1 \phi_2 d\zeta = \int_0^t k \phi_2 \exp\left\{ (1 - \phi_2) + \chi_{23} (1 - \phi_2)^2 \right\} (V_2 / V_{2g})(P_2^0 / P) dt \tag{5}$$

where ϕ_2 is the volume fraction of diluent, D_{23} is the mutual diffusion coefficient of the diluent and k is the gas-side mass transfer coefficient. P and P_2^0 represent the total pressure and the pure diluent vapor pressure, respectively. V_{2g} and V_2 are the partial specific volume of the diluent in the gas phase and the partial specific volume of the diluent in the liquid phase. χ_{23} is the diluent-polymer interaction parameter. Subscript 0 denotes the initial value.

The conservation equation of the polymer in the membrane solution is given by:

$$L(t) = L_0 \phi_{30} / \int_0^1 \phi_3 d\zeta \tag{6}$$

where ϕ_3 is the volume fraction of polymer.

The equation describing the heat transfer process with coordination transformation described by equation 1 is given by:

$$\frac{\partial T}{\partial t} = \frac{\zeta}{L(t)} \frac{dL(t)}{dt} \frac{\partial T}{\partial \zeta} + \frac{\alpha}{L(t)^2} \frac{\partial^2 T}{\partial \zeta^2} \tag{7}$$

$$t = 0; \ T = T_0 \tag{8}$$

$$\zeta = 0; \ T = T_0; \ T/\zeta = 0 \tag{9}$$

$$\zeta = 1; \ K/L(t)\partial T/\partial \zeta = h\left(T_{gas} - T\right) + \sigma\varepsilon\left(T_{gas}^4 - T^4\right) - N_2 \Delta H_2 \tag{10}$$

where α is the thermal diffusivity of the polymer solution. K and h are the thermal conductivity of the polymer solution and the gas-side heat transfer coefficient, respectively. T_{gas} is the temperature in the gas bulk phase, σ and ε are the Stefan-Boltzmann constant and emissivity of the polymer solution, and ΔH_2 is the latent heat of evaporation of the diluent. Here, the temperature at the glass-facing surface is assumed to be equal to the initial temperature.

To simulate the evaporation process by using equations 2-10, several parameters must be determined. The diffusion coefficient based on the free volume theory by Vrentas and Duda (40,41) was used as D_{23} in equation 2. All parameters except for χ_{23} were estimated according to Zielinsky and Duda's method (42). The interaction parameter χ_{23} for the iPP-DPE system is given by McGuire (23). Therefore, no fitting parameters were used to calculate D_{23}. The gas-side mass transfer coefficient k and gas-side heat transfer coefficient h were estimated by correlations applicable to free convection conditions (43).

Experimental

The polymer and diluent used are isotactic polypropylene (iPP, Himont Co., 6824PM) and diphenyl ether (DPE, Aldrich Chemical Co., 99% purity). All chemicals were used without further purification. The preparation of the homogeneous polymer-diluent sample was similar to that described earlier (*31*). The resulting solid sample had an approximate constant thickness of 600 μm. After putting the solid polymer-diluent sample in a glass bottle shown in Figure 2, the sample was sealed and heated to 433.2 K to cause melt-blending. Then, the glass cap was taken off and diluent was allowed to evaporate from the air-facing surface of the nascent membrane, thereby creating a polymer concentration gradient in the sample. After a certain evaporation period, the glass cap was replaced and the glass bottle including the sample was removed from the oven and slowly cooled.

Two cooling procedures were used in order to test their effect on structure. In one procedure, samples were cooled in air at 298 K, 333.2 K, or 373.2 K by moving the sample to an oven set to the desired temperature. In another procedure, a temperature gradient was imposed on the sample after the evaporation period. In this procedure, the top (air-solution) surface of the sample was immersed in ice-water for 5 seconds, while the bottom portion of the melt-blend maintained contact with the glass bottle. The glass cap was then replaced, and the glass bottle with the sample was placed at room temperature (about 298 K). It should be noted that in all these experiments, the polymer concentration gradients caused by the evaporation are the same, although the cooling conditions were different.

Diphenyl ether was extracted from the membrane by immersing it in methanol overnight. Methanol was then allowed to evaporate from the resulting microporous structure. The final sample was immersed in liquid nitrogen, fractured, and coated with gold-palladium using a sputtering coater (Commonwealth Model 3, Pelco). The cross-section of the membrane was viewed using a scanning electron microscope (JOEL JSN-35C) under an accelerating voltage of 25 W.

A DSC (Perkin Elmer DSC-7) was used to determine the dynamic crystallization temperature, T_c, of polymer-diluent samples at different concentrations. A 3 to 6 mg sample was sealed in an aluminum DSC pan, melted at 433.2 K, then cooled at several controlled rates (2. 0, 10. 0, or 20. 0 K/min) to 298.2 K. The onset of the exothermic peak during the cooling was taken as the dynamic crystallization temperature.

Simulation Result of Evaporation Process.   Figure 3 shows the calculated polymer concentration profiles across the thickness of the membrane for evaporation periods ranging from zero to six minutes. The polymer volume fraction at the top surface increases as evaporation continues; the concentration at the bottom surface remains unaffected for at least four minutes. Evaporation leads to a concentration gradient in the nascent membrane and a decrease in membrane thickness with increased evaporation time.

A comparison of the calculated results and the experimental data for the dimensionless membrane weight and membrane thickness are shown in Figure 4. The dimensionless membrane weight is defined as the membrane weight after evaporation divided by the initial membrane weight. The membrane thicknesses

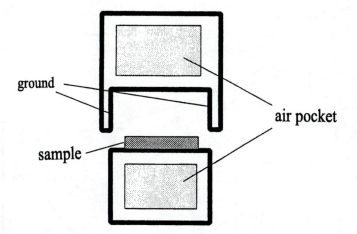

Figure 2. Schematic diagram of glass container used for preparation of thermally-induced phase separation (TIPS) membranes.

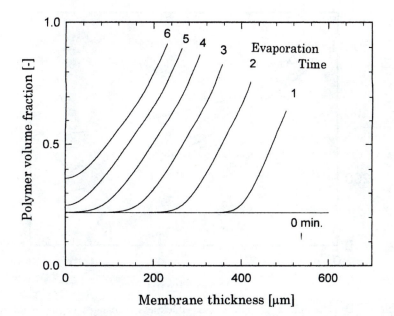

Figure 3. Calculated relationship between the polymer volume fraction across the melt-blended solution and the length of the evaporation period. The right sides in these lines correspond to the top surface, while the left sides correspond to the bottom surface. Sample: 20 wt% iPP

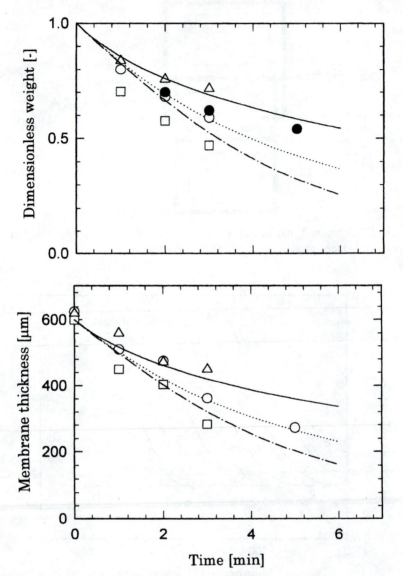

Figure 4. Dimensionless weight and thickness changes over time. Experimental data: □: 15 wt% iPP, ○●: 20 wt% iPP, Δ: 30 wt% iPP Calculated results: (— - —): 15 wt% iPP, (----): 20 wt% iPP, (——): 30 wt% iPP.

were measured after cooling the solution to cause phase separation and extracting the diluent with methanol. The open circles (○) and filled circles (●) are data from different experiments with the same sample (20 wt%). The agreement between data indicates the good reproducibility of these experiments. For lower initial polymer concentration, the diffusion coefficient and the vapor pressure of the diluent will be higher, which leads to a higher evaporation rate and a higher weight and thickness loss over the same time period. The solid lines are the calculated results. The calculated results for both the dimensionless weights and the membrane thicknesses show reasonable agreement with the corresponding experimental data.

Membrane Structure Resulting from Samples Cooled at Room Temperature. Figure 5 shows the cross-sections of the membranes obtained for the different evaporation times. Figure 6 shows the corresponding top and bottom surfaces of the membranes at a higher magnification. In this case, 20 wt% iPP-DPE samples were used. In all cases, cellular pore structures were observed, which is characteristic of a polymer mixture undergoing liquid-liquid TIPS. Also, spherulites were formed due to the crystallization of iPP. Therefore, the phase separation process in this case is liquid-liquid phase separation with subsequent polymer crystallization.

In the case of no evaporation, the membrane structure was almost isotropic (Figure 5(a)) with a pores size at the bottom surface (~5.0 μm) nearly equal to that at the top surface (Figure 6(a)). This result indicates clearly that the cooling rates at both surfaces were nearly equal in this experiment.

On the other hand, as the evaporation time increases, anisotropic structures were formed (Figure 5(b)-(d)). The gradient from smaller pores at the top surface to larger pores at the bottom surface is caused not by the difference in the cooling rates, but by the polymer concentration gradient induced by the evaporation. The difference in the pore sizes at both surfaces is clearly shown in Figure 6(b)-(d). These experimental results indicate that anisotropic structures can be successfully obtained by introducing an evaporation process before the temperature quenching step. As shown in Figure 5(b)-(d), the depth to which the smaller pores extends into the membrane cross-section increases with increased evaporation time. Careful visual inspection of the micrographs allows one to estimate a boundary line between the larger pores (that is, those unaffected by evaporation) and the smaller pores (that is, those affected by evaporation). The observed dimensionless boundary positions from the bottom surface are 0.90, 0.68 and 0.43 for the 1 min, 2 min, and 3 min evaporation cases, respectively. The calculated results are 0.74, 0.55, and 0.36 for the 1 min, 2 min, and 3 min cases. These calculated positions were obtained from Figure 3. The observed boundary positions are approximately in agreement with the calculated positions, which also shows the validity of the simulation results. The pore size at the bottom surface remained virtually unchanged over the range of 1 min evaporation time to the case of 3 min evaporation time, as shown in Figure 6(b)-(d). This observation agrees with the calculated results in Figure 3 where the polymer

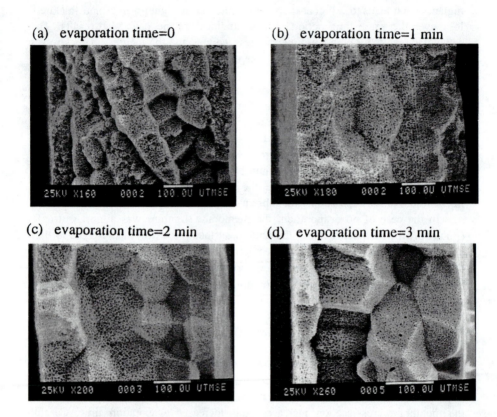

(a) evaporation time=0

(b) evaporation time=1 min

(c) evaporation time=2 min

(d) evaporation time=3 min

Figure 5. Micrographs of the cross-sections of 20 wt% sample cooled in room temperature air. (a) evaporation time = 0, (b) evaporation time = 1 min, (c) evaporation time − 2 min, (d) evaporation time = 3 min. The right and left sides correspond to the top and bottom surfaces, respectively.

concentration at the bottom surface is equal to the initial concentration for evaporation times of up to 4 minutes. The pore size at the top surface decreases as the evaporation time increases (Figure 6(b)-(d)). This observation corresponds to the increase in polymer concentration at the top surface predicted by the model, as shown in Figure 3.

Effects of Thermal History on Membrane Structures

Phase Diagram Considerations. Figure 7 shows the binodal curve for the iPP-DPE system (*23*), the experimental equilibrium crystallization curve (*23*), and dynamic crystallization curves determined in this study at cooling rates of 2.0, 10.0, and 20.0 K/min. Based on the calculated polymer volume fractions shown in Figure 3, the polymer concentration at the top surface is shown for evaporation periods of 0, 1, and 2 minutes. The polymer concentration at the top surface increases with a slight decrease in temperature during the evaporation process. As indicated in Figure 3, the concentration of the bottom portion of the melt-blend remains unchanged during the evaporation for up to 4 minutes. The various quench temperatures (373.2, 333.2, 298K) are also indicated. The quench paths, which are represented by vertical lines, are discussed below in association with the membrane structures.

Non-isothermal Cooling. A series of membranes were formed on a hot-stage at controlled cooling rates (1, 2.5, 10, and 100 K/min) and with no evaporation. The resulting membranes were prepared for SEM analysis, and the average cell sizes of these isotropic membranes were determined. Thus, the relation between the cooling rate and the average cell diameter was obtained at 20 wt% initial polymer concentration. This correlation was then used to estimate cooling rates for the samples prepared by quenched in air as described below; that is, by measuring average cell size and comparing to the correlation, it was possible to estimate the cooling rate experienced by the membrane when it was quenched in air. These values are summarized in Table I.

Table I. Estimated cooling rates for isothermal quench conditions studied.

Quench condition	Estimated cooling rate (K/min)
Air at 298K	9.0
Air at 333.2K	3.5
Air at 372.2K	1.5

Isothermal Cooling in Air. Figure 8 shows the cross-sections near the top and bottom surfaces of membranes obtained via evaporation for a specified period followed by a quench in 333.2 K air. The average cell size with no evaporation in Figure 8(a) is larger than that in Figure 6(a) (~6.2 μm versus ~5.0 μm, respectively).

34

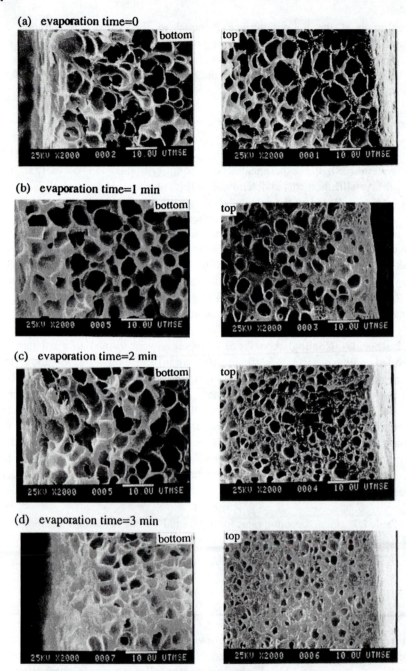

Figure 6.   Structures at top and bottom surfaces for 20 wt% sample cooled in room temperature air. (a) evaporation time = 0, (b) evaporation time = 1 min, (c) evaporation time = 2 min, (d) evaporation time = 3 min.

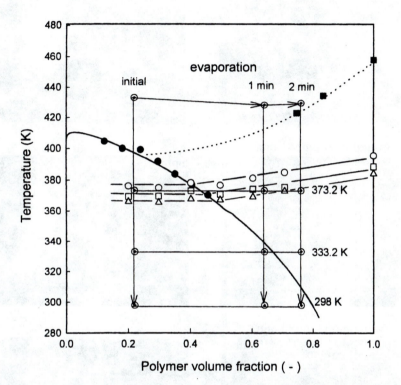

Figure 7. Phase diagram for iPP/DPE system. (○) dynamic crystallization temperature (cooling rate = 2 K/min); (□) dynamic crystallization temperature (cooling rate = 10 K/min); (△) dynamic crystallization temperature (cooling rate = 20 K/min); (⊙) polymer solution conditions. Evaporation path and quench paths are included in this figure. Other symbols and lines are the same as those described in Figure 1. Reproduced with permission from (*44*) Copyright 1998 Elsevier.

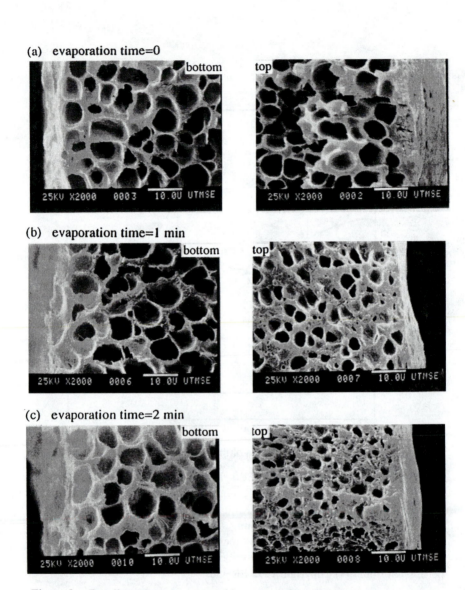

Figure 8. Detail structures at top and bottom surfaces of samples at a quench temperature of 333.2 K. (a) evaporation time = 0, (b) evaporation time = 1 min, (c) evaporation time = 2 min. Reproduced with permission from (*44*). Copyright 1998 Elsevier.

This difference in the cell size results from a difference in cooling rate. When the sample is quenched to 333.2 K, the temperature difference between the initial hot temperature (433.2 K) and the quench temperature is smaller than when the sample is quenched to 298 K. Therefore, the cooling rate in the former case is slower. By decreasing the cooling rate, the cell size increases because more time is allowed for coarsening of the droplets generated by the liquid-liquid phase separation (*20,24,25*).

In contrast to the isotropic membranes formed with no evaporation, anisotropic membranes were formed by evaporation of diluent from the nascent membrane (Figure 8(b)). The process path for the bottom surface is represented in Figure 7 by a vertical line from the "initial" point to the point at 333.2 K. On the other hand, the process path for the top surface is shown in Figure 7 as a vertical line labeled according to the evaporation time. After 1 minute of evaporation the iPP volume fraction at the solution-air interface is 0.64, and the quench condition lies below both the binodal and the dynamic crystallization curve. The estimated cooling rate in the case of quenching in 333.2 K air was 3.5 K/min, which is close to the dynamic crystallization curve determined at 2.0 K/min and shown in Figure 7. After 2 minutes of evaporation, the polymer volume fraction at the top surface is 0.75, and the quench condition lies below the dynamic crystallization curve but above the binodal. Consequently, the top surface of the membrane formed with 1 and 2 minute evaporation time undergoes different phase separation and structure formation mechanisms. For the top surface that has experienced 1-minute evaporation, the solution near the top surface crosses the dynamic crystallization curve prior to crossing the binodal curve during the cooling. However, because the crystallization rate is slower than the liquid-liquid phase separation rate (*27*), liquid-liquid phase separation probably occurs after crystallization starts. This process leads to the cell structure at the top surface shown in Figure 8(b). For the top surface that has experienced 2 minutes of evaporation, the solution near the top surface crosses the dynamic crystallization curve but does not cross the binodal. Consequently, the solution near the surface undergoes solid-liquid TIPS without any liquid-liquid TIPS, and a skin layer is formed (Figure 8(c)).

The results of the 333.2 K quench can again be compared to the results of the 298 K air quench. The cooling rate for the 298 K quench is 9 K/minute, which corresponds closely with the dynamic crystallization curve determined at 10 K/min (Figure 7). Therefore, the condition at the top surface even after 2 minutes of evaporation is inside the binodal, which leads to cells at the top surface (Figure 6(c)) instead of a skin layer as seen in the 333.2 K quench.

Figure 9 shows cross-sections of the membranes obtained in the case of quenching in air at 373.2 K. The average cell size at the bottom surface of 9.2 μm is larger than that shown in Figure 6 and Figure 8 because of the decreased cooling rate (1.5 K/min versus 3.5 K/min and 9 K/min, respectively). Near the top surface, the cells are much smaller than those at the bottom, and a skin layer is formed at the

38

(a) evaporation time=0

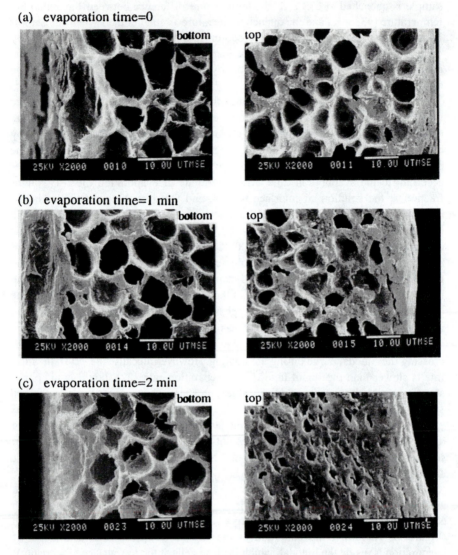

(b) evaporation time=1 min

(c) evaporation time=2 min

Figure 9. Detail structures at top and bottom surfaces of samples at a quench temperature of 373.2 K. (a) evaporation time = 0, (b) evaporation time = 1 min, (c) evaporation time = 2 min. Reproduced with permission from (*44*). Copyright 1998 Elsevier.

very top even for 1-minute evaporation time. As shown in Figure 7, the quench conditions at the top surface for evaporation times of 1 and 2 minutes are far from the cloud point curve and, therefore, liquid-liquid TIPS is not possible; only solid-liquid TIPS occurs, which leads to a skin layer in both cases (Figure 9(b) and (c)).

Use of a Temperature Gradient as well as a Concentration Gradient. Figure 10 shows cross-sections of the membranes produced when only the top surface of the membrane was immersed in ice-water. Even in the case of no evaporation, the difference in cooling rates experienced by the two surfaces results in an anisotropic structure (Figure 10(a)). The 3-minute evaporation resulted in a thick skin of about 45 μm (Figure 10(b)). Thus, the combined use of a temperature gradient and a concentration gradient can result in a pronounced asymmetric structure.

Conclusions

Anisotropic and asymmetric membrane structures were obtained by the TIPS process by introducing an evaporation step before the temperature quench. Evaporation from one side of the melt-blended solution induced a polymer concentration gradient in the membrane, which led to smaller pores or a skin at the top surface. The effects of the cooling rate and the quench temperature on the membrane structures were investigated. When cooled in air, decreasing the quench temperature resulted in an increase in the cooling rate, which led to a decrease in the average cell diameter at the bottom surface. On the other hand, a skin layer was formed at the top surface when the quench temperature was so high that the quench condition for the top surface of the sample solution was outside the binodal curve. Furthermore, it was found that the combined use of a temperature gradient and concentration gradient produced a pronounced asymmetric structure. The evaporation process was simulated by solving the appropriate mass transfer and heat transfer equations, and the changes of the polymer volume fractions during the evaporation were calculated.

Literature Cited

1. Castro, A.J. *U.S. Patent* 4,247,498 (1981).
2. Lipps, B. *U.S. Patent* 3,546,209 (1970).
3. Mahoney, R. *U.S. Patent* 4,020,230 (1977).
4. Mahoney, R. *U.S. Patent* 4,115,492 (1978).
5. Nohmi,T. *U.S. Patent* 4,229,297 (1980).
6. Tanzawa, H. *U.S. Patent* 3,896,061 (1975).
7. Shipman, G.H. *U.S. Patent* 4,539,256 (1985).
8. Vitzthum, G.H.; Davis, M.A. *U.S. Patent* 4,490,431 (1984).
9. Kinzer, K.E. *U.S. Patent* 4,867,881 (1989).
10. McAllister, J.W.; Kinzer, K.E.; Mrozinski, J.E.; Johnson, E.J.; Dyrud, J.F. *U.S. Patent* 4,957,943 (1990).

40

(a) evaporation time=0

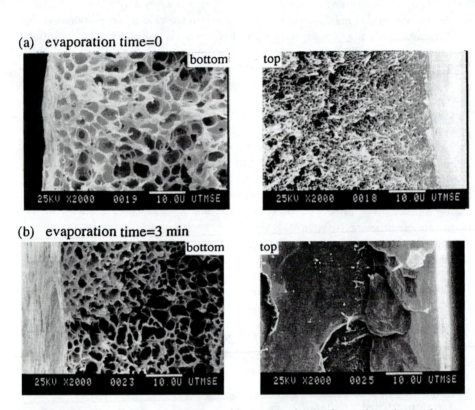

(b) evaporation time=3 min

Figure 10. Detail structures at top and bottom surfaces of samples when only the top surface of the membrane was immersed in ice-water. (a) evaporation time = 0, (b) evaporation time = 3 min. Reproduced with permission from (*44*). Copyright 1998 Elsevier.

11. Mrozinski, J.S. *U.S. Patent* 4,726,989 (1988).
12. Mrozinski, J. S. *U.S. Patent* 4,863,792 (1988).
13. Caneba, G.T.; Soong, D. S. *Macromolecules* **1985**, *18*, 2538.
14. Caneba, G.T.; Soong, D. S. *Macromolecules* **1985**, *18*, 2545.
15. Caneba, G.T.; Soong, D.S. *U.S. Patent* 4,659,470 (1987).
16. Tsay, F.-J.; Torkelson, J.M. *Macromolecules* **1990, 23, 775.**
17. Kesting, R.E. *ACS Symp. Ser.* **1985**, *269*, 131.
18. Yen, L.Y.; Lopatin, G. *U.S. Patent* 4,906,377 (1990).
19. Yen, L.Y.; Lopatin G. *U.S. Patent* 5,032,274 (1991).
20. Berghmans, S.; Berghmans, H.; Meijer, H.E.H. *J Membr. Sci.* **1996**, *116*, 171.
21. Berghmans, S.; Mewis, J.; Berghmans, H.; Meijer, H.E.H. *Polymer* **1995,** *36*, 3085.
22. Caplan, M.R.; Chiang, C.Y.; Lloyd, D.R.; Yen, L.Y. *J Membr. Sci.* **1997,** *130*, 219.
23. McGuire, K. S. *Ph.D. Dissertation,* The University of Texas at Austin, 1995.
24. McGuire, K. S.; Laxminarayan, A.; Lloyd, D.R. *Polymer* **1995,** *36, 495 1.*
25. McGuire, K.S.; Laxminarayan, A.; Martula, D.S.; Lloyd, D.R. *J. Colloid Interface Sci.* **1996,** *182*, 46.
26. Laxminarayan, A. *Ph.D. Dissertation,* The University of Texas at Austin, 1994.
27. Laxminarayan, A.; McGuire, K. S.; Kim, S. S.; Lloyd, D.R. *Polymer* **1994,** *35*, 3060.
28. Chiang, C.-Y. *Ph.D. Dissertation,* The University of Texas at Austin, 1995.
29. Chiang, C.-Y.; Starov, V.M.; Lloyd, D.R. *Colloid J. of Russian Academy of Sci.* **1995,** *57*, 715.
30. Chiang, C.-Y.; Lloyd, D.R. *J Porous Materials* **1996,** *2*, 273.
31. Kim, S. S.; Lloyd, D.R. *J Membr. Sci.* **1991,** *64*, 13.
32. Smolders, C.A.; van Aartsen, J.J.; Steenbergen, A. *Kolloid-Z Z Polymere* **1971,** *243*, 14.
33. Aubert, J.H. *Macromolecules* **1990,** *23*, 1446.
34. Laxminarayan, A. *MS Thesis,* Michigan Technical University, 1990.
35. Lal, J.; Brasil, R. *Macromolecules* **1991,** *24*, 290.
36. Song, S.-W.; Torkelson, J.M. *Macromolecules* **1994,** *27*, 6389.
37. Song, S.-W.; Torkelson, J.M. *J Membr. Sci.* **1995,** *98, 209.*
38. Zeman, L,J.; Zydney, A.L. Microfiltration and Ultrafiltration; Marcel Dekker, Inc.; New York, 1996, pp. 108-114.
39. Tsay, C. S.; McHugh, A.J. *J. Membr. Sci.* **1991,** *64*, 81.
40. Vrentas, J. S.; Duda, J. L. *AIChE J.* **1979,** *25*, 1.
41. Duda, J.L.; Vrentas, J. S.; Ju, S. T.; Liu, H. T. *AIChE J.* **1982,** *28*, 279.
42. Zielinski, J.M.; Duda, J.L. *AIChE J.* **1992,** *38*, 405.
43. Chapman, AJ. Heat Transfer; Macmillan Publishing Co.: New York, 1984, pp. 317-318
44. Matsuyama, H.; Berghmans, S.; Batarseh, M.T.; Lloyd, D.R. *J Membrane Sci.* **1998,** *142*, 27.

Chapter 3

Thermally-Induced Phase Separation Mechanism Study for Membrane Formation

Sung Soo Kim, Min-Oh Yeom, and In-Sok Cho

Department of Chemical Engineering, Institute of Material Science and Technology, Kyung Hee University, Yongin City, Kyunggido 449–701, Korea

The thermally-induced phase separation mechanism was investigated for the system isotactic polypropylene and diphenyl ether. Droplet growth was monitored by using a thermo-optical microscope. The degree of supercooling for liquid-liquid phase separation affected the size and growth rate of the droplets. Increasing the polymer content affected the solution viscosity and interaction parameter which resulted in a decrease of the size and growth rate of the droplets. Crystallization interfered with the droplet growth at lower quench temperatures from the onset of crystallization. The droplet growth rate followed the equation proposed by Furukawa, and the droplet growth was well described by the theories for coarsening and coalescence. Equilibrium droplet size was estimated by the Laplace equation and was in good agreement with the experimental data. The cell size of the membrane was about half of the droplet size of the melt sample due to shrinkage of the sample during extraction of the diluent and the drying process.

Development of semicrystalline polymer membranes is important for many membrane processes because they can withstand high-temperature operating conditions and are resistant to aggressive chemicals (*1*). Moreover, microporous semicrystalline polymer membranes can be used in other applications such as battery separators (*2*). The thermally-induced phase separation (TIPS) process is one of the most promising fabrication methods of making microporous semicrystalline membranes (*3*). Controlling the structure of the membrane made via TIPS requires an understanding of the phase separation mechanisms. Specifically, the liquid-liquid phase separation process induced by the thermodynamic instability

© 2000 American Chemical Society

of the system must be examined in detail for the interpretation of the microcellular structure formation (*4*).

As shown in Figure 1, when the homogeneous melt solution (D) enters the liquid-liquid phase separation region (C), it undergoes spinodal decomposition to form a polymer-rich (R) and a polymer-lean (L) phase (*5,6*). If the sample is held at T, the phase-separated domain (liquid droplet) grows via a coarsening process followed by coalescence of the droplets to form an even greater droplet (*7-9*). The sample is then quenched in ice water to induce crystallization of the polymer, which freezes the structure. Following extraction of the diluent, a microcellular structure is obtained (*10,11*).

In this work, the system isotactic polypropylene (PP) and diphenyl ether (DPE) was investigated for membrane fabrication via TIPS. The PP/DPE system forms a homogeneous melt solution above the melting temperature of the mixture and undergoes liquid-liquid or solid-liquid phase separation depending on the initial polymer concentration. Several factors affecting the phase separation mechanisms were examined, that is, (i) phase separation temperature, (ii) crystallization temperature, and (iii) initial composition of the melt solution. Droplet formation and growth were monitored for interpretation of the phase separation mechanism. The droplet growth kinetics were analyzed and compared with the results of McGuire and Laxminarayan who also studied the TIPS process for the PP/DPE system (*12-15*). The interfacial tension between the phase-separated domains was determined to estimate the equilibrium domain size. The cell size of the membrane was compared with the droplet size of the melt solution to study the droplet growth behavior to the real cell growth behavior.

Experimental

Isotactic PP (H730F) was supplied by SK Co., Korea, and DPE was obtained from Aldrich Co. PP is one of the most preferred materials in this application, because it has advantages over other materials in terms of processability, thermal and chemical resistance, and price. DPE was selected as a diluent in this work, because it forms a homogeneous melt solution with PP at high temperature, which undergoes liquid-liquid phase separation by lowering the temperature. The refractive index of DPE (n_D^{20}=1.579) is different from that of PP (n_D^{20}=1.490), which enabled us to observe the sharp interfaces between the polymer-lean and polymer–rich phases formed by the liquid-liquid phase separation process via optical microscopy. Methanol (Aldrich Co.) was used to extract the DPE after phase separation was completed. PP and DPE were melt blended at 210°C to make homogeneous melt solutions. The sample was quenched into ice water to form solid samples.

Perkin Elmer DSC 7 was used for the thermal analyses of the samples and a Hitachi S-4200 scanning electron microscope (SEM) was used to examine the structure of the membranes. A thermo-optical microscope (TOM) system was assembled as shown in Figure 2, which consisted of a Zeiss Jenaval microscope, a

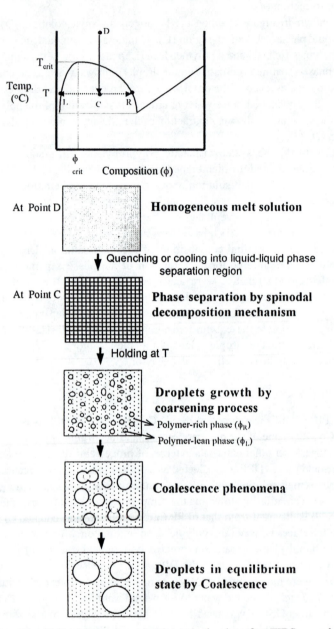

Figure 1.   Liquid-liquid phase separation mechanism for TIPS membrane formation.

Mettler hot-stage (FP-82) with central processor (FP-90), and an image analyzer system (IP lab) equipped with CCD camera, VCR, and video printer. The Laplace equation was used to estimate the equilibrium droplet size. In order to determine the interfacial tension, the surface tensions of the melt solution and diluent were measured by using a modified TOM system, as shown in Figure 3. Each shape of the drop was captured for the calculation of the surface tension by using the pendant drop method proposed by Andreas *et al.* (*16*).

Results and Discussion

The phase diagram of the PP/DPE system was determined experimentally, as shown in Figure 4. Cloud points were determined using the TOM system, and the melting temperatures (T_m) and crystallization temperatures (T_c) were obtained by DSC at a scanning rate of 10°C/min. The melt-blended samples were held at 210°C for 10 min to eliminate the thermal history of the sample before scanning. Figure 4 shows a typical phase diagram for a liquid-liquid phase separation system of a semicrystalline polymer, whose monotectic point, the intersection point of the cloud point curve and crystallization curve, is located at 94°C and 40 wt.% PP at a 10°C/min cooling rate. McGuire *et al.* also generated a phase diagram for the same system which showed similar cloud points as reported here (*12*). However, McGuire *et al.* determined the equilibrium melting temperature, which is much higher than the dynamic T_m and T_c determined in this work. Therefore, the location of the monotectic points should be different from each other due to the different scanning rates. The location of the sample relative to the equilibrium melting temperature in the phase diagram had great influence on the crystallization of PP. When the PP content was lowered to less than 10 wt.%, the sample had poor integrity and could not be used for the determination of the phase diagram.

A homogeneous melt solution of a 10 wt.% PP/90 wt.% DPE sample at 210°C was placed onto the hot-stage preset at 110°C. Placing the 210°C sample on the hot-stage at 110°C made the hot-stage temporarily unstable and it took about 3 seconds for the sample and the hot-stage to reach equilibrium. It was difficult to quantify the cooling rate for this quenching step. The sample was prepared between the cover slips and the edges were sealed with silicon sealant. There was little diluent loss during the experiment, and the weight change before and after the test was less than 3% of the initial weight. The phase separation images at 110°C were captured at different times, as shown in Figure 5. The number labeled in each picture represents the elapsed time after the hot-stage reached equilibrium in hr:min:sec mode. Images of phase separation by spinodal decomposition were observed immediately; however, the onset of spinodal decomposition could not be observed. After 5 minutes, phase-separated droplet growth via coarsening was observed in the optical microscope. The droplets continued to grow until 1 hr and 30 min. In its later stages of liquid-liquid phase separation, coalescence of droplets

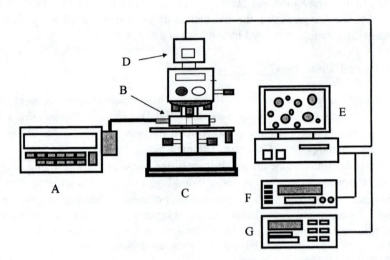

Figure 2   A schematic diagram of a thermo-optical microscope system. A: central processor, B: hot-stage, C: optical microscope, D: CCD camera E: image analyzer system, F: VCR, G: video printer.

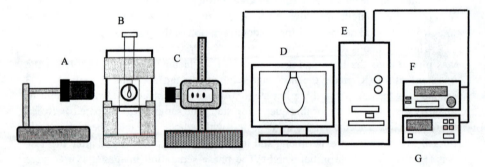

Figure 3.   A schematic diagram of a surface tension measurement system. A: light source, B: pendant drop assembly, C: CCD camera, D: monitor, E: image analyzer system, F: VCR, G: video printer.

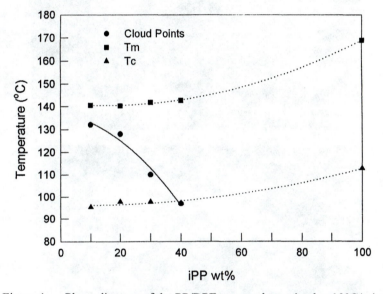

Figure 4.   Phase diagram of the PP/DPE system determined at 10°C/min.

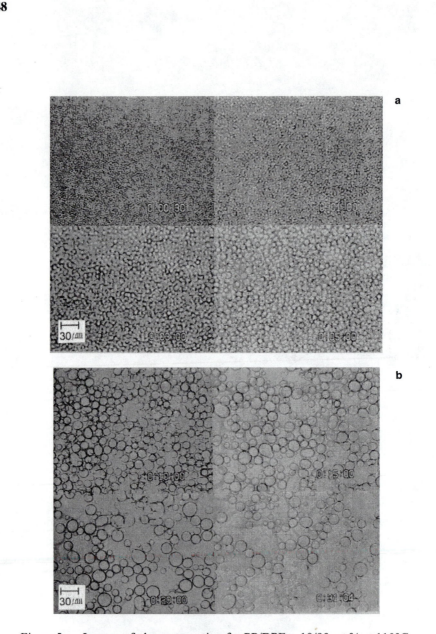

Figure 5.   Images of phase separation for PP/DPE = 10/90 wt.% at 110°C
(numbers in each image represents hr:min:sec).

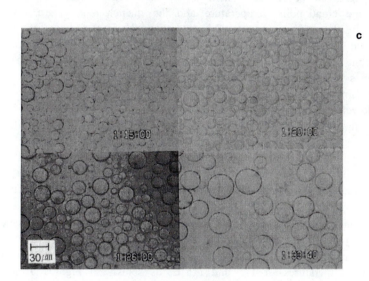

c

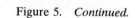

Figure 5. *Continued.*

was observed, as shown in the upper left corner of Figure 6. Coalescence of droplets resulted in large cells after extraction of the diluent.

Similar observations were obtained for other samples with different compositions and at other quench temperatures. Each droplet size was obtained by averaging the sizes of 50 to 60 droplets, and each experiment was repeated three times to ensure the reproducibility of the data. Temperature dependence of the droplet growth was examined in more detail by selecting various quench temperatures from 105°C to 125°C. The temperature dependence of the growth rate is shown in Figures 7 and 8. Our results are similar to those of McGuire *et al.* (*12*). The growth rate and the size of the droplets increased as the quench temperature was lowered. A decrease in quench temperature caused a greater supercooling for liquid-liquid phase separation, which is defined as the temperature difference between the cloud point temperature and the quench temperature. Greater supercooling changed the viscosity of the matrix and the droplet phases, the volume fractions of the two phases, and the interfacial tension. Therefore, increased supercooling accelerated the droplet growth (*17-19*) and the size and growth rate of the droplets increased as the holding temperature was lowered for the 10 and 20 wt.% PP samples.

At lower temperature PP crystallization interfered with the droplet growth. Although the holding temperature was located above the dynamic crystallization curve at a scanning rate of 10°C/min, it is located under the equilibrium melting temperature reported by McGuire *et al.* (*12*). Consequently, there was a thermodynamic driving force for crystallization (*20,21*). The degree of supercooling for crystallization is defined as the temperature difference between the equilibrium melting temperature and the holding temperature. Greater supercooling also accelerated crystallization, and there was no more droplet growth after a certain degree of crystallization, because the crystallization process froze the structure.

The dependence of droplet growth on composition was also explored for phase separation temperatures of 115 and 125°C, as shown in Figures 9 and 10, respectively. The droplet growth behavior depended strongly on the composition. Both the growth rate and the size of the droplet decreased with an increase of PP content at each quench temperature. An increase of polymer concentration increased the solution viscosity, which limited the droplet growth. The interaction parameter between PP and DPE depended on the composition and it also had influence on the droplet growth (*22*). With an increase in polymer concentration the nucleation density for crystallization increased (*20,21*). For a 20 wt.% PP sample the droplet growth was suspended at 20 min (115°C) and 40 min (125°C) by the interference of PP crystallization. Crystallization of PP at 105°C began at 6 minutes and it interfered with the droplet growth beyond this point, as shown in Figure 11. The droplet could not reach the equilibrium size, when the PP composition was high and the crystallization temperature was low.

In order to estimate the equilibrium droplet size we used the Laplace equation

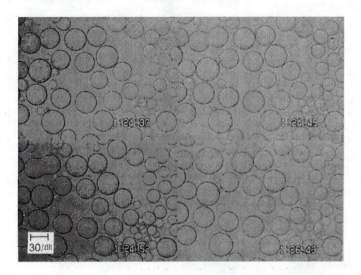

Figure 6.   Coalescence images at the later stage of phase separation.

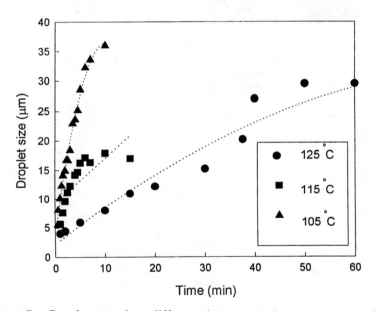

Figure 7.   Droplet growth at different phase separation temperatures for PP/DPE=10/90 wt.%.

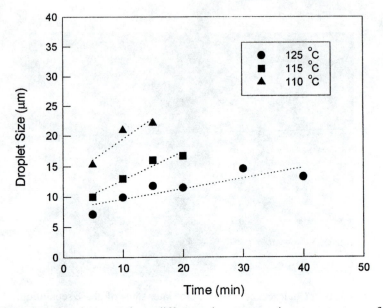

Figure 8. Droplet growth at different phase separation temperatures for PP/DPE=20/80 wt.%.

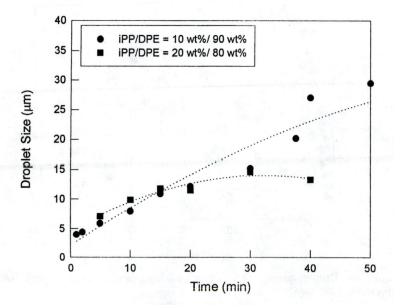

Figure 9. Droplet growth for different PP/DPE compositions at 115°C.

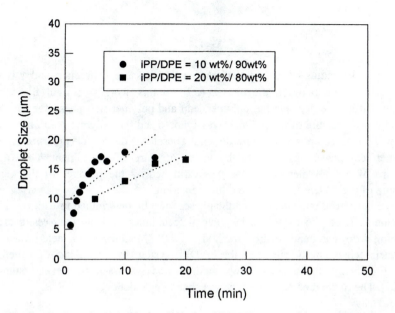

Figure 10. Droplet growth for different PP/DPE compositions at 125°C.

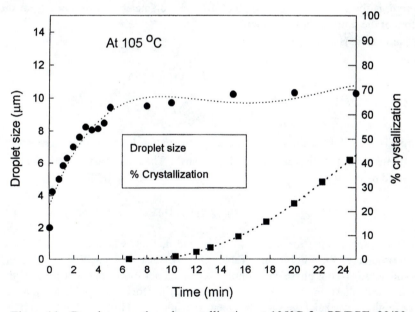

Figure 11. Droplet growth and crystallization at 105°C for PP/DPE=20/80 wt.%.

for a spherical drop:

$$r = 2\gamma / \Delta p \tag{1}$$

where r is the radius of the droplet, Δp is the pressure difference, and γ is the interfacial tension between the phases (23). In this work, we used the vapor pressure difference between the polymer-lean and polymer-rich phase for equation 1. The vapor pressure of the polymer was ignored and the polymer-lean phase was assumed to be composed of pure diluent. Therefore, the vapor pressure of the polymer lean-phase was equal to the vapor pressure of the diluent. The vapor pressure of the polymer-rich phase was estimated by the product of the vapor pressure of the diluent (24) and the diluent fraction.

The interfacial tension between the phases must be determined to estimate r by equation 1. It can be measured by several techniques such as the sessile drop, spinning drop, and pendant drop method (16,23). McGuire et al. determined the interfacial tension of the phases in the PP/DPE system by a spinning drop method (12). Previously, the pendant drop method was discussed by several authors (23,25). The method of Andreas defines a shape-dependent quantity S as:

$$S = d_s / d_e \tag{2}$$

where d_s is the equatorial diameter, and d_e is the diameter at a distance from the bottom of the drop as shown in Figure 12.

The parameters H and β are defined as:

$$H = -\beta (d_e / b)^2 \tag{3}$$

$$\beta = -(\Delta\rho\, g\, b^2) / \gamma \tag{4}$$

where b is the radius of the drop at the apex, $\Delta\rho$ is the density difference between the phases and g is the gravitational constant. Then the surface tension, γ, is expressed by equation 5.

$$\gamma = (\Delta\rho\, g\, d_e^2) / H \tag{5}$$

From the shape of the drop we measured d_e and d_s from which S was calculated. 1/H was obtained from Fordham's Table (25). Because the densities of polymer-rich and polymer-poor phase were similar, we could not form a pendant drop of the polymer-lean phase in the polymer-rich phase. Therefore, we could not directly measure the interfacial tension between the phases. However, for nonpolar compounds Fowke's relationship can be applied, which is expressed as

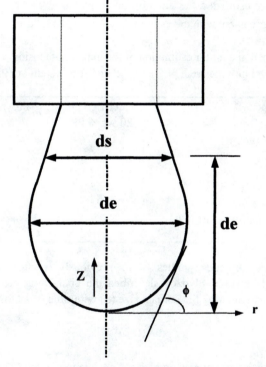

Figure 12. Geometry of the pendant drop in a selected plane.

$$\gamma_{12} = (\gamma_1^{1/2} - \gamma_2^{1/2})^2 \qquad (6)$$

where γ_{12} is the interfacial tension between phases 1 and 2, and γ_1 and γ_2 are the surface tensions of phases 1 and 2, respectively (26).

The PP/DPE system phase separated into a polymer-lean phase ($\phi_p=0$) and a polymer-rich phase ($\phi_p=0.2$) at 125°C. We measured the surface tensions of pure diluent and the 20 wt.% PP sample at 125°C. The interfacial tension between the phases at 125°C was 0.59 dyne/cm, which was higher than that determined by McGuire et al. (12). The experimental data for obtaining the interfacial tension are summarized in Table I. With the interfacial tension value and the vapor pressure of the diluent (24), we estimated the equilibrium droplet size as 12.6 μm at 125°C, which is in good agreement with the experimental value of 13.3 μm.

Table I. Experimental data for estimation of the interfacial tension of polymer-lean phase and polymer-rich phase of the PP/DPE system at 125°C

Property	Polymer-rich phase (20 wt.% PP)	Polymer-lean phase (pure diluent)
Density (g/cm^3)	0.877	0.984
d_e	1.85	1.70
d_s	1.20	1.23
S	0.65	0.72
1/H	0.95	0.73
Surface tension(dyn/cm)	27.9	20.4
Interfacial tension (dyn/cm)	0.59	

Droplet growth kinetics studies were previously performed by several investigators. Lifshitz and Slyozov described the coarsening process using an asymptotic power law relating the domain size (d) with time (t) as follows:

$$d \sim (D\,\xi)^{1/3}\, t^{1/3} \qquad (7)$$

where D is the diffusion coefficient and ξ is the correlation length (27,28).

When two droplets overlap and dissolve at their interface, a larger drop will be formed. This process is referred to as coalescence or collision-combination mechanism. In this case, the size of the dispersed phase (d) is expressed as,

$$d \sim [k_B T/\eta]^{1/3}\, t^{1/3} \qquad (8)$$

where k_B is the Boltzmann constant, T is the phase separation temperature, η is the solution viscosity, and t is time (29-31).

In the bicontinuous phase after phase separation, the pressure gradient between the regions with different curvature causes the hydrodynamic flow to form larger droplets. In this case, the droplet diameter (d) can be expressed as a function of surface tension (σ), viscosity of the continuous phase (η), and time (t) (7,28):

$$d \sim (\sigma / \eta)\, t \tag{9}$$

Recently, Furukawa proposed a theory in which the single length scale was set as a function of time. This is a more general form of the coarsening process, and the length scale (d) is expressed as:

$$d = \kappa \cdot t^{\lambda} \tag{10}$$

where κ is a proportional constant and λ is a scaling exponent or growth exponent which is related to the microscopic particle growth (32). This mechanism can be applied to a phase-separated polymer solution system regardless of whether the system undergoes phase separation by spinodal decomposition or by nucleation and growth.

Furukawa's theory was applied to the systems studied in this work, and the results are shown in Table II.

Table II. Proportional constants and scaling exponents of Furukawa equation for PP/DPE systems at different composition and temperature.

(a) 10 wt.% PP/DPE			
Quench temperature	125°C	115°C	105°C
Proportional constant (κ)	3.54	6.91	9.77
Scaling exponent (λ)	0.38	0.43	0.51

(b) 20 wt% PP/DPE			
Quench temperature	125°C	115°C	110°C
Proportional constant (κ)	4.47	5.37	8.91
Scaling exponent (λ)	0.31	0.36	0.35

As the quench temperature decreased, κ increased, which indicates an increase of the droplet size and its growth rate. The scaling exponent, λ, increased as the PP content decreased from 20 to 10 wt.%. For the 10 wt.% PP sample, λ increased as the quench temperature decreased. On the other hand, λ was essentially constant (~0.3) for the 20 wt.% PP sample independent of the quench temperature, as proposed by Furukawa (32). McGuire et al. also determined the λ value for the PP/DPE system, and showed a similar trend to our data. However, for the 20 wt.% PP sample we obtained a greater value of λ than McGuire (12).

Droplets formed during liquid-liquid phase separation consisted mainly of diluent and the sites occupied by the droplets formed cells after extraction of the diluent. Therefore, the cell size of the membrane was expected to be directly related to the droplet size. The samples on the hot-stage for the droplet growth observation were saved and the diluent was extracted. SEM images of the samples are shown in Figures 13 to 15. As the holding time of the 20 wt.% PP sample at 110°C increased, the cell size increased (Figure 13), which is the same trend for the droplet growth, as shown in Figures 7 to 10. However, there was not much difference between the 10 min and 30 min samples, because PP crystallization interfered with the droplet growth of the sample after 10 minutes.

The effect of quench temperature on the cell size was also examined, as shown in Figure 14. By lowering the holding temperature, the cell size increased as well as the droplet size. The cell size increased until the quench temperature reached 110°C. However, at 105°C the cell size decreased due to crystallization of the polymer, although it had the greatest supercooling for liquid-liquid phase separation of the samples. An increase of polymer concentration reduced the cell size, as shown in Figure 15.

From the images of droplets and SEM images of cells we can conclude that the cell size is directly related to the droplet size. The absolute size of the droplet and the cell of the same sample are compared in Figure 16. The cell size was about half of the droplet size. The smaller cell size resulted from shrinkage of the sample during extraction of the diluent and the subsequent drying process. Factors that determine the degree of shrinkage are: (i) the type of extractant, (ii) the polymer concentration of the sample, and (iii) the evaporation rate of the extractant. A more systematic study on the sample shrinkage is currently being performed and will be reported in a future paper.

Conclusions

As the degree of supercooling for liquid-liquid phase separation increased in the PP/DPE system, the size and growth rate of the droplets increased. However, the degree of supercooling for crystallization also increased and crystallization interfered with the droplet growth at a lower quench temperature. An increase of polymer concentration affected the solution viscosity and interaction parameter, which resulted in a decrease of the size and growth rate of the droplets. The

 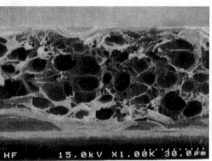

(a) holding time :1min   (c) holding time :10min

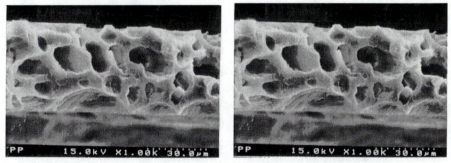

(b) holding time : 5min   (d) holding time :30min

Figure 13. SEM images of membranes held at 110°C for different intervals (PP/DPE = 20/80 wt.%).

60

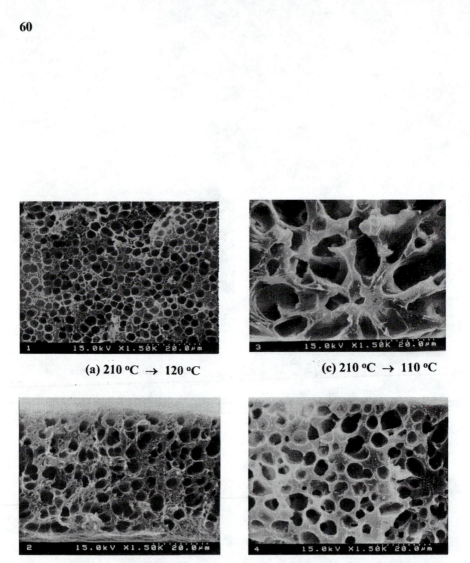

(a) 210 ºC → 120 ºC

(c) 210 ºC → 110 ºC

(b) 210 ºC → 115 ºC

(d) 210 ºC → 105 ºC

Figure 14. SEM images of membranes held at different temperatures for 10 minutes (PP/DPE = 20/80 wt.%).

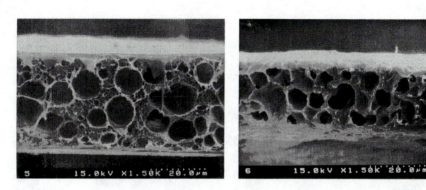

(a) 210 ºC → 115 ºC   **(c) 210 ºC → 115 ºC**

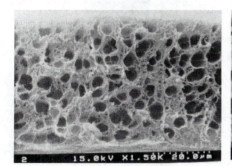 

(b) 210 ºC → 105 ºC   **(d) 210 ºC → 105 ºC**

iPP/DPE = 10/90 wt% sample   **iPP/DPE = 20/80 wt% sample**

Figure 15. SEM images of membranes prepared from different PP/DPE compositions.

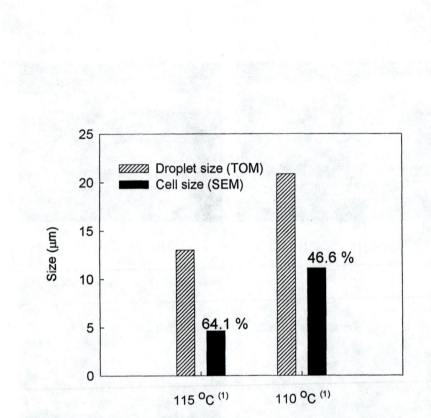

Figure 16. Comparison of droplet and cell sizes of the sample PP/DPE=20/80 wt.% held for 10 min.

equilibrium droplet size was estimated by the Laplace equation. The estimated droplet size was in good agreement with the experimental value. The droplet growth kinetics were examined in terms of the equation proposed by Furukawa. κ increased as quench temperature decreased, and λ increased with an increase of PP content. For a 10 wt.% PP sample, λ increased as the quench temperature decreased. For a 20 wt.% PP sample, λ was essentially constant ($\sim$0.3) for all quench temperatures, as previously proposed by Furukawa. The cell size in the membrane was about half of the droplet size because of shrinkage induced by extraction of the diluent and the subsequent drying process.

Acknowledgements

The authors gratefully acknowledge the financial support of Korea Science and Engineering Foundation (#961-0803-025-2 and #97K3-1005-02-09-3).

Literature Cited

1. Howell, J.A.; Sanchez, V.; Field, R.W. *Membranes in Bioprocessing*; Chapman & Hall: London, 1993.
2. Degen, P.J.; Lee, J. *U.S. Patent 5,492,781* (1996).
3. Kesting, R. E. *Synthetic Polymeric Membranes;* John Wiley & Sons: New York, 1985.
4. Kim, S.S.; Lloyd, D.R. *J. Membrane Sci.* **1991**, *64*, 13.
5. Cahn, J.W. *ACTA Metallurgica* **1961**, *9*, 795.
6. Smolders, C.A.; van Aartsen, J.J.; Steenbergen, A. *Kolloid-Z. u. Z. Polymere* **1971**, *243*, 14.
7. Song, S.W.; Torkelson, J.M. *Macromolecules* **1994**, *27*, 6389.
8. Tsai, F.J.; Torkelson, J.M. *Macromolecules* **1990**, *23*, 775.
9. Nojima, S.; Shiroshita K.; Nose T. *Polymer J.* **1982**, *14*, 289.
10. Kim, S.S.; Lloyd, D.R.; Kinzer, K.E. *J. Membrane Sci.* **1991**, *64*, 1.
11. Lloyd, D.R.; Kinzer, K.E.; Tseng, H.S. *J. Membrane. Sci.* **1990**, 52, 239.
12. McGuire, K.; Laxminarayan, A.; Lloyd, D.R. *Polymer* **1995**, *36*, 4951.
13. McGuire, K.; Laxminarayan, A.; Martula, D.S.; Lloyd, D.R. *J. Colloid & Interface Sci.* **1996**, *182*, 46.
14. Laxminarayan, A. Ph.D. Dissertation, The University of Texas at Austin, 1994.
15. Laxminarayan, A.; McGuire, K.S.; Kim, S.S.; Lloyd, D.R. *Polymer* **1994**, *35*, 3060.
16. Andreas, J.M.; Hauser, E.A.; Tucker, W.B. *J. Phys. Chem.* **1938**, *42*, 1001.
17. Lee, H.K.; Myerson, A.S.; Levon, K. *Macromolecules* **1992**, *25*, 4002.
18. Zryd, I.L.; Burghardt, W.R. *J. Appl. Polym. Sci.* **1995**, *57*, 1525.

19. McGuire, K.S.; Lloyd, D.R.; Lim, G.B.A. *J. Membrane. Sci.* **1993**, *79*, 27.
20. Wang, Y.F.; Lloyd, D.R. *Polymer* **1993**, *34*, 2324.
21. Wang, Y.F.; Lloyd, D.R. *Polymer* **1993**, *34*, 4740.
22. Cha, B.J.; Char, K.; Kim, J.-J.; Kim, S.S.; Kim, C.K. *J. Membrane Sci.* **1995**, *108*, 219.
23. Miller, C.M.; Neogi, P. *Interfacial Phenomena;* Marcel Dekker, Inc., New York, 1985; pp 1-44.
24. *CRC Handbook of Chemistry and Physics, 75th Ed.;* Lide, D.R. Ed.; CRC Press, Boca Raton, 1995.
25. Adamson, A. W. *Physical Chemistry of Surfaces, 5th Ed;* Wiley-Interscience: New York, 1990.
26. Van Krevelen, D.W. *Properties of Polymers, 2nd Ed.*; Elsevier, Amsterdam, 1976.
27. Lifshitz, I.M.; Slyozov, V.V. *J. Phys. Chem. Solids* **1961**, *19*, 35.
28. Siggia, E. D. *Phys. Rev.* **1979**, *A20*, 59.
29. Binder, K.; Stauffer, D. *Phys. Rev. Lett.* **1974**, *33*, 1006.
30. Wong, N.C.; Knobler, C.M. *J. Chem. Phys.* **1978**, 69, 725.
31. Chou, Y.C.; Goldburg, W.I. *Phys. Rev.* **1979**, *A20*, 2105.
32. Furukawa, H. *Adv. Phys.* **1985**, *34*, 703.

Chapter 4

A New Type of Asymmetric Polyimide Gas Separation Membrane Having Ultrathin Skin Layer

Hisao Hachisuka, Tomomi Ohara, and Kenichi Ikeda

Product Development Center, Membrane Sector, Nitto Denko Corporation, 1–1–2 Shimohozumi, Ibaraki, Osaka 567–8680, Japan

A new type of asymmetric gas separation membrane having an ultrathin, essentially defect-free skin layer using diethylene glycol dimethylether (DGDE) as a dope solvent was developed. The asymmetric membrane made of 6FDA-BAAF polyimide consisted of an ultrathin skin layer supported by a sponge-like porous substructure. The skin layer was essentially defect-free and had a thickness of 40-60 nm. The formation of such ultrathin skin layer was contributed to the mixing properties of the dope solvent and water as the quench medium. Solvents such as DGDE are miscible with water; however, a sharp interface between DGDE and water forms during the initial stages of the quench step. The ultrathin skin layer was possibly formed by solidification at such interface.

The preparation of membranes having a thin skin layer is one of the most significant factors to achieving high gas fluxes. In addition, the skin layer of gas separation membranes must be essentially defect-free to obtain a high selectivity. Recently, thick, isotropic films of polyimides containing 2,2-bis(3,4-dicarboxyphenyl) hexafluoropropane dianhydride (6FDA) were reported to exhibit both higher gas selectivity and permeability than previous gas separation membrane materials (1-9). However, the gas fluxes of thick, isotropic films are far too low for industrial applications. For any practical use, membranes with a thin, selective layer must be developed. In particular, thin-skinned asymmetric membranes can have high gas fluxes and mechanical strength due to their structure (10-16). However, membranes with a skin layer thickness of less than 100 nm are very difficult to prepare by conventional methods (10-16). The preparation of a high-performance asymmetric membrane having a skin layer thickness of less than 100 nm was investigated using a new casting solution method. The formation mechanism of such asymmetric membrane was examined in terms of the interaction between the polymer, solvent, and water by using a phase diagram. The gas separation performance of the thin-skinned asymmetric polyimide membrane was determined in pure-gas permeation experiments with nitrogen, oxygen, methane, and carbon dioxide.

Experimental

Preparation and Analysis of Asymmetric 6FDA-BAAF Membranes. The polyimide used in this study was synthesized from 6FDA (Hoechst AG Co., 99% purity) and 2,2-bis(4-amino phenyl) hexafluoropropane (BAAF) (Central Glass Co.,>99% purity) (17). The structure of 6FDA-BAAF is shown in Figure 1. The casting solution was prepared by dissolving the 6FDA-BAAF polyimide in diethylene glycol dimethylether (DGDE). The polymer concentration of the casting solution was 18 wt%. Asymmetric membranes were prepared by casting the 6FDA-BAAF/DGDE dope on a polyester non-woven and were then solidified in water at 30 and 38°C, respectively. Thereafter, the water-wet membranes were dried at 60°C.

The cross-section structure of the membranes prepared at 30°C using 6FDA-BAAF/DGDE dope was studied by field-emission electron microscopy (UHR FE-SEM), as shown in Figure 2. The morphology of the asymmetric 6FDA-BAAF membrane consisted of an ultrathin skin layer supported by a sponge-like porous substructure. The interface between the two layers appears more distinct than that of other asymmetric membranes (10-16). The thickness of the skin layer was about 60 nm and there were no finger voids in the sponge-like porous layer. Asymmetric polyimide membranes were also prepared from a 6FDA-BAAF solution in N-methyl-2-pyrrolidone (NMP), a conventional solvent for membrane preparation. The cross-section structure of the resulting asymmetric membrane consisted of a sponge-like porous layer containing finger voids and without a skin layer, as shown in Figure 3. Previous work showed that asymmetric membranes prepared from low polymer concentration tend to form finger voids or very porous structures (18). In our study, both casting dopes had the same polymer concentration and viscosity. The formation mechanism of their different structures was examined using a phase diagram.

Determination of the Phase Diagram. The phase diagram was determined by titrating water into the 6FDA-BAAF/DGDE dope until the first permanent turbidity was obtained (18). This method gave the boundary of the one-phase and the two-phase region. A small quantity of water was then added and the two phases were separated and analyzed. The 6FDA-BAAF content of the upper phase was determined from the weight after evaporation to dryness. The DGDE and water concentrations were determined using gas chromatography. The 6FDA-BAAF, DGDE, and water concentration of the bottom phase were determined from the mass balance. The 6FDA-BAAF content of the final membranes was determined measuring the weight of the dried sample.

The phase diagram of the 6FDA-BAAF/DGDE/water system is shown in Figure 4. It was previously shown that the 6FDA-BAAF/DGDE dope gelled at a

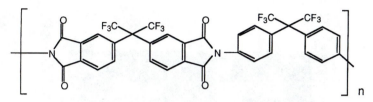

Figure 1. Repeat unit of 6FDA-BAAF polyimide.

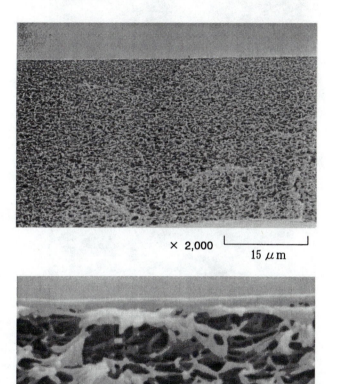

× 2,000

15 μm

×50,000

600 nm

Figure 2. FE-SEM photographs of a cross section of an asymmetric 6FDA-BAAF membrane prepared from DGDE dope at 30°C. (Reproduced with permission from reference 17. Copyright 1996.)

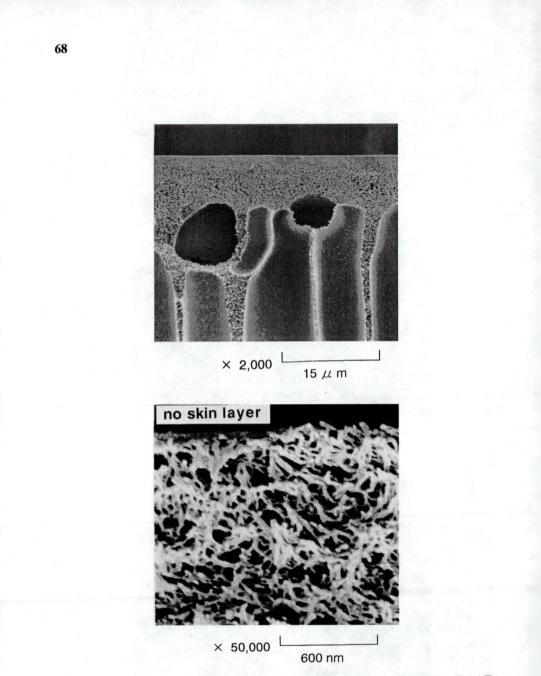

× 2,000

15 μ m

× 50,000

600 nm

Figure 3. FE-SEM photographs of a cross section of an asymmetric 6FDA-BAAF membrane prepared from a NMP-based dope. (Reproduced with permission from reference 17. Copyright 1996.)

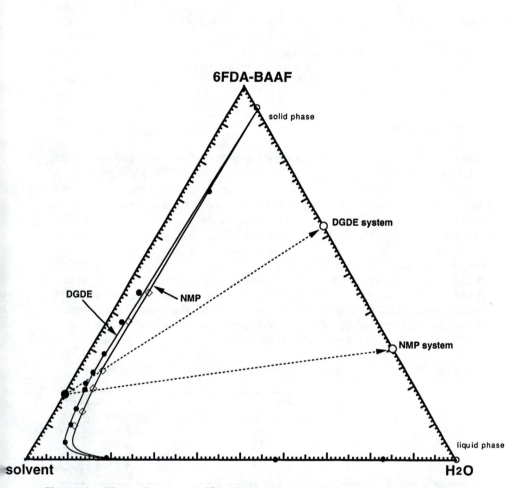

Figure 4. Phase diagram of (●) 6FDA-BAAF/DGDE/water and (◇) 6FDA-BAAF/NMP/water. The broken lines show the influence of solvent on the precipitation paths. (Reproduced with permission from reference 17. Copyright 1996.)

lower water content than that of the conventional cellulose acetate system (*18*). The more hydrophobic nature of the 6FDA-BAAF polyimide may contribute to this behavior. The phase diagram of the 6FDA-BAAF/NMP was also determined by the same method. The phase diagram was almost identical to that of the 6FDA-BAAF/DGDE system, as shown in Figure 4. The composition of the membrane, which is given by the position on the 6FDA-BAAF-water line, consists of a mixture of solid polymer and a liquid phase. The composition of the asymmetric 6FDA-BAAF membrane obtained using DGDE was different from that obtained using NMP as shown in Figure 4. This difference can depend on the path from the 6FDA-BAAF/DGDE dope or 6FDA-BAAF/NMP dope to the final membrane composition.

Results and Discussion

Influence of Solvent Properties on Membrane Structure. It is well known that polar solvents such as NMP or N,N-dimethyl formamide (DMF) can generally be used as a suitable polymer solvent to form asymmetric membranes. Such solvent dissolves in water easily and mixing occurs immediately. On the other hand, mixing of DGDE and water occurs after a delay period. A sharp interface between DGDE and water is formed for a short period when DGDE is poured slowly on water, as shown in Figure 5(a). Such sharp interface is not formed in the case of the NMP/water system, as shown in Figure 5(b). This result suggests that 6FDA-BAAF may solidify at the sharp interface formed between the 6FDA-BAAF/DGDE dope and water before mixing of DGDE and water occurs. The difference between DGDE and NMP may be correlated with the dielectric constant and/or dipole moment of the solvents. The dielectric constant and dipole moment of DGDE and other solvents are listed in Table I.

Table I. Comparison of dielectric constant and/or dipole moment of solvents.

Solvent	Dielectric constant (-)	Dipole moment (debye)
DGDE	5.97	1.97
NMP	32.0	4.00
water	78.3	1.94
n-hexane	1.98	0.08

The dielectric constant of DGDE is 13-fold lower than that of water but is only 3-fold higher than that of n-hexane which is a very hydrophobic solvent. On the other hand, the dipole moment of DGDE is much larger than that of n-hexane. Although DGDE is infinitely miscible with water it is thermodynamically more difficult to dissolve DGDE in water as compared with NMP. As a result, a sharp interface between DGDE and water is formed during the initial stages of the solvent/nonsolvent exchange. It was also confirmed that similar solvents, such as diethylene glycol diethylether, can also form asymmetric membranes having an

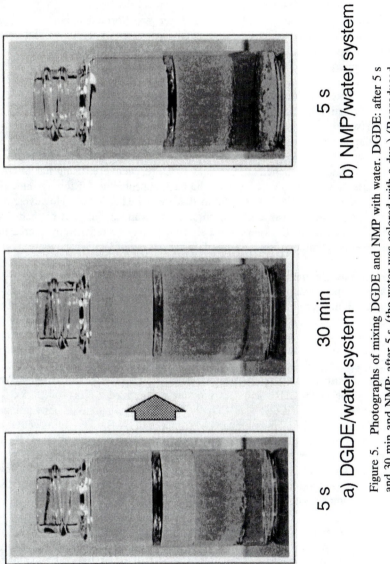

5 s 30 min 5 s

a) DGDE/water system b) NMP/water system

Figure 5. Photographs of mixing DGDE and NMP with water. DGDE: after 5 s and 30 min and NMP: after 5 s. (the water was colored with a dye.) (Reproduced with permission from reference 17. Copyright 1996.)

ultrathin skin layer. The solidification of the interfacial layer of the 6FDA-BAAF/DGDE dope on a water surface is compared with that of the 6FDA-BAAF/NMP dope in Figure 6. The interfacial layer of the 6FDA-BAAF/DGDE dope in water was clear and transparent suggesting that an ultrathin and dense layer was formed. On the other hand, the interfacial layer of the 6FDA-BAAF/NMP dope in water was opaque suggesting that a porous layer was formed.

The sharp interface between DGDE and water was formed only for a short period and then the DGDE dissolved completely in water. After the initial delay period and formation of the dense skin layer, liquid-liquid phase separation occurred in the bulk membrane forming the porous substructure.

Influence of Quench Medium Temperature on Skin Layer. As mentioned above, the formation of an ultrathin skin layer depends on the slow mixing of DGDE and water. In general, the solubility and mixing kinetics increase by increasing the temperature. Therefore, the thickness of the skin layer can be controlled by the quench medium temperature. The cross-section structures of polyimide membranes quenched at 30°C and 38°C, respectively, using the 6FDA-BAAF/DGDE dope are shown Figure 7. The basic morphology of the asymmetric membrane quenched at 38°C was similar to that quenched at 30°C. However, the thickness of the skin layer was about 40 nm and was thinner than that quenched at 30°C (~60 nm). Thus, the thickness of the skin layer may be reduced by increasing the quench medium temperature, because of faster mixing of DGDE and water.

Gas Permeation Properties of Asymmetric 6FDA-BAAF Membranes. The steady-state gas permeation properties of the membranes were measured by using the constant volume/variable pressure method at 25°C. A membrane was placed in a permeation cell and the permeation apparatus was evacuated by a vacuum pump. The feed pressure was 3 kgf/cm^2 and the permeate pressure was always less than 10 mmHg. The permeation rate was determined using a Baratron pressure transducer (Type 122A) and digital equipment with a recorder.

The asymmetric 6FDA-BAAF membranes mentioned above had a selectivity (α) of 30 for CO_2/CH_4 and a pressure-normalized flux of 4.8×10^{-4} cm^3(STP)/cm^2·s·cmHg for CO_2. This CO_2 flux is higher than that of conventional commercial gas separation membranes. Isotropic 6FDA-BAAF films have a selectivity of about 50 for CO_2/CH_4 and a CO_2 permeability coefficient, P, of 50×10^{-10} cm^3(STP)·cm/cm^2·s·cmHg. The asymmetric membranes had a selectivity of less than 50 for CO_2/CH_4. It is evident, therefore, that the skin layer contained a few defects. Recent simulations showed that the defect area in the skin layer of the polyimide membranes is about 10^{-6}% (19).

Casting of 6FDA-BAAF Membranes on Commercial-Sized Equipment. The asymmetric 6FDA-BAAF membrane was also prepared by a continuous casting process on commercial-sized equipment (1 m wide and 60 m long). The quench temperature was 30°C. The performance of the resulting membrane (sample size: 1m x 1m) is shown in Figure 8. The carbon dioxide/methane selectivity was at least 20 and the average CO_2/CH_4 selectivity was 33. The performance of a machine-cast membrane made at a quench temperature of 38°C is shown in Figure 9. The pressure-normalized fluxes were higher than those of polyimide membranes

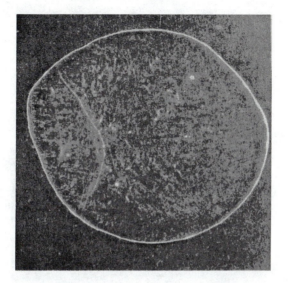

5 s
a) DGDE dope in water

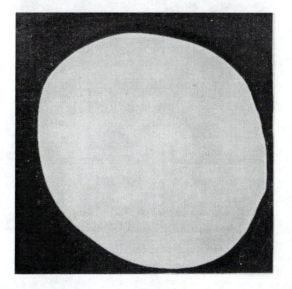

5 s
b) NMP dope in water

Figure 6. Surface photographs of a 6FDA-BAAF/DGDE dope and a 6FDA-BAAF/NMP dope in water after 5 s. (Reproduced with permission from reference 17. Copyright 1996.)

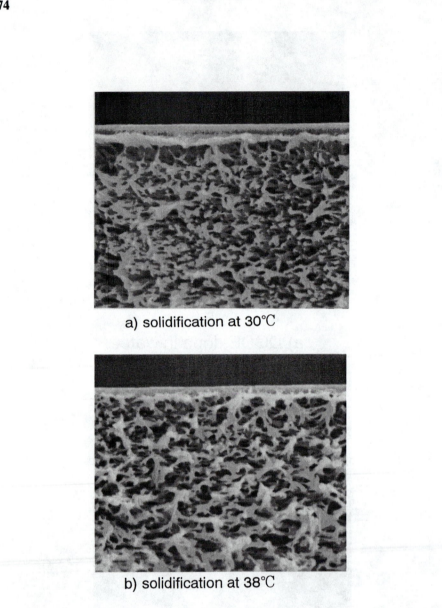

a) solidification at 30°C

b) solidification at 38°C

× 20,000   1.2 μm

Figure 7. FE-SEM photographs of cross sections of asymmetric 6FDA-BAAF membranes quenched in water at 38°C and at 30°C.

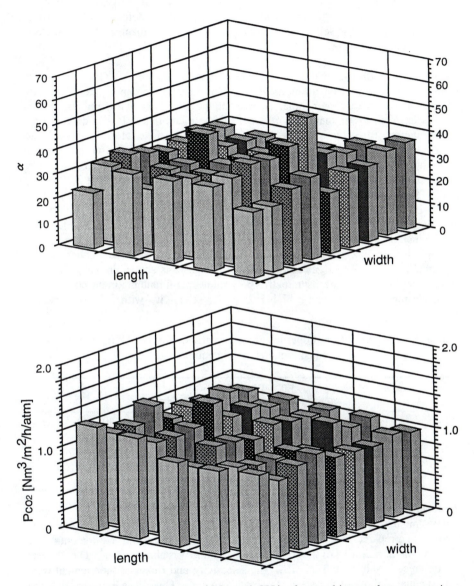

Figure 8. Permeation properties (CO_2 and CH_4) of a machine-made asymmetric 6FDA-BAAF membrane (quenched in water at 30°C). Sample size: 1m × 1m.

quenched at 30°C. This increase in flux can be attributed to a thinner skin layer as shown in Figure 7. On the other hand, the average CO_2/CH_4 selectivity decreased to 22. This decrease in selectivity can be attributed to a slight increase of skin layer defects. However, large-sized and/or more numerous defects, which would drastically decrease the selectivity, were not formed by the machine-casting method on a manufacturing scale.

Influence of Silicone Coating on Membrane Properties. Gases permeate orders of magnitude faster through a defect than through the dense part of the skin layer of an asymmetric membrane. As a result, the selectivity of a membrane having defects is much lower than that of a defect-free membrane. If the defects of a membrane are filled by a highly permeable polymer material it is possible to improve the selectivity of the asymmetric membrane (*20*). The asymmetric membrane of this study had defects as mentioned above. A silicone rubber layer was applied to fill the small number of defects. As silicone rubber ($P = 2,000 \times 10^{-10}$ $cm^3(STP) \cdot cm/cm^2 \cdot s \cdot cmHg$ for CO_2) has a much higher gas permeability than 6FDA-BAAF, the decrease in carbon dioxide flux caused by the filling of defects may be small. A simulation using the resistance model suggested that the selectivity can be increased by filling the defects with silicone rubber without decreasing the carbon dioxide flux drastically. The selectivity of a silicone rubber coated asymmetric 6FDA-BAAF membrane for various gases is shown in Table II. The results confirmed that the asymmetric polyimide membrane made on commercial-sized casting equipment had a higher CO_2/CH_4 selectivity with a silicone rubber coating.

Table II. Gas permeation properties of asymmetric 6FDA-BAAF membrane made on continuous casting equipment.

Gas	Pressure-normalized flux ($10^{-6} cm^3(STP)/cm^2 \cdot s \cdot cmHg$)	Selectivity (α)
CO_2	370	40 (CO_2/CH_4)
		23 (CO_2/N_2)
CH_4	9.1	-
O_2	66	4.2 (O_2/N_2)
N_2	16	-

Performance of a Polyimide Spiral-Wound Membrane Element. A spiral-wound membrane element was fabricated using the asymmetric 6FDA-BAAF membrane prepared on continuous, commercial-sized casting equipment. The effective membrane area of the spiral-wound module was about 1.0 m^2. The element performance is shown in Table III. The selectivity and fluxes of the element were almost equal to those of flat-sheet membrane samples. Thus, it is expected that the asymmetric 6FDA-BAAF gas separation membrane can be utilized for various gas separation applications at the industrial scale in spiral-wound module configuration.

Conclusions

Asymmetric 6FDA-BAAF polyimide membranes with an ultrathin (~ 40-60 nm), essentially defect-free skin layer and a macrovoid-free, sponge-like substructure

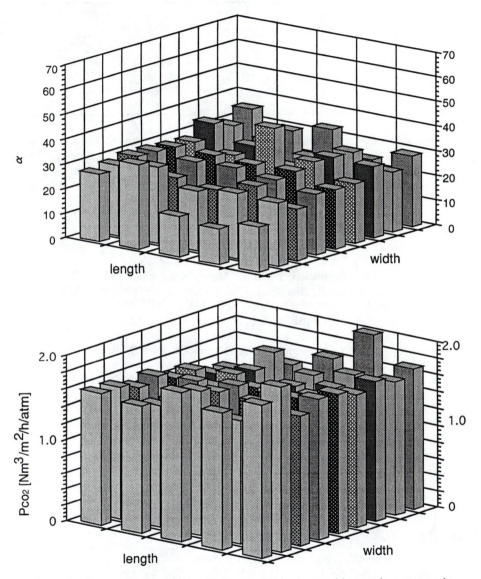

Figure 9. Permeation properties (CO_2 and CH_4) of a machine-made asymmetric 6FDA-BAAF membrane (quenched in water at 38°C). Sample size: 1m × 1m.

Table III. Gas permeation properties of a 6FDA-BAAF spiral-wound
membrane element for various Gases.

Gas	Gas flux $(m^3(STP)/h \cdot atm)$	Selectivity (α)
CO_2	0.96	38 (CO_2/CH_4)
		21 (CO_2/N_2)
CH_4	0.025	-
O_2	0.2	4.3 (O_2/N_2)
N_2	0.046	-

were made by a phase inversion process using diethylene glycol dimethylether
(DGDE) as the polymer solvent and water as the quench medium. The formation of
the ultrathin skin layer can be explained by the slow mixing of DGDE and water
during the initial stages of the quench process. The ultrathin skin layer was formed
by polymer solidification without phase separation at such interface.

Literature Cited

1.   Matsumoto, K.; Xu, P. *J. Appl. Polym. Sci.* **1993**, *47*, 1961.
2.   Matsumoto, K; Minamizaki, Y.; Xu,P. *Maku (Membrane)* **1992**, *17*, 395.
3.   Matsumoto, K.; Xu, P.; Nishikimi, T. *J. Membrane Sci.* **1993**, *81*, 15.
4.   Matsumoto, K.; Xu, P. *J. Membrane Sci.* **1993**, *81*, 23.
5.   Kim, T.H.; Koros, W.J.; Husk, G.R.; O'Brien, K.C. *J. Membrane Sci.* **1988**, *37*, 45.
6.   Stern, S.A.; Mi, Y.; Yamamoto, H. *J. Polym. Sci., Part B: Polym. Phys. Ed.* **1989**, *27*, 1887.
7.   Kim, T.H.; Koros, W.J.; Husk, G.R. *J. Membrane Sci.* **1989**, *46*, 43.
8.   Coleman, M.R; Koros, W.J. *J. Membrane Sci.* **1990**, *50*, 285.
9.   Tanaka, K.; Kita, H.; Okano, M.; Okamoto, K. *Polymer* **1992**, *33*, 585.
10. Chung, T.S.; Kafchinski, E.R.; Vora, R. *J. Membrane Sci.* **1994**, *88*, 21.
11. Loeb.S.; Sourirajan S. *Adv. Chem Ser.* **1963**, *38*, 117.
12. Fritzsche, A.K.; Murphy, M.K.; Cruse, C.A.; Malon, R.F.; Kesting, R.E.; *Gas Sep Purif.* **1989**, *3*, 106.
13. Kesting, R.E.; Fritzsche, A.K.; Cruse, C.A.; Moore, M.D. *J. Appl. Polym. Sci.* **1990**, *40*, 1557.
14. Fritzsche, A.K.; Cruse, C.A.; Kesting, R.E.; Murphy, M.K.; *J. Appl. Polym. Sci.* **1990**, *41*, 713.
15. Kesting, R.E.; Fritzsche, A.K.; Murphy, M.K.; Handermann, A.C.; Cruse, C.A.; Malon, R.F. *U.S. Patent* 4,871,494, 1989.
16. Chung, T.S.; Kafchinski, E.R.; Foley, P. *J. Membrane Sci.* **1993**, *75*, 181.
17. Hachisuka, H.; Ohara T.; Ikeda, K. *J. Membrane Sci.* **1996**, *116*, 265.
18. Strathmann, H.; Scheible, P.; Baker, R.W. *J. Appl. Polym. Sci.* **1971**, *15*, 811.
19. Hachisuka, H.; Ohara T.; Ikeda, K. *J. Appl. Polym. Sci.* **1996**, *61*, 1615.
20. Henis, J.M.S; Tripodi, M.K. *Sep.Sci.Techn.* **1980**, *15*, 1059.

Chapter 5

Influence of Surface Skin Layer of Asymmetric Polyimide Membrane on Gas Permselectivity

Hiroyoshi Kawakami and Shoji Nagaoka

Department of Applied Chemistry, Tokyo Metropolitan University, Hachioji, Tokyo 192–0397, Japan

Asymmetric 6FDA-APPS polyimide membranes were made by a dry/wet phase inversion technique. The membranes had high gas selectivity without any additional coating of the skin layer. The apparent skin layer thickness depended strongly on evaporation time, which was of great importance for the formation of an asymmetric membrane with high gas permeance. The surface roughness of the membranes, analyzed from atomic force microscopy (AFM) micrographs, increased with decreasing the apparent skin layer thickness. The permeances of the asymmetric membranes for CO_2, O_2, N_2, and CH_4 were measured at 35°C and 76 cmHg. The asymmetric membranes exhibited gas selectivities equal to or even higher than those determined for dense, isotropic 6FDA-APPS membranes.

Recently, aromatic polyimides have been identified as membrane materials with high gas selectivity for gas pairs such as H_2/N_2, O_2/N_2, and CO_2/CH_4 (*1-3*). There have been many studies of novel, structurally modified polyimides to enhance gas permeance and selectivity. However, the permeance of most polyimides is low; therefore, it is highly desirable to develop a polyimide membrane that provides higher gas permeance. The most important factor for enhancing the gas permeance is to make the membrane as thin as possible without introducing defects or pinholes which will reduce the selectivity significantly.

To minimize thickness, an asymmetric membrane is preferred with a defect-free ultrathin skin layer and a porous substructure to support the skin layer during high-pressure operation (*4,5*). The phase separation method is the most widely used technique for fabricating asymmetric membranes, and such membranes are prepared by controlled phase separation of polymer solutions into two phases with high and low polymer concentrations (*6,7*). In recent years, Pinnau and Koros reported that defect-free asymmetric membranes with skin layer thicknesses as thin as 20 nm can be formed by a dry/wet phase inversion process and that phase separation is induced in the outermost region of the cast membrane during the evaporation step (8,9). In this method, an evaporation step prior to immersion in the coagulation medium is necessary for the formation of a defect-free skin layer.

In this paper, we describe the effect of the surface skin layer of an asymmetric polyimide membrane on its gas transport properties. A number of asymmetric membranes were prepared by dry/wet phase inversion, and the permeances of the membranes to CO_2, O_2, N_2, and CH_4 were determined at 35°C and 76 cmHg. To investigate the influence of the evaporation kinetics on the surface skin layer properties, a series of asymmetric polyimide membranes were prepared using different evaporation times.

Experimental

Preparation of Asymmetric Membranes. 6FDA-APPS was synthesized from 2,2'-bis(3,4-dicarboxyphenyl)hexafluoropropane dianhydride (6FDA) and bis[4-(4-aminophenoxy)phenyl]sulfone (APPS) (Figure 1). Asymmetric 6FDA-APPS membranes were made by a dry/wet phase inversion technique (10,11). The composition of the casting solution used for preparation of asymmetric membranes was 12 wt% polyimide, 55 wt% methylene chloride, 23 wt% 1,1,2-trichloroethane, and 10 wt% butanol. High volatility methylene chloride and low volatility 1,1,2-trichloroethane were good solvents for 6FDA-APPS and butanol was used as a nonsolvent additive. The polymer solutions were filtered and subsequently degassed. The solutions were cast on glass plates with a knife gap of 250 μm and then air-dried for 15-600 sec. After evaporation, the membranes were coagulated in methanol, washed in methanol for 12 hours, air-dried for 24 hours, and finally dried in a vacuum oven at 150°C for 15 hours to remove any remaining solvent.

Structure of Asymmetric Membrane. The cross-sections of the asymmetric membranes were observed with a scanning electron microscope (SEM: JXP-6100P, JEOL). The membranes were cryogenically fractured in liquid nitrogen and then coated with Pt/Pd.

The surface morphology of asymmetric membranes was visualized using an atomic force microscope (AFM: Seiko SPI3700) in air at room temperature. The cantilevers (Seiko SN-AF01), with a spring constant of 0.38 N/m, were microfabricated from silicon nitride. The surface was continuously imaged in the feedback mode with a scan area of 500 nm x 500 nm and a constant scan speed of 2 Hz. The surface roughness (R_z) of the membranes was characterized by the average vertical peak to valley distance.

Permeability Measurements. The gases used in this study (carbon dioxide, oxygen, nitrogen, and methane) had a purity of at least 99.99% and were used without further purification.

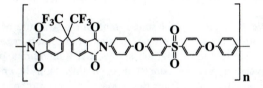

Figure 1.   Structure of 6FDA-APPS polyimide.

Asymmetric 6FDA-APPS membranes were mounted in a permeation cell with a surface area of 1.0 cm^2. Gas permeances at 76 cmHg were determined with a high vacuum apparatus (Rika Seiki Inc., K-315-H). The pressures on the upstream and downstream sides were measured using a Baratron absolute pressure gauge. The error in the permeance was estimated to be ±0.1 to 0.5%. The apparent skin layer thickness of a defect-free asymmetric membrane was calculated from the known permeability coefficient and permeance using

$$L = P / Q \tag{1}$$

where L is the apparent skin layer thickness, P is the gas permeability coefficient as determined for a dense, isotropic film of known thickness $(cm^3(STP) \cdot cm/cm^2 \cdot s \cdot cmHg)$, and Q is the gas permeance of the asymmetric membrane $(cm^3(STP)/cm^2 \cdot s \cdot cmHg)$.

Results and Discussion

Thermogravimetric analysis (TGA) was used to test the membranes for residual solvent. The membranes showed no weight loss below 450°C in nitrogen. This result indicates that the membrane preparation protocol completely removed all solvent.

A representative scanning electron micrograph of the structure of an asymmetric 6FDA-APPS polyimide membrane prepared after an evaporation time of 15 sec is shown in Figure 2. The photomicrograph shows that a membrane prepared by the dry/wet phase inversion process consisted of a thin skin layer and a porous substructure. The apparent skin layer thickness of the asymmetric membrane was calculated from equation 1. The thickness was determined based on the oxygen permeability coefficient of a dense 6FDA-APPS polyimide film. The thickness of membranes coagulated after an evaporation time of 15-600 sec increased with an increase in evaporation time.

Pinnau *et al.* proposed a mechanism for the formation of the surface skin layers of asymmetric membranes made by dry/wet phase inversion (*12*). It was suggested that the outermost region of the membrane undergoes phase separation by spinodal decomposition during the initial stages of the evaporation process. Furthermore, the evaporation step prior to immersion of the membrane in the nonsolvent strongly influences the surface skin layer thickness and is also necessary for the formation of the defect-free skin (*12*).

Differences in the morphology were evaluated by the surface roughness (R_z), which is the mean difference between the ten highest peaks and the ten lowest valleys as measured by AFM. Figure 3 shows the relationship between the apparent skin layer thickness and R_z for asymmetric membranes prepared by dry/wet phase inversion. R_z decreased as the skin layer thickness increased. Interestingly, the surface of a dense membrane showed a uniform nodular structure ($R_z = 0.89$ nm). The average nodule size and surface morphology of the dense membrane were similar to that of the asymmetric membranes coagulated after

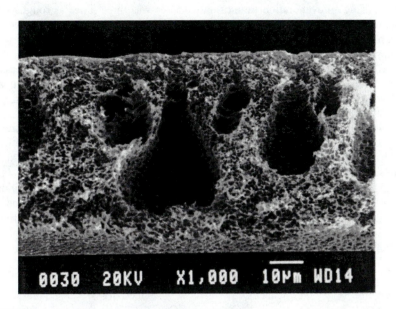

Figure 2.   Cross-section of an asymmetric 6FDA-APPS membrane made by dry/wet phase inversion. Evaporation time: 15 sec.

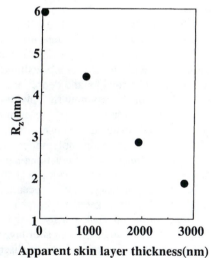

Apparent skin layer thickness(nm)

Figure 3.   Dependence of R_z on apparent skin layer thickness for asymmetric 6FDA-APPS membranes.

evaporation times longer than 120 sec. The surface of the asymmetric membrane prepared using an evaporation time of only 15 sec was three times as rough as the membrane surface with an evaporation time of 600 sec. A faster exchange rate between solvent and nonsolvent at the membrane surface with shorter evaporation time appears to be responsible for the higher roughness of the membrane. We believe that the wet process in the nonsolvent determines the surface roughness of the membrane. However, the influence of the exchange rate between solvent and nonsolvent on the surface morphology of the membrane must be elucidated in a future study.

Table I shows gas permeances and selectivities of the 6FDA-APPS membranes. The selectivities for O_2/N_2 and CO_2/CH_4 in the asymmetric membrane with an apparent skin layer thickness of 25 nm were 6.3 with an O_2 permeance of 7.7 x 10^{-5} $[cm^3(STP)/(cm^2 \cdot s \cdot cmHg)]$ and 40 with a CO_2 permeance of 3.5 x 10^{-4} $[cm^3(STP)/(cm^2 \cdot s \cdot cmHg)]$, respectively, without the necessity of an additional coating to seal any possible defects. These gas permeances were approximately 2000 times higher than those determined for the dense, isotropic 6FDA-APPS film.

Table I. Gas permeance and selectivity of 6FDA-APPS polyimide membranes at 35°C and 76 cmHg.

Membrane type	Permeance Q*		Selectivity		Thickness (nm)
	CO_2	O_2	CO_2/CH_4	O_2/N_2	
Asymmetric	350	77	40	6.3	25
Dense film	0.182	0.038	36	6.1	50,000

*Q in 10^{-6} $cm^3(STP)/cm^2 \cdot s \cdot cmHg$

Figure 4 shows the relationship between the CO_2/CH_4 selectivity and the apparent skin layer thickness. The gas selectivity and permeance of the asymmetric membranes is presented in Figure 5. The selectivity of the asymmetric membranes increased with decreasing thickness, and the membranes exhibited gas selectivities equal to or even higher than those determined for the dense membrane. This result indicates that the surface skin layer of the asymmetric polyimide membrane was essentially defect-free. Therefore, the gas transport of the asymmetric membranes occurred predominantly by a solution/diffusion mechanism.

Both structure and dynamics of aromatic polyimides are influenced by intermolecular and intramolecular charge transfer interactions between the polymer chains, because aromatic polyimides contain an alternating sequence of electron-rich donor and electron-deficient acceptor moieties (13,14). Recently, we reported that asymmetric polyimide membranes prepared by dry/wet phase inversion exhibit stronger charge transfer complex formation than a dense polyimide film (10,11). This means that the packed polyimide structure in the skin layer is formed by the charge transfer complex and provides high size and shape discrimination for gas molecules. The skin layer of the asymmetric membrane coagulated after a short evaporation time might form a more efficiently packed structure relative to that of membranes prepared with longer evaporation times. However, these arguments

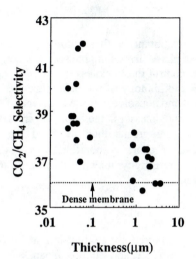

Figure 4. Relationship between CO_2/CH_4 selectivity and apparent skin layer thickness for asymmetric 6FDA-APPS membranes at 35°C and 76 cmHg.

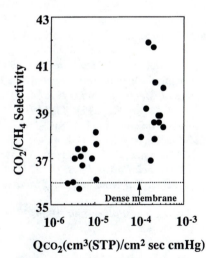

Figure 5. Relationship between CO_2/CH_4 selectivity and CO_2 permeance for asymmetric 6FDA-APPS membranes at 35°C and 76 cmHg.

require validation by UV-visible or fluorescence spectroscopy of the surface skin layer.

Conclusions

Asymmetric polyimide membranes with an ultrathin, defect-free skin layer were prepared by the dry/wet phase inversion process. By controlling the evaporation time, it was possible to control the thickness of the skin layer of the asymmetric membrane. The surface morphology of the membrane was investigated by AFM. The exchange of solvent and nonsolvent at the interface of the coagulation medium determined the surface roughness of the skin layer. Gas selectivities of the asymmetric membranes were greater than those of the dense polyimide film. Membranes coagulated after short evaporation times exhibited the highest gas selectivities. The higher selectivity may result from a more efficiently packed structure in the skin layer.

Acknowledgements

This work was partially supported by a Grant-in Aid for Scientific Research from the Ministry of Education, Science, and Culture, Japan.

Literature Cited

1. Kim, T.H.; Koros,W.J.; Husk, G.B.; O'Brien, K.C. *J. Membrane Sci.* **1988**, *37*, 45.
2. Stern, S.A.; Mi, Y.; Yamamoto, H. *J. Polym. Sci., Polym. Phys.* **1989**, *27*, 1887.
3. Kawakami, H.; Mikawa, M.; Nagaoka, S. *J. Appl. Polym. Sci.* **1996**, *62*, 965.
4. Kimura, S. G. *U. S. Patent* 3,709,774 (1973).
5. Chung, T.S.; Kafchinski, E.R.; Foley, P. *J. Membrane Sci.* **1992**, *75*, 181.
6. Pinnau, I.; Wind, J.; Peinemann, K.-V. *Ind. Eng. Chem. Res.* **1990**, *29*, 2028.
7. Lai, J.Y.; Liu, M.J.; Lee, K.R. *J. Membrane Sci.* **1994**, *86*, 103.
8. Pinnau, I.; Koros, W.J. *J. Appl. Polym. Sci.* **1991**, *43*, 1491.
9. Pinnau, I.; Koros, W.J. *J. Membrane Sci.* **1992**, *7*, 81.
10. Kawakami, H.; Anzai, J.; Nagaoka, S. *J. Appl. Polym. Sci.* **1995**, *57*, 789.
11. Kawakami, H.; Mikawa, M.; Nagaoka, S. *J. Membrane Sci.* **1996**, *118*, 223.
12. Pinnau, I.; Koros, W.J. *J. Polym. Sci. Polym. Phys.* **1993**, *31*, 419.
13. Ishida, H.; Wellinghoff, S.T.; Baer E.; Koening, J.L. *Macromolecules* **1980**, *13*, 826.
14. Hasegawa, M.; Mita, I. *J. Polym. Sci., Polym. Lett.* **1989**, *27*, 263.

Chapter 6

Effect of Surfactant as an Additive on the Formation of Asymmetric Polysulfone Membranes for Gas Separation

A. Yamasaki[1], R. K. Tyagi[2], A. E. Fouda[2], K. Jonnason[2], and T. Matsuura[3]

[1]National Institute of Materials and Chemical Research,
1–1 Higashi Tsukuba 305, Japan
[2]Institute of Chemical Process and Environmental Technology, National
Research Council of Canada, Ottawa K1A 0R6, Canada
[3]Industrial Membrane Research Institute, Department of Chemical
Engineering, University of Ottawa, Ottawa K1N 6N5, Canada

The addition of a surfactant, sodium dodecyl sulfate (SDS), to the casting solution increased the oxygen permeance through asymmetric polysulfone membranes significantly, while maintaining relatively high selectivity of oxygen over nitrogen. A phase inversion technique with partial solvent evaporation (0 to 15 min at 95°C) and a dual coagulation step using 2-propanol and water was used to prepare the asymmetric polysulfone membranes. The content of SDS in the solution was varied from 0 to 2.0 wt.%. The oxygen permeance showed a maximum at 1.0 wt.% of SDS for a given solvent evaporation time. Scanning electron microscopy (SEM) indicated that the skin layer growth was very limited with an increase in the solvent evaporation time for the membranes prepared with SDS.

The phase inversion technique is a common method for the preparation of asymmetric membranes (*1-4*). Phase inversion membranes for gas separation are generally comprised of an ultrathin, dense skin layer and a porous substructure. The gas permeation properties of an asymmetric membrane are determined primarily the skin layer because most of the permeation resistance is concentrated in the skin layer. For an ideal phase inversion membrane, the skin layer should be as thin as possible to obtain a high flux. In addition, the skin layer must be defect-free to obtain the intrinsic selectivity of the membrane material. Optimization of phase inversion membranes is very difficult because many parameters are involved in the membrane formation process (*1-4*).

In previous work (*5*), we studied the growth of the skin layer during the phase inversion process. A two-step coagulation process using 2-propanol and water, respectively, and partial solvent evaporation before coagulation, were used for membrane formation. The two-step coagulation process is based on the concept proposed by van't Hof et al. (6), except for the additional solvent evaporation step. The growth of the skin layer was observed with scanning electron microscopy (SEM). It was found that the skin layer thickness increased with increased coagulation time in the 2-propanol bath or increased solvent evaporation time. With an increase in the skin layer thickness, the oxygen permeance decreased and the oxygen/nitrogen selectivity increased. This result suggested that the density of the skin layer as well as its thickness increased with an increase in the coagulation time and/or the solvent evaporation time.

One option for controlling the membrane formation process may be using an additive in the casting solution. For example, the addition of an inorganic salt such as $CaCl_2$ (7), a polymer such as poly(vinyl pyrrolidone) (8), or an organic acid such as propionic acid (9,10), can change the structure of phase inversion membranes, especially the density of the skin layer. These additives are assumed to change the polymer configuration in the casting solution, which could result in morphological changes in the skin layer. It is known that some polymers form a polymer-surfactant complex in aqueous solution (11-15) through hydrophobic interactions between polymer and surfactant. Because of the complex formation, the properties of the polymer solution can be changed significantly. Therefore, the addition of a surfactant to the casting solution for a phase inversion membrane should greatly influence the polymer configuration, and consequently, the final membrane structure.

In this study, a typical surfactant, sodium dodecyl sulfate (SDS), was used as an additive in the casting solution for the preparation of asymmetric polysulfone membranes and its effect on the membrane structure and gas permeation properties was investigated.

Experimental

Membrane Preparation. Polysulfone (Udel-P1700, Amoco Performance Products) was used as the membrane-forming polymer, N,N-dimethylacetamide (DMAc) was used as a solvent, and sodium dodecyl sulfate (SDS) was used as a surfactant. The initial polymer concentration in the casting solution was 27.5 wt.% for all membranes in this study. The SDS concentration was varied from 0 to 2.0 wt.% in the casting solution. When the SDS content was larger than 2.0 wt.%, the surfactant precipitated in the casting solution. The membranes were cast on a glass plate with a thickness of 200 μm (8 mils). The casting process was carried out at ambient conditions and the relative humidity in the casting chamber was controlled to be less than 10%. After casting, the solvent was evaporated in a constant temperature oven at 95°C for a given period (0-15 min) Then, the film was immersed in a 2-propanol bath for 5 s at room temperature where the initial solidification of the membrane took place. Thereafter, the membrane was immersed in a water bath for more than 12 h, and subsequently washed with fresh water. The membrane was dried in a desiccator for several days before the gas permeation measurements were carried out. X-ray photoelectron spectroscopy (XPS) was used to detect any remaining SDS (sodium atom) in the membranes; no sodium was detected for all membranes studied within the detection limit; thus, SDS was completely extracted from the membranes.

The asymmetric polysulfone membranes were coated with a thin layer of polydimethylsiloxane (thickness 1-3 μm) to eliminate any skin layer defects. The permeance of defect-free, silicone-coated polysulfone membranes was calculated based on the series-resistance model (5).

Gas Permeation Measurements. A constant-pressure/variable volume system was used for the gas permeation experiments. The gas permeation experiments were carried out with oxygen and nitrogen at 25°C at a feed pressure of 100 psig and atmospheric permeate pressure (0 psig).

Results and Discussion

The effect of the evaporation time on the oxygen permeance of the membranes is shown in Figure 1. For a given SDS content, the oxygen permeance decreased with an increase in the solvent evaporation time. When the SDS content was 1.0 wt.%, the oxygen permeance showed a maximum value for a given evaporation time, that is, nearly ten times larger than the permeance of the membrane prepared without SDS. The oxygen permeance of the membrane prepared with 2.0 wt.% SDS was generally lower than that made from 1.0 wt.% SDS, but still much larger than that made without SDS. The oxygen permeance of the membrane prepared with 0.5 wt.% SDS was almost equal to that made without SDS.

The effect of the solvent evaporation time on the oxygen/nitrogen selectivity of the various membranes is shown in Figure 2. The oxygen/nitrogen selectivity increased with an increase in the solvent evaporation time for the membranes made without SDS up to about 10 min, where the selectivity was about 6, and leveled off thereafter. The selectivity of a dense, homogeneous polysulfone film was reported to be 6.2 (2). The oxygen/nitrogen selectivity of the membrane made from 1.0 wt.% SDS was slightly lower that made without SDS. For the membranes made from 0.5 and 2.0 wt.% SDS, on the other hand, the oxygen/nitrogen selectivity of 3 was essentially constant, which was always less than that made without SDS.

Figures 3(a)-3(d) show the SEM images of the cross-sections of the membranes prepared without SDS at different evaporation times. The skin layer thickness increased with an increase in the solvent evaporation time from less than 1 μm (0 min.) to about 3-4 μm (15 min.). The increase in skin layer thickness was consistent with the decrease in the oxygen permeance as the solvent evaporation time increased.

Figures 4(a)-4(d) show the change of the membrane structures with an increase in the solvent evaporation time for the membranes prepared with 1.0 wt.% SDS. The skin layer thickness was less than 1 μm for all solvent evaporation times up to 10 min, and about 1 μm when the solvent evaporation time was 15 min. This result suggests that the gas permeance of the membranes increased because the growth of the skin layer was reduced with the addition of SDS.

The growth of the skin layer with the solvent evaporation time can be explained as follows. During evaporation of the solvent, a higher concentration of polymer is formed in the topmost part of the cast film. This region is converted into the skin layer during the solidification process when the polymer concentration exceeds a

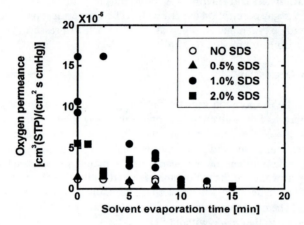

Figure 1. Effect of sodium dodecyl sulfate (SDS) concentration on the oxygen permeance of asymmetric polysulfone membranes.

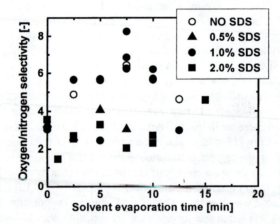

Figure 2. Effect of sodium dodecyl sulfate (SDS) concentration on the oxygen/nitrogen selectivity of asymmetric polysulfone membranes.

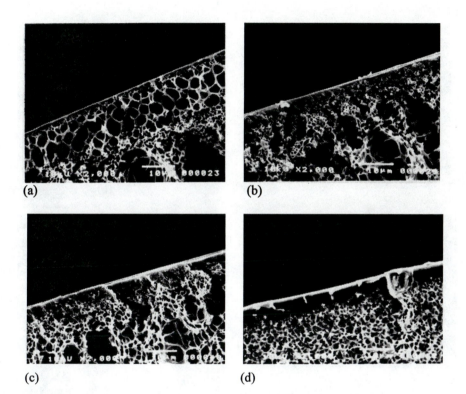

(a)

(b)

(c)

(d)

Figure 3. SEM images of polysulfone membrane cross-sections (made without sodium dodecyl sulfate): evaporation time (a) 0 min (b) 5 min (c) 10 min (d) 15 min.

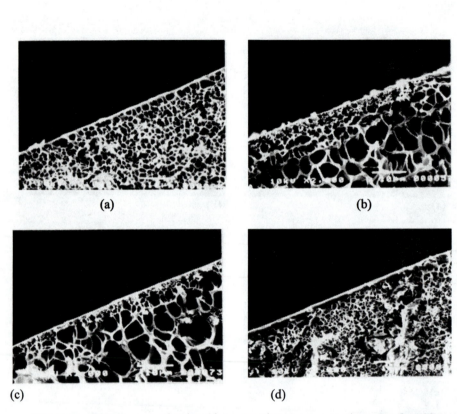

Figure 4.   SEM images of polysulfone membrane cross-sections (made with
1.0 wt.% sodium dodecyl sulfate): evaporation time (a) 0 min (b) 5 min (c) 10
min (d) 15 min.

critical value. A longer solvent evaporation time results in a thicker region of high polymer concentration; hence, the skin layer thickness increases with evaporation time (Figure 5). The change in the oxygen/nitrogen selectivity can be explained in terms of the configurational change of polymer chains in the skin layer. When the evaporation time is short, the density of the skin layer will be small because the packing density of the polymer chains is small due to a large distance among the polymer chains in the cast film. Thus, the selectivity is relatively low for membranes made at short evaporation time. An increase in the solvent evaporation time results in a more closely packed structure. Thus, the density of the skin layer increases and, consequently, the selectivity increases.

The addition of a surfactant to the casting solution may have two effects on the membrane formation process as shown schematically in Figure 6: (i) a reduction of the solvent evaporation rate and (ii) a decrease of the interactions among polymer chains due to polymer-surfactant complex formation. Because SDS is amphiphilic and DMAc is hydrophilic, a layer of SDS molecules will be formed on the surface of the cast film which decreases the solvent evaporation rate. The reduced solvent evaporation rate results in a slower growth of the highly concentrated polymer region in the cast film. Because DMAc is hydrophilic and polysulfone is hydrophobic, the SDS molecules and the polymer are likely to form a micelle-like complex in the solution. This complex will reduce the interaction among the polymer chains; thus, the coagulation of polymer chains will be delayed with SDS, and, consequently, the growth of the skin layer will be delayed. The effect of SDS on the configurational change of the polymer solution was supported by the following experimental observations: (i) the viscosity of the casting solution was significantly reduced by the addition of SDS and (ii) the solution rate of polysulfone in DMAc was greatly increased when SDS was added to the solvent. Both results suggest that the polymer-polymer interactions decreased by the addition of SDS. Previously, these phenomena were observed for aqueous solutions of polymer and surfactant (11-15) where formation of polymer-surfactant complexes occurred.

The configurational change of the polymer solution depends on the SDS content. When the SDS content was only 0.5 wt.%, most of the SDS molecules were freely distributed in the solvent phase (DMAc) because the SDS concentration was too small for the formation of a polymer-surfactant complex. Thus, the coagulation rate of polysulfone chains was not much affected by the addition of 0.5 wt.% of SDS. However, the free SDS molecules in the solvent phase (DMAc) may leave defects in the skin layer after being leached out; consequently, the oxygen/nitrogen selectivity decreased. When the SDS content was 2.0 wt.%, SDS molecules formed free micelles (without polymer) in the solvent phase in addition to the polymer-surfactant complex. These free micelles may also leave defects in the skin layer and decrease the oxygen/nitrogen selectivity. For the case of 1.0 wt.% SDS additive in the casting solution, most of the SDS molecules can form a polymer-surfactant complex, which leads to a delay in the coagulation of the polymer. This delay in coagulation resulted in a decrease in the skin layer thickness. However, the coagulation delay did not affect the packing density of the polymer chains in the

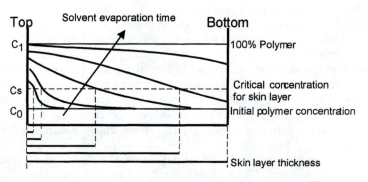

Figure 5.   Schematic concentration profile in the cast film.

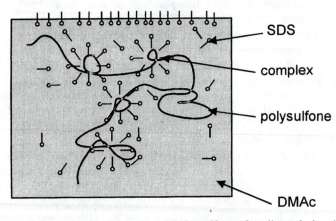

Figure 6.   Schematic representation for the effect of sodium dodecyl sulfate on the structure formation of phase inversion membranes.

skin layer of the membrane. Hence, the oxygen/nitrogen selectivity was not affected by addition of 1.0 wt% SDS.

In this paper, the addition of a surfactant, SDS, was found to decrease the skin layer growth rate and, therefore, led to a reduction in skin layer thickness. Further detailed investigations will be necessary to elucidate the morphological changes of the polymer solution and kinetics of the membrane formation process.

Conclusions

Asymmetric polysulfone membranes made with an SDS additive showed a significant increase in the oxygen permeance when the SDS content in the casting solution was 1.0 wt.%. The oxygen/nitrogen selectivity of these membranes was almost equal to that made without SDS. The growth of the skin layer was reduced significantly by the addition of SDS. These effects can be attributed to the surface-active properties of SDS.

Literature Cited

1. Kesting, R. E. *Synthetic Polymeric Membranes*, Wiley Interscience: New York, NY, 1985.
2. Koros, W. J.; Pinnau, I.. In *Polymeric Gas Separation Membranes*; Paul, D. R. and Yampolskii, Y. P., Ed.; CRC Press, Boca Raton, FL, 1994; pp 210-271.
3. Mulder, M. *Basic Principles of Membrane Technology*, Kluwer Academic Publishers, Dordrecht, The Netherlands, 1996.
4. Matsuura, T. *Synthetic Membranes and Membrane Separation Processes*, CRC Press, Boca Raton, FL, 1994.
5. Yamasaki, A.; Tyagi, R. K.; Fouda, A. E.; Jonnason, K.; Matsuura, T. *J. Membrane Sci.* **1997**, *123*, 89.
6. van't Hof, J.; Reuters, A. J.; Boom, R. M.; Rolevink, H. H. M.; Smolders, C. A. *J. Membrane Sci.* **1992**, *70*, 17.
7. Nguyen, T.D.; Chan, K.; Matsuura, T; Sourirajan, S. *Ind. Eng. Chem. Prod. Res. Dev.* **1985**, *24*, 655.
8. Miyano, T.; Matsuura, T; Carlsson, D.J.; Sourirajan, S. *J. Appl. Polym. Sci.* **1990**, *41*, 407.
9. Fritzsche, A. K.; Cruse, C. A.; Kesting, R. E.; Murphy, M. K. *J. Appl. Polym. Sci.* **1990**, *39*, 1915.
10. Fritzsche, A. K.; Cruse, C. A.; Kesting, R. E.; Murphy, M. K. *J. Appl. Polym. Sci.* **1990**, *39*, 1949.
11. Saito, S. *J. Colloid Sci.* **1960**, *15*, 283.
12. Isemura, T.; Imanishi, A. *J. Polym. Sci.* **1958**, *33*, 37.
13. Jones, M. N. *J. Colloid Interface Sci.* **1967**, *23*, 36.
14. Lewis, K. E.; Robinson, C. P. *J. Colloid Interface Sci.* **1970**, *32*, 539.
15. Francois, J.; Dayantis, J.; Sabbadin, *J. Eur. Polym. J.* **1985**, *21*, 165.

Chapter 7

Preparation of Poly(ether sulfone) and Poly(ether imide) Hollow Fiber Membranes for Gas Separation: Effect of Internal Coagulant

D. Wang, K. Li, and W. K. Teo

Department of Chemical and Environmental Engineering, National University of Singapore, 10 Kent Ridge Crescent, Singapore 119260

The internal coagulant plays a very important role in the preparation of hollow fiber membranes. In this paper, the influence of the internal coagulant on hollow fiber formation, structures, and gas separation performance of polyethersulfone (PESf) and polyetherimide (PEI) membranes are presented. The internal coagulants used include water, methanol, ethanol, 2-propanol, and mixtures of the alcohols and water. Water was always used as the external coagulant. Polymer dopes containing a polymer and a solvent mixture of N-methyl-2-pyrrolidone (NMP) and ethanol (EtOH) were formulated to give low coagulation values. The correlation of coagulant properties with gas permeation properties, membrane structure, and membrane integrity is addressed.

Polymeric membranes have been applied successfully in some industrial gas separation processes over the past two decades. Development of highly permeable asymmetric hollow fiber membranes from highly selective polymers such as polysulfone (PSf), polyethersulfone (PESf), polyetherimide (PEI), and polyimide (PI) has been a subject of intense research (1). The fabrication of hollow fiber membranes is a complicated process involving complicated phase equilibria, rapid phase inversion kinetics, and interfacial mass transfer during the spinning process. In addition, the nascent fiber emerging from the tip of the spinnerette must simultaneously experience internal and external coagulation processes during membrane formation. It has been reported that the drying process during the external coagulation is very important in forming high-performance gas separation membranes from polymer solutions containing nonsolvent-additives (2,3). In the dry-wet phase inversion process, coagulation of the internal surface of the nascent fiber starts immediately after extrusion from the spinnerette, whereas the external surface experiences coalescence and orientation of polymer aggregates before solidification in the coagulation bath. It has been shown that the nature of the coagulant is important in determining the morphology of asymmetric membranes made by the phase inversion process (4-6).

In hollow fiber spinning, water is usually used as the external coagulant because of economic and environmental reasons; however, other nonsolvents for the polymer, besides water, may be used as the internal coagulant because of the small amount of internal coagulant involved. Although the internal coagulant may play an important role in controlling membrane structure and separation performance in fabricating integral-asymmetric hollow fibers, only a few detailed studies have been reported to elucidate its effect (7,8).

In our previous studies (2,3,9), PESf and PEI hollow fiber membranes with good gas separation performance were fabricated from a NMP/EtOH solvent system with water as the internal and external coagulant. However, some of the membranes prepared contained big macrovoids adjacent to the internal skin layer which are not desirable for high-pressure operation. Moreover, in some cases the external skin layers were damaged due to the rapid internal coagulation process and the fast movement of the coagulation front. In this study, the effects of internal coagulant on hollow fiber integrity, structure, and gas separation properties were examined for PESf and PEI hollow fiber membranes.

Materials and Experimental Methods

Materials. Polyetherimide (Ultem 1000) and polyethersulfone (Radel A-300) were purchased from General Electric Company and Amoco Performance Products, respectively. The polymers were first dissolved in NMP/EtOH solvent mixtures. The polymer dopes were formulated to be very close to incipient phase separation by the addition of a predetermined amount of ethanol. The coagulation values (G_v) of the PEI dope and PESf dope were 1 g and 1.5 g, respectively, using water as the coagulant. The coagulation value has been defined as the quantity of coagulant (water) in grams required to induce phase separation in a 2 wt% polymer dope (10). The mass ratios of the solvent (NMP) and the nonsolvent-additive (EtOH) in these polymer dopes were determined from the measured precipitation values of the polymer/solvent/nonsolvent systems used according to the method reported elsewhere (11). The precipitation value is defined as the grams of a nonsolvent required to cause visual turbidity of a binary solution containing 100g of solvent and 2g of polymer (12). The precipitation values of ethanol and water in PESf/NMP and PEI/NMP systems at 25°C are shown in Table I. The compositions and properties of the polymer spinning dopes are given in Table II.

Table I.   Precipitation values of six nonsolvents in PEI/NMP
and PESf/NMP at 25°C.

	H_2O	CH_3OH	C_2H_5OH	2-C_3H_7OH	$CH_3OH/$ H_2O (1:1)[a]	$C_2H_5OH/$ H_2O (1:1)[a]
PV-1	6.1	22.7	25.4	24.8	9.2	9.7
PV-2	15.4	51.5	51.4	-	23.0	26.0

PV-1: precipitation values in PEI/NMP system;
PV-2: precipitation values in PESf/NMP system. [a] by weight.

Water, methanol, ethanol, 2-propanol, and mixtures of 50 wt% MeOH/50 wt% H_2O, and 50 wt% EtOH/50 wt% H_2O were used as the internal coagulants in this study. Their precipitation values in PESf/NMP and PEI/NMP systems were also measured at 25°C and are shown in Table I. Polydimethylsiloxane (Sylgard-184) supplied by the Dow Corning Co. was used as the coating material for the spun hollow fibers using n-pentane as a solvent. Silicone-coated membranes were prepared through external surface coating in a solution containing 3 wt% Sylgard-184 in n-pentane.

Table II.   Composition and properties of polymer spinning dopes at 25°C.

Spinning solution	Spinning solution composition (wt%)			Solution properties	
	NMP	C_2H_5OH	polymer	η (cp)	G_v (g)
PESf-S	48.78	21.22	30.0	55,700	1.5
PEI-S	62.36	12.64	25.0	13,340	1.0

η: viscosity of polymer dope at 25°C; G_v: coagulation value

Hollow Fiber Spinning.   All hollow fiber membranes were prepared at ambient conditions (25±1°C and 60-65% RH) in our laboratory using a spinnerette with an orifice diameter/inner tube diameter of 0.60/0.15 mm. The spinning apparatus used and procedure have been described in details elsewhere (2,3). The spinning conditions are summarized in Table III.

Characterisation of Hollow Fiber Membranes.   The hollow fiber membranes were characterised by measuring their gas permeation properties and examining the cross-section structure of the membranes using scanning electronic microscopy (SEM). The spun hollow fiber membranes were kept in a spray of water for at least three days and then dried at ambient conditions (25±1°C and RH = 60%-65%). Test modules containing 30-40 hollow fibers of about 20 cm in length were made. The fibers were immersed in the coating solution for two minutes under vacuum. Thereafter, the silicone coating was cured at room temperature for 24 h. The apparatus used for the measurement of gas fluxes through the silicone-coated membranes was the same as previously described (*2,3*). The steady-state volumetric gas fluxes were determined using a soap flowmeter at ambient conditions. The pressure-normalized fluxes of He and N_2 for all hollow fiber membranes were determined at a pressure difference of 5 bar and 25°C.

Table III. Spinning conditions for preparation of hollow fiber membranes.

Spinning conditions	PEI-S dope	PESf-S dope
Spinning temperature	25±1°C	25±1°C
Room humidity	60%-65%	60%-65%
Coagulant temperature	25±1°C	25±1°C
Solution flow rate	2.5 ml/min	2.6 ml/min
Bore fluid flow rate	0.5-0.9 ml/min	1.2-1.5ml/min
Take-up velocity	6.6 m/min	8 m/min

Results and Discussion

Polyetherimide Hollow Fiber Membranes. Polyetherimide (PEI) hollow fiber membranes were spun from the formulated polymer dopes using different internal coagulants including H_2O, EtOH, 2-PrOH, and mixtures of 50 wt% MeOH/50 wt% H_2O and 50 wt% EtOH/50 wt% H_2O at various air-gap lengths. Other spinning conditions are given in Table III. Pure-gas permeation properties of the silicone-coated PEI membranes for He and N_2 were then determined at 25°C and a pressure difference of 5 bar. Figures 1 and 2 show the pressure-normalized He flux and ideal selectivity of He/N_2 as a function of the air-gap. The best combination of selectivity and gas flux appears to occur for membranes prepared at an air-gap of 2 cm. When the length of the air-gap was zero, rapid coagulation of the external surface took place immediately after leaving the spinnerette resulting in a porous external skin with low selectivity (*9*). With increasing the length of the air-gap beyond 2 cm, the He/N_2 selectivity generally decreased. The extent of influence of the air-gap on the selectivity for different internal coagulants is different and follows the order H_2O>MeOH/H_2O>EtOH/H_2O>alcohols.

With water as the internal coagulant, the membrane exhibited the highest flux but also the lowest selectivity among all the internal coagulants examined and over the air-gap range studied. This result suggests that fast diffusion of water might have occurred and reached the outer surface which caused damage to the outer skin layer. Membranes spun using ethanol and 2-propanol as the internal coagulant displayed excellent selectivity, especially for the latter; however, the membranes had only moderate gas fluxes. The membrane prepared using a mixture of 50 wt% MeOH/50 wt% H_2O as coagulant also showed very good selectivity but rather low fluxes due to a thicker membrane wall as depicted in Figure 3c. EtOH and a mixture of 50 wt%

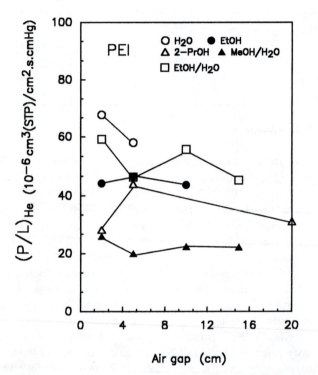

Figure 1.   Pressure-normalized He flux of silicone-coated polyetherimide
(PEI) membranes spun using five internal coagulants and at various air-gaps.

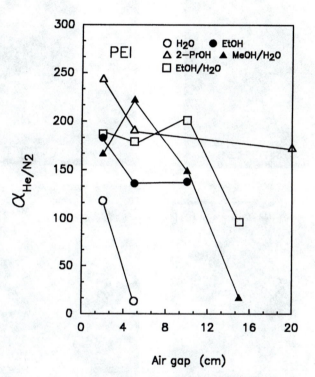

Figure 2.   Ideal selectivity for He/N$_2$ of silicone-coated PEI membranes spun using five internal coagulants and at various air-gaps.

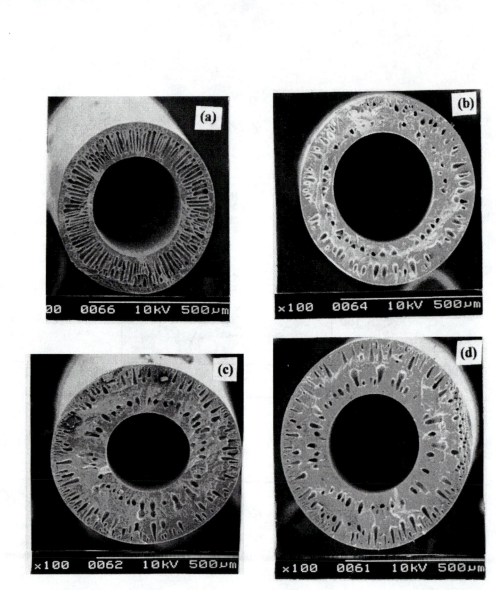

Figure 3.   Photomicrographs of cross-sections of PEI membranes spun using four internal coagulants and at an air-gap of 5 cm.: (a) H_2O; (b) EtOH; (c) 50 wt% MeOH/50 wt% H_2O; (d) 50 wt% EtOH/50 wt% H_2O.

EtOH/50 wt% H_2O appeared to be the "ideal" internal coagulants for the PEI/NMP/EtOH dope system yielding high gas fluxes and good selectivity. The presence of EtOH as additive in the internal coagulant medium seems to reduce the extent of diffusion of coagulant such that there is no effect of the coagulant on the external skin layer which remained thin and dense.

The features of cross-section structures of the PEI hollow fibers prepared from the four internal coagulants are shown in Figures 3a-d. These photomicrographs clearly elucidate the influence of internal coagulant and coagulation strength on the membrane structures. As water is a very strong nonsolvent for PEI, strong polymer-polymer contacts led to a rapid growth of the polymer-lean phase during internal coagulation. Internal water quickly diffused in the polymer-lean phase along the fiber wall. Because of the slower process of coagulation in the external skin, the extended time of progressive solidification from the internal skin to the external skin tended to accelerate the growth of the polymer-lean phase and the formation of large macrovoids. The resulting membrane exhibits long internal finger-like structures (Fig. 3a), whereas the layer adjacent to the external skin displays short and small macrovoids.

The membranes prepared using the other three internal coagulants exhibit much smaller internal macrovoids and the number of macrovoids was also reduced. Figures 3b, 3c, and 3d show cross-section structures of the membranes prepared using EtOH, MeOH/H_2O, and EtOH/H_2O as the internal coagulant, respectively. The internal macrovoids are smaller and the external macrovoids are larger in these membranes as compared to that prepared with water as the internal coagulant (Fig.3a). When weak nonsolvents such as EtOH and 2-PrOH were used as the internal coagulant, the coagulant diffusion rate along the fiber wall was reduced. For a given air-gap, the diffusion distance of the internal coagulant is shorter when the overall fiber wall is rapidly solidified by the external wet coagulation process. In this case, internal macrovoids are smaller, whereas the external macrovoids are larger. Because the nonsolvent strength of the alcohol/water mixture was significantly reduced in comparison with water, similar structures are displayed by the membranes spun using mixtures of EtOH/H_2O and MeOH/H_2O as internal coagulant. For gas separation membranes employed at high pressure, the membrane should have good mechanical strength and, therefore, a sponge-like substructure is highly desirable. The use of an internal coagulant with suitable nonsolvent strength can efficiently reduce large macrovoids and maintain good gas separation performance

Polyethersulfone Hollow Fiber Membranes.   In spinning polyethersulfone (PESf) hollow fiber membranes, it was observed that hollow fibers with circular lumen could not be prepared using methanol and ethanol as the internal coagulant even at a very high flow rate of the internal coagulant (Fig. 6b). Thus, very weak nonsolvents for the polymer, such as alcohols, are not suitable as internal coagulants in preparing PESf hollow fibers. On the other hand, PESf hollow fibers with circular lumen could easily be prepared using water, or a mixture of 50 wt% MeOH/50 wt% H_2O or a mixture of 50 wt% EtOH/50 wt% H_2O as the internal coagulant.

The influence of the air-gap on the permeation properties of the PESf hollow fibers is illustrated in Figures 4 and 5 which show the pressure-normalized He flux and ideal He/N_2 selectivity, respectively. These data were determined at a pressure difference of 5 bar and 25°C for silicone-coated membranes. In general, membranes with high selectivity were produced at longer air-gaps as compared to the PEI membrane-forming system. An air-gap of 10 cm to 15 cm appeared to be optimum. Over this air-gap range, He/N_2 selectivities of the membranes prepared from the different coagulants, except EtOH, were very similar, in the range of 95 to 115. This result suggests that the internal coagulation medium had little influence on the outer skin layer formation. The gas flux generally decreased with decreasing coagulant strength, that is, $H_2O>MeOH/H_2O>EtOH/H_2O>MeOH>EtOH$. The low gas flux of the membranes prepared from methanol and ethanol may be caused by a denser substructure and/or a dense inner skin layer. Membranes spun using mixtures of the alcohols and water (50 wt% MeOH/50 wt% H_2O and 50 wt% EtOH/50 wt% H_2O) exhibited both high flux and high selectivity. With increasing air-gaps, the gas flux of the membranes prepared from both these mixtures increased.

The photomicrographs of the fibers prepared using water, ethanol, and mixtures of 50 wt% MeOH/50 wt% H_2O and 50 wt% EtOH/50 wt% H_2O as internal coagulants are shown in Figure 6a-d. The membrane prepared using MeOH as coagulant displayed the same square lumen as that of ethanol. Both of these fibers exhibited very dense, sponge-like structure. On the other hand, many small finger-like voids appeared in the inner edge of the membrane prepared using water as the internal coagulant. The macrovoids near the inner skin of the membranes prepared from both the $EtOH/H_2O$ and $MeOH/H_2O$ mixtures were substantially reduced compared to that prepared using water as the internal coagulant alone. Like the PEI hollow fibers, the use of a suitable weak internal coagulant can effectively reduce the large macrovoids in the substrate.

The above results illustrate clearly that the internal coagulant plays a very important role in the control of the membrane structure and integrity. In general, the internal coagulant had a more pronounced effect on PEI membranes than on PESf membranes. It appears that polymer-coagulant interaction is a very important variable in the control of the phase inversion process besides the affinity and diffusivity of the solvent-coagulant and the coagulant-additive systems. For a given polymer/solvent system, the degree of interaction between a polymer and a coagulant may be indicated by the precipitation value of the coagulant in the polymer/solvent system (12). A small precipitation value of a coagulant, such as water, indicates that the coagulant is a strong nonsolvent for the polymer. Therefore, weak interaction of coagulant-polymer and strong interaction of polymer-polymer existed in the phase inversion process. Alcohols, such as

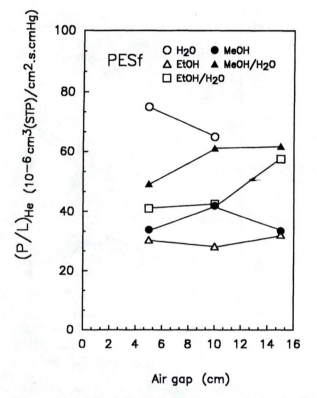

Figure 4. Pressure-normalized He flux of silicone-coated polyethersulfone (PESf) membranes spun using five internal coagulants and at various air-gaps.

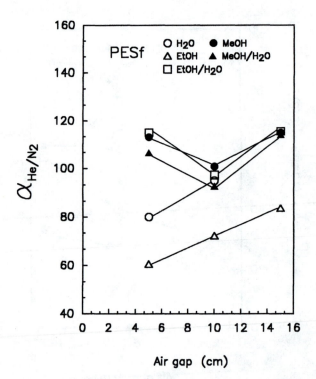

Figure 5. Ideal selectivity for He/N$_2$ of silicone-coated PESf membranes spun using five internal coagulants and at various air-gaps.

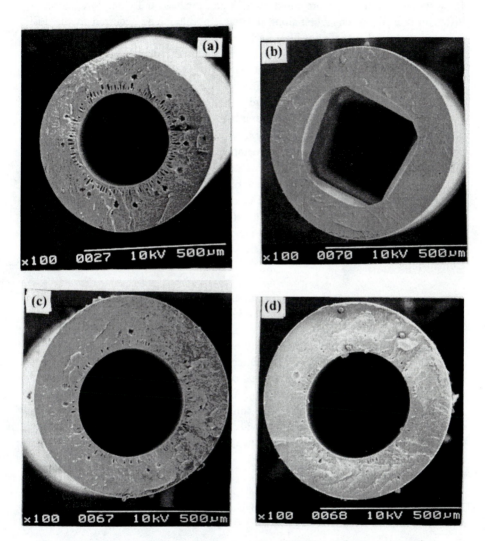

Figure 6.   Photomicrographs of cross-section of PESf membranes spun using four internal coagulants and at an air-gap of 10 cm.: (a) H_2O; (b) EtOH; (c) 50 wt% MeOH/50 wt%H_2O; (d) 50 wt% EtOH/50 wt% H_2O.

methanol, ethanol, and propanol are weak nonsolvents for the polymers and may be useful to reduce the coagulation rate. For a given coagulant, the precipitation value in the PEI/NMP system is generally much lower than that in the PESf/NMP system. It is anticipated that a more rapid coagulation process and, therefore, a faster diffusion rate of the coagulant along the nascent membrane occurred in the PEI spinning system. Large (finger-like) macovoids were formed in this case. It should be noted that the coagulation rate is closely related to the polymer solution state. In this study, both polymer dopes were close to incipient phase separation as indicated by their small coagulation values. The experimental results also showed that hollow fiber with a desired structure may be prepared by selecting an internal coagulant with a proper nonsolvent strength as indicated by its precipitation value in the polymer/solvent system. A mixture of a strong nonsolvent and a weak nonsolvent which gives a moderate precipitation value appears to be a suitable internal coagulant for the polymer systems examined. The coagulant strength can be adjusted by using a mixture of the proper ratio of a strong nonsolvent and a weak nonsolvent. In this study, a mixture of alcohol to water (1:1 by weight) was used for simplicity. It is likely that there is an optimal mass ratio of a strong nonsolvent and a weak nonsolvent for different polymer solution systems.

Conclusions

The internal coagulant plays an important role in the preparation of high-performance hollow fiber membranes. The hollow fiber structure, integrity, and gas separation properties are highly influenced by the internal coagulant. Polyethersulfone and polyetherimide hollow fiber gas separation membranes with good gas separation properties and fiber integrity can be prepared using mixtures of ethanol/water or methanol/water as internal coagulant. For the spinning of a desired hollow fiber, an internal coagulant medium should have an appropriate nonsolvent strength for a polymer, which can be adjusted by using a mixture of a strong nonsolvent and a weak nonsolvent with the proper ratio according to its precipitation value for the polymer/solvent system.

Acknowledgements

The financial support from the National University of Singapore in the form of research grants (RP940621/A and RP970634) is gratefully acknowledged.

Literature Cited

1. Membrane Separation Systems: Recent Developments and Future Directions, Baker, R W.; Cussler, E. L.; Koros, W. J.; Strathmann, H.; (Eds.), Noyes Data Corp., Park Ridge, NJ, 1991.

2. Wang, D. L., Ph.D. Thesis, National University of Singapore, 1995.
3. Wang, D. L.; Li, K.; Teo, W. K. *J. Membrane Sci.* **1996**, *115*, 8.
4. Strathmann, H. In *Materials Science of Synthetic Membranes*, ACS Symp. Ser. No. 269, Lloyd, D. R. (Ed); American Chemical Society, Washington, DC, 1985.
5. Bottino, A.; Capanelli; G.; Munari, S. *J. Appl. Polym. Sci.* **1985**, *30*, 3009.
6. Chung, T. S.; Kafchinski, E. R.; Foley, P. *J. Membrane Sci.* **1994**, *21*, 88.
7. Pinnau, I.; Koros, W. J. *J. Membrane Sci.* **1992**, *71*, 81.
8. Pesek, S. C.; Koros, W. J. *J. Membrane Sci.* **1994**, *88*, 1.
9. Wang, D. L; Li, K.; Teo, W. K. *J. Membrane Sci.* **1998**, *138*, 193.
10. Wang, D. L; Li, K.;Teo, W. K. *J. Membrane Sci.* **1995**, *98*, 233.
11. Kai, M.; Ishii, K.; Tsugaya, H.; Miyano, T. In *Reverse Osmosis and Ultrafiltration*, ACS Symposium Series No. 281, Matsuura, T., and Sourirajan, S., (Eds.), American Chemical Society, Washington, DC, 1985.
12. Wang, D. L; Li, K.; Sourirajan, S.; Teo, W. K. *J. Appl. Polym. Sci.* **1993**, *50*, 1693.

Chapter 8

Effect of Solvent Exchange on the Morphology of Asymmetric Membranes

H. C. Park[1], Y. S. Moon[2], H. W. Rhee[2], J. Won[1], Y. S. Kang[1], and U. Y. Kim[1]

[1]Division of Polymer Science and Engineering, Korea Institute of Science and Technology, P.O. Box 131, Cheongryang, Seoul, Korea
[2]Department of Chemical Engineering, Sogang University, Seoul, Korea

The effect of solvent exchange and the subsequent drying process on membrane morphology was investigated for microporous and integrally-skinned asymmetric membranes by using scanning electron microscopy (SEM) and atomic force microscopy (AFM). For the microporous membranes, significant changes in morphology and transport properties were observed due to the collapse of micropores in the top skin layer. In the case of integrally-skinned asymmetric membranes, only negligible differences in membrane structure were observed upon solvent exchange. The effects of solvent exchange depend strongly on the capillary forces imposed on the membrane matrix by a liquid present in the membrane pores. A membrane dried using supercritical CO_2 experienced no capillary forces upon drying, and, hence, the membrane morphology observed by SEM showed the nascent, original membrane structure formed by the phase inversion process.

When a membrane is prepared by the nonsolvent-induced phase inversion process by precipitating a polymer solution in a coagulation medium, its pores are filled with the coagulation medium (*1-3*). Thereafter, the wet membranes are typically dried. The nascent membrane morphology can change due to the capillary forces exerted by the coagulation medium on the membrane matrix during the drying process. The capillary pressure arises from the curvature of liquid in the interstitial capillaries. It provides a force that can deform the membrane matrix whereas the mechanical modulus of the material resists the deformation. This concept was previously studied in the formation of homogeneous films from disperse latex particles (*4*) and adopted to qualitatively explain the morphological changes of phase inversion membranes on drying by Pinnau (5). A liquid in the interstitial capillary system develops a different capillary pressure depending on the pore radius (6). The capillary pressure exerts in a direction normal to the water-polymer interface and tends to deform the polymer particles in that direction (4-6).

Brown developed a simple mathematical model to calculate the capillary pressure of a close-packed sphere structure using the Young-Laplace equation (4). Here, the capillary pressure was expressed by 12.9 σ/r, where σ and r are the surface tension of the liquid and the radius of particles, respectively. For a closely packed system of polymer particles with a diameter of 0.01 μm in a liquid with a surface tension of 70 dyne/cm, the capillary pressure is calculated to be 1.8×10^3 kg/cm^2. Because of this extremely high capillary pressure, the nascent membrane can be deformed during the drying process.

A typical morphological change in the membrane is the collapse of micropores in the top skin layer. In an asymmetric membrane, the pores are usually increasing in size from the top (skin layer) to bottom structure and the capillary force will, then, be the highest in the top skin layer. Therefore, in most cases, the skin morphology of a membrane can be changed significantly on drying, and its transport properties will be altered markedly. For controlling membrane morphology during drying, the solvent exchange method has been commonly employed. The purpose of the solvent exchange is to lower the surface tension of the liquid present in the membrane pores and, consequently, preventing the collapse of the micropores in the top skin layer (7-10).

In this study, the effect of solvent exchange was investigated on the morphological changes of microporous and integrally-skinned asymmetric membranes during drying. In addition, the supercritical fluid drying method was utilized to minimize the morphological changes. The supercritical fluid does not undergo any phase transition and, thereby, imposes negligible capillary pressure on the membrane matrix during drying. Using this procedure, the original nascent membrane morphology can be observed even after drying.

Experimental

Membrane Preparation. Two types of asymmetric phase inversion membranes were prepared from polysulfone (PSf, Amoco Co., P3500, Mw 35,000 g/mol). The casting solutions were prepared by dissolving PSf (25 wt.%) in N-methyl-2-pyrrolidone (NMP, Aldrich Chemical, Inc.) or a mixture (6:4 by weight) of NMP and tetrahydrofuran (THF, Merck Co.). The solution was cast onto a glass plate using a doctor blade of 250 μm clearance. The PSf/NMP solution was immersed into a water/NMP (1:1 by weight) coagulation bath, whereas the PSf/NMP/THF solution was quenched into a water coagulation bath. The former is a microporous membrane and will be designated as T1 in this paper. The latter is an integrally-skinned asymmetric membrane and will be designated as T2. After the phase inversion process was completed, membranes were kept in a water bath for 2 days at room temperature to replace the residual solvent, NMP and/or THF, with H_2O and then were subject to the solvent exchange. For the solvent exchange, water in a membrane was first replaced with ethanol by soaking the membrane in an ethanol bath for 2 days, and then ethanol was replaced with hexane for 2 days. Membrane samples were taken at each step and dried in air for 2 days and then in a vacuum oven for another 2 days at room temperature. The samples dried after being kept in the water bath are denoted as T1W and T2W, the samples dried after ethanol-exchange as T1E and T2E, and the samples dried after hexane-exchange as T1H and T2H, respectively. Membrane samples which were dried using supercritical CO_2 after ethanol-exchange are denoted as T1SC. The surface tension of water, ethanol, and hexane are 72.5, 22.7, and 17.9 dyne/cm, respectively.

Supercritical CO_2 Drying. To investigate the effect of solvent exchange on the membrane morphology, the supercritical CO_2 drying method was used. After the replacement of the casting solvent NMP with water, the membrane sample was soaked in an ethanol bath for 5 days to ensure the complete replacement of water and the residual NMP with ethanol. The ethanol-soaked membrane was placed in the reactor cell filled with ethanol. The pressure and temperature in the cell were increased up to 100 bar and 40°C, and then CO_2 was introduced and maintained for 12 hours under the same conditions (the critical pressure and temperature of CO_2 are 100 bar and 30°C, respectively). After complete replacement of ethanol with CO_2, the pressure and temperature inside the reactor cell were lowered to atmospheric conditions. The drying path of the membrane is shown schematically in Figure 1.

Investigation of Membrane Morphology and Performance. The cross-sectional morphology of a membrane fractured in liquid nitrogen was investigated by field emission electron microscopy (FE-SEM, Hitachi S-4200). The SEM images were taken for membrane samples coated with Pt-Pd at an accelerating voltage of 15 kV. The surface morphology of a membrane was investigated with an Atomic Force Microscope (AFM, CP, Park Sci. Inst.). The permeance of pure O_2 and N_2 was measured by using a constant-pressure gas permeation apparatus with a bubble flowmeter at a feed gas pressure of 70 psig at room temperature (-25°C).

Results and Discussion

Membrane with a Porous Skin Layer (T1). T1 type membranes which were produced by coagulation of PSf (25 wt.%)/NMP solution in a NMP/water (1/1 by weight) mixture had a rather thick porous skin layer on top of a sponge-like substrate. These membranes showed very interesting features after the solvent

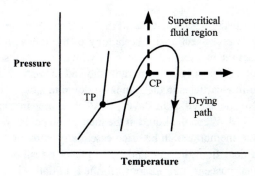

Figure 1.   Supercritical CO$_2$ drying path in a P-T diagram.

exchange. The top skin layer contained many micropores and, therefore, it was affected strongly by the capillary forces during the drying step.

Figure 2(a) shows the FE-SEM image of the cross-section of a T1W membrane that was dried in air directly after removing the residual solvent from the membrane for 2 days in the water bath. Because the membrane pores were filled with water whose surface tension was the highest among the liquids used in this study, the largest capillary pressure was exerted during the drying process for a given pore size. The nodule-like particles in the top skin layer seem to be fused together and the skin layer is compact and thick.

Figure 2 (b) shows the FE-SEM image of the membrane cross-section dried after ethanol exchange. It also shows a compact and thick skin structure, which is, however, thinner than that of T1W. The cross-section of a membrane dried after ethanol-hexane exchange is shown in Figure 2(c). The hexane-exchanged membrane shows a much more open structure than the former two membranes. The less compact structure resulted from the decreased capillary force imposed on the pore walls due to the lower surface tension of hexane present in the membrane pores during drying.

Figure 2(d) shows the cross-section of a T1SC membrane dried by using supercritical CO_2. The cross-sectional morphology of the top skin layer was quite different from those of the membranes dried in air with or without solvent exchange. The FE-SEM image of the T1SC membrane showed a very uniform structure composed of nodule-like particles in the skin layer down to a depth of 500 nm from the top surface. No densified skin layer was observed in the cross-section investigated by FE-SEM. When the liquid in the membrane pores was replaced by a supercritical fluid, the membrane can be dried essentially without a capillary force because the supercritical fluid did not undergo a phase transition. Therefore, the membrane retains its nascent morphology formed initially by the immersion precipitation process. The region under the top skin of the T1SC membrane showed a seemingly uncollapsed and loose structure. From the cross-section structure of the T1SC membrane, it can be speculated that the top skin layer of an asymmetric membrane initially composed of nodule-like particles was compacted to a dense layer during drying due to the capillary action of the liquid.

AFM is a very useful method for studying the surface structure of polymeric membranes (11-16). The changes in the surface structure of the skin layers investigated by the AFM images (Figures 3 and 4) coincide with the FE-SEM images shown in Figure 2. The collapse of micropores during drying results in the change of the cross-sectional morphology as well as the surface of the skin layer. The skin surface of the T1W membrane appears much rougher than that of the T1SC membrane, which can be confirmed and quantified by the surface roughness parameters: (i) average roughness, (ii) root-mean-square roughness, and (iii) lateral mean diameter. The average roughness, R_{avg}, is given by the average deviation from the average of the height,

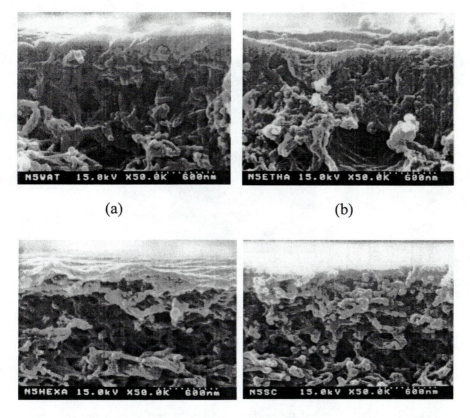

(a)   (b)

(c)   (d)

Figure 2. FE-SEM images of microporous polysulfone membranes (T1 type) dried under different conditions: (a) T1W (water), (b) T1E (ethanol), (c)T1H (hexane), and (d) T1SC (supercritical CO_2).

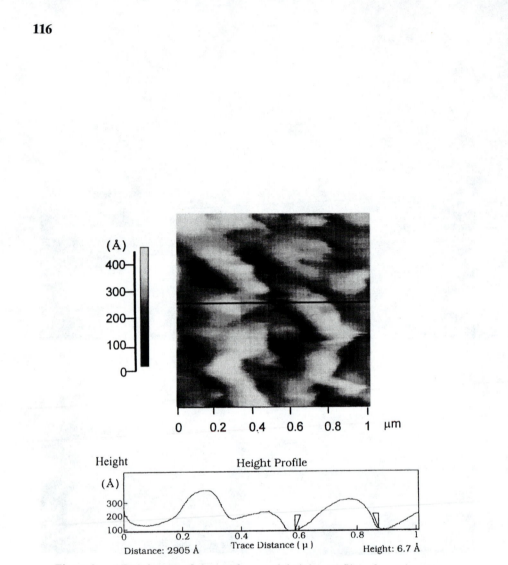

Figure 3. AFM image of the surface and height profile of a microporous polysulfone membrane (T1W) dried in air without solvent exchange.

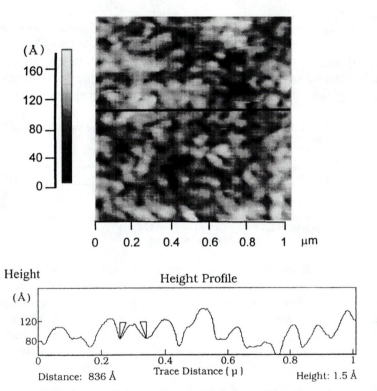

Figure 4. AFM image of the surface and height profile of a microporous polysulfone membrane (T1SC) dried using supercritical CO_2.

$$R_{avg} = \sum_{n=1}^{N} \left| Z_n - \overline{Z} \right| / N \qquad (1)$$

where Z_n is a current Z (height) value, $\overline{Z}$ is a mean height, and N is the number of points within a given area. The root-mean-square roughness, R_{rms}, is given by the standard deviation of the data,

$$R_{rms} = \sqrt{\sum_{n=1}^{N} \left(Z_n - \overline{Z} \right)^2 / (N-1)} \qquad (2)$$

Because R_{rms} contains square terms, the large deviation from the average Z height is weighted more heavily than that in R_{avg}. The lateral mean diameter, P_L, is a convenient parameter to compare the various sizes together with both R_{rms} and R_{avg}. It can be obtained by fitting the ellipse to the height profile of AFM images as shown in Figure 5. The calculated R_{avg}, R_{rms}, and P_L values are given in Table I.

Table I. Surface roughness of microporous polysulfone membranes (T1) dried under different conditions.

Drying conditions	R_{avg} (Å)	R_{rms} (Å)	P_L (Å)
T1W	55	68	1,420
T1SC	18	23	592

R_{avg}: average roughness; R_{rms}: root-mean-square roughness;
P_L: lateral mean diameter of hemispherical structure.

As expected, all values of R_{avg}, R_{rms}, and P_L are smaller in T1SC than in T1W. This result demonstrates clearly that the surface roughness decreases with a decrease in the surface tension of liquid present in the membrane pores during drying. This might suggest that the capillary action caused the aggregation of smaller particles to bigger ones and eventually to dense film formation leading to a thick skin layer. The morphological changes of the top skin layer during drying affected the gas permeances of the membranes, as shown in Table II.

Table II. The effect of solvent exchange on the gas transport properties of microporous polysulfone (T1) membranes.

Membrane	Surface tension (dyne/cm)	Permeance (P/l) (10^{-6} cm^3(STP)/cm^2·s·cmHg)		Selectivity (O$_2$/N$_2$)
		O$_2$	N$_2$	
T1W	72.5	11.6	11.9	0.98
T1E	22.7	24.7	25.2	0.98
T1H	17.9	25.8	26.2	0.99

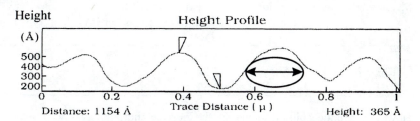

Figure 5. Lateral mean diameter P_L calculated from the height profile of the AFM image.

The gas permeance decreased with an increase in the liquid surface tension. The decrease in permeance can be explained by an increase in the skin layer thickness. In addition, although the T1W membrane appeared to have a gas-tight top skin layer by FE-SEM, it did not show any O_2/N_2 selectivity.

Membrane with a Thin Gas-Tight Skin Layer (T2). T2 type membranes which were produced from a 25 wt.% solution of PSf in a NMP/THF (6/4 by weight) solvent mixture by coagulation in pure water had a gas-tight skin layer and a finger-like substructure. This membrane type was not much affected by the solvent exchange treatment. As can be seen in Figure 6, morphological differences are hardly observed in the cross-sections of the top skin layers from SEM images. The permeance of O_2 and N_2 as well as the O_2/N_2 selectivity were also hardly affected by the solvent exchange conditions, as shown in Table III.

Table III. The effect of solvent exchange on the gas transport properties of integrally-skinned asymmetric polysulfone (T2) membranes.

Membrane	Surface tension (dyne/cm)	Permeance (P/l) (10^{-6} cm^3(STP)/cm$^2 \cdot$s$\cdot$cmHg)		Selectivity (O_2/N_2)
		O_2	N_2	
T2W	72.5	2.41	0.79	3.1
T2E	22.7	2.52	0.83	3.0
T2H	17.9	2.53	0.83	3.0

However, a noticeable difference in the surface morphology was observed. Figures 7 and 8 show the AFM surface images and the height profile of the T2 type membranes. As given in Table IV, the surface morphology and height profile from AFM images show that the surface becomes smoother by replacing water with ethanol and then with hexane.

Table IV. The effect of liquid surface tension on the membrane surface roughness of integrally-skinned asymmetric polysulfone (T2) membranes.

Liquid	Surface tension (dyne/cm)	R_{avg} (Å)	R_{rms} (Å)	P_L (Å)
water	72.5	81	103	1,150
ethanol	22.5	37	47	924
hexane	17.9	37	47	880

Conclusions

For asymmetric membranes made by the phase inversion process, the drying step can cause additional changes in the membrane morphology. The top skin layer of a

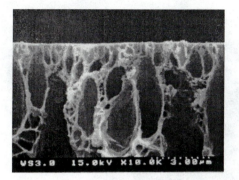

(a)

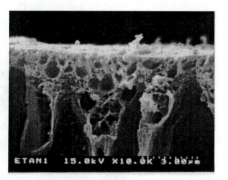

(b)

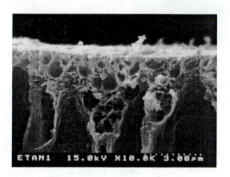

(c)

Figure 6.   FE-SEM images of integrally-skinned asymmetric polysulfone membranes dried under different conditions: (a) T2W (water), (b) T2E (ethanol), and (c) T2H (hexane).

122

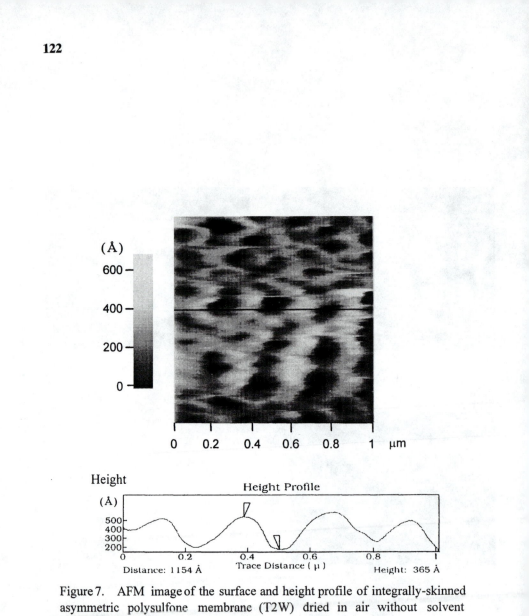

Figure 7. AFM image of the surface and height profile of integrally-skinned asymmetric polysulfone membrane (T2W) dried in air without solvent exchange.

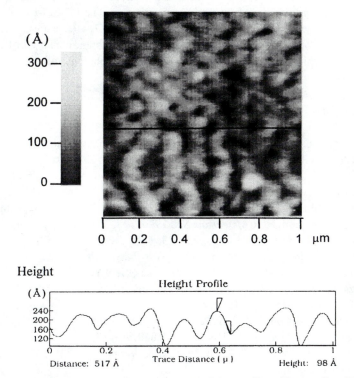

Figure 8.   AFM image of the surface and height profile of integrally-skinned asymmetric polysulfone membrane (T2H) dried in air after ethanol-hexane solvent exchange.

microporous asymmetric membrane densified on drying due to the collapse of micropores. Such a trend became less severe as the surface tension of the solvent decreased in the order of water, ethanol, and hexane. It was also observed from AFM analysis that the surface became smoother with a decrease in the liquid surface tension. Such a smooth and loose skin layer resulted in an increase of the gas permeance of the membrane. However, integrally-skinned asymmetric membranes with a very thin gas-tight skin layer on top of a finger-like substrate showed negligible change in the cross-sections and only a small change in the surface roughness. The solvent exchange did not show any effect on the gas transport properties of integrally-skinned asymmetric membranes. The FE-SEM image of the cross-section of the membrane dried using supercritical CO_2 after exchange of water with ethanol showed that the top skin layer of the membrane consisted of nodule-like particles, which might represent the nascent morphology of the membrane before drying.

Literature Cited

1. Kesting, R.E. *Synthetic Polymeric Membranes - A Structural Perspective*; 2nd Ed.; John Wiley and Sons, New York, NY, 1985; pp. 237-286.

2. Mulder, M. *Basic Principles of Membrane Technology*; 2nd Ed.; Kluwer Academic Publishers, Dordrecht, The Netherlands, 1996; pp. 75-140.

3. Wienk, I.M.; Boom, R.M.; Beerlage, M.A.M.; Bulte, A.M.W.; Smolders, C.A.; Strathmann, H. *J. Membrane Sci.* **1996**, *113*, 361.

4. Brown, G.L. *J. Polym. Sci.* **1956**, *22*, 423.

5. Pinnau, I.; Koros, W.J. *J. Polym. Sci: Part B: Polym. Phys.* **1993**, *31*, 419.

6. Brinker, C.J. *Sol-Gel Science*; Academic Press, New York, NY, 1990.

7. Fritzsche, A.K.; Arevalo, A.R.; Connolly, A.F.; Moore, M.D.; Elings, V.; Wu, C.M. *J. Appl. Polym. Sci.* **1992**, *45*, 1945.

8. Merten, U.; Gantzel, P.K. *U.S. Patent* 3,415,038, 1968.

9. MacDonald, W.; Pan, C.-Y. *U.S. Patent* 3,842,515, 1974.

10. Manos, P. *U.S. Patent* 4,080,743, 1978; *U.S. Patent* 4,080,744, 1978; *and U.S. Patent* 4,120,098, 1978.

11. Ohya. H.; Konuma, H. *J. Polym. Sci.* **1977**, *21*, 2515.

12. Dietz, P.; Hansuma, P.K.; Herrmann, K.H.; Inacker, O.; Lehmann, H.D. *Ultramicroscopy* **1991**, *35*, 155.

13. Kim, J.; Fane, A.G.; Fell, C.J.D.; Suzuki, T.; Dickson, M.R. *J. Membrane Sci.* **1990**, *54*, 89.

14. Bowen, W.R.; Robert, N.H.; Lovitt, W.; Williams, P.M. *J. Membrane Sci.* **1996**, *110*, 229.

15. Krausch, G.; Patterson, D. *Macromolecules* **1994**, *27*, 6768.

16. Bowen, W.R.; Robert, N.H.; Lovitt, W.; Williams, P.M. *J. Membrane Sci.* **1996**, *110*, 233.

Chapter 9

Thin-Film Composite Membranes Prepared from Sulfonated Poly(phenylene oxide): Preparation, Characterization, and Performance

G. Chowdhury, S. Singh, C. Tsang, and T. Matsuura

Department of Chemical Engineering, Industrial Membrane Research Institute, University of Ottawa, Ottawa, Ontario K1N 6N5, Canada

Thin-film composite (TFC) membranes were prepared by coating a thin layer of sulfonated poly(2,6-dimethyl-1,4-phenylene oxide) (SPPO) on the top surface of porous polyethersulfone support membranes of various pore sizes. Both thin-film composite membranes and porous support membranes were characterized by the mean pore size determined using solute transport data and atomic force microscopy. The effects of solvents used in the preparation of the coating solution on the surface structure of the thin-film composite membranes and their performance were also studied. The SPPO thin-film composite membrane in hydrogen form was ion-exchanged to a metal cation form, and the effects of ion-exchange on the solute rejection and flux were studied.

A promising membrane material for reverse osmosis (RO) applications is sulfonated poly(2,6-dimethyl-1,4-phenylene oxide) (SPPO) (*1*). Membranes prepared from SPPO possess a number of advantages when compared to other commercially available membranes with electric charges. SPPO can be cast into thin films having excellent chemical and physical stability along with good water flux and salt rejection characteristics. Asymmetric membranes prepared from SPPO were initially developed for RO treatment of brackish water (*2*). Characterizations of this material as a barrier layer in thin-film composite (TFC) membranes were carried out in the 1980s (*3-5*). Kubota and Yanase (*6*) prepared thin-film composite membranes from SPPO coated onto microporous polysulfone hollow-fibers for high-pressure RO applications.

SPPO-based TFC membranes for low pressure RO applications have been developed since the early 1990s in our laboratory (*7-9*). Preliminary experiments indicated that the solvent used in the coating solution had a marked effect on the composite membrane performance. Specifically, the water flux was affected significantly, whereas solute rejection changed to a much smaller extent. The SPPO-based membranes were characterized in our laboratory using solute transport measurements as well as by atomic force microscopy (10). Reverse osmosis performance of SPPO-based TFC membranes in different quarternary ammonium forms were also studied (11). Separation of inorganic and organic solutes by these membranes was controlled primarily by the membrane charge and the hydrophobic alkyl hydrocarbon chain of the quarternary ammonium.

This paper is focused on SPPO-based membranes for reverse osmosis and nanofiltration applications. Characterization of pore size distribution, surface roughness by atomic force microscope (AFM), effect of membrane material, membrane preparation, and testing conditions on membrane performance are discussed. Understanding these effects and properties of SPPO membranes may help predict their utility in potential applications.

Experimental

Membrane Material and Membrane Preparation. SPPO of the desired ion-exchange capacity (IEC) was prepared by direct sulfonation of PPO (available from General Electric Co.) in chloroform with a stoichiometric amount of chlorosulfonic acid following the method outlined by Plummer et al. (2). The exact IEC value of the polymer was determined using acid-base titration. The resulting modified PPO was in the hydrogen form (SPPOH).

Porous support membranes were prepared from polyethersulfone (PES, Victrex 4100P) by the phase inversion technique using casting solutions of different PES concentrations (10, 12, 15, and 20 wt. %) in N-methyl-2-pyrrolidone (NMP). Polyvinylpyrrolidone (PVP) was added to the casting solution in a ratio of PES:PVP=1:1 (by weight) according to the method described by Lafréniere et al.(12). Some membranes were stored in distilled water without being dried and they are hereafter referred to as 10U, 12U, 15U, and 20U membranes, depending on the concentration of PES. Some of the membranes were immersed in a 30 wt.% glycerol solution for 24 hours prior to air drying at room temperature for several days. The glycerol-treated membranes are hereafter referred to as 10UD, 12UD, 15UD, and 20UD membranes.

The SPPO composite membranes were prepared as follows: a dilute solution of SPPO (1-3 wt.%) in a solvent or a solvent mixture was poured and spread over the skin side of a microporous substrate. The excess solution was drained by holding the membrane vertically, leaving a thin layer of SPPO solution on the substrate surface. The SPPO-coated membrane was then dried. Composite membranes made from a 1 wt.% SPPO solution in methanol are designated as 10S, 12S, 15S, and 20S membranes. The reproducibility of the thin-film composite membranes so prepared was ± 3 % in solute rejection and ± 6 % in water flux. To obtain SPPO in metal ion form (SPPOMe$^+$), the membrane in the SPPOH form was equilibrated in an aqueous solution of the hydroxide of the respective metal.

Membrane Characterization. The membranes were tested for ultrafiltration of polyethylene glycol (PEG, molecular weight up to 35,000 g/mol) and polyethylene oxide (PEO, molecular weights of 100,000 to 200,000 g/mol) at 345 kPa (gauge) after compaction at 552 kPa (gauge) for 5 hours. AFM images were obtained by using Nanoscope III from Digital Instruments Inc., USA. A non-contact mode of AFM in air was used to obtain the membrane pore sizes and roughness parameters.

Results and Discussion

Surface Morphology of Microporous Polyethersulfone and Composite SPPO Membranes. Differences in the membrane surface morphology can be expressed in terms of various roughness parameters, such as the mean roughness, R_a, the root-mean-square of Z data, R_q, and the mean difference in the height between the five highest peaks and the five lowest valleys, R_z. All these parameters can be determined by AFM.

The mean roughness is the mean value of surface relative to the center plain, and is calculated as:

$$R_a = \frac{1}{L_x L_y} \int_0^{L_x} \int_0^{L_y} |f(x,y)| dx dy \tag{1}$$

where f(x,y) is the surface relative to the center plane and L_x and L_y are the dimensions of the surface.

The root-mean-square of Z values (R_q) is the standard deviation of the Z values within a given area and is calculated as:

$$R_q = \sqrt{\frac{\sum \left(Z_i - Z_{avg}\right)^2}{N_p}} \tag{2}$$

where Z_i is the current Z value and Z_{avg} is the average of the Z value within the given area and N_p is the number of points within the given area. The average difference in height, R_z, between the five highest peaks and five lowest valleys is calculated relative to the mean plane, which is the plane about which the image data has a minimum value.

The solute separation of PEG and PEO of the microporous PES and SPPO/PES composite membranes versus the solute diameters is shown in Figure 1. The values for mean pore size, μ_p, and standard deviation, σ_p, are summarized in Table I.

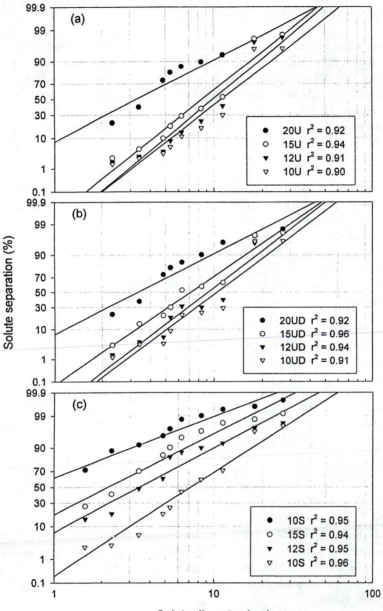

Figure 1. Solute separation curves (solute diameter versus their separation) plotted on log-normal probability paper for (a) U (uncoated without drying) membranes, (b) UD (uncoated with drying in air) membranes and (c) S (SPPOH-coated) membranes.

Table I. Geometric mean pore size (μ_p) and geometric standard deviation (σ_p) of various membranes calculated from separation data and from AFM images. Molecular-weight cut-off values (MWCO) obtained from solute transport measurements are also shown.

Membrane[a]	From solute transport			From AFM images	
	MWCO (g/mol)	Mean pore size μ_p (nm)	Geometric Std Dev. σ_p	Mean pore size μ_p (nm)	Geometric Std Dev. σ_p
20U	21,000	3.2	2.4		
20UD	20,000	3.4	2.3		
20S	3,500	0.7	3.3		
15U	84,000	8.4	1.7		
15UD	75,000	7.2	1.8	25	1.6
15S	11,000	2.2	2.4		
12U	94,000	9.9	1.7		
12UD	91,000	9.1	1.7	32	1.5
12S	20,000	3.4	2.3		
10U	98,000	11.1	1.8		
10UD	94,000	10.4	1.8	38	1.4
10S	71,000	7.1	2.0	31	1.5

[a] First two digits indicate the polymer concentration in the casting solution of the porous polyethersulfone support membrane; U indicates uncoated membranes without drying; UD indicates uncoated membranes with drying; S indicates membranes coated with SPPO.

In general, there was a tendency that the mean pore size, μ_p, increased, whereas σ_p decreased, as the PES concentration, (indicated by the first two digits in the membrane code) decreased. The mean pore size for the 10U membrane (molecular weight cut-off (MWCO) = 98,000 g/mol) was found to be 11 nm, which is comparable to the mean size of 15 nm of a sulfonated polysulfone membrane with a MWCO of 100,000 g/mol obtained from solute transport data (16). 20U, 20UD and 12S membranes which showed a MWCO of about 20,000 g/mol had similar mean pore sizes of 3.2, 3.4 and 3.4 nm, respectively. When a thin layer of SPPO was coated on the surface of UD membranes, the pore sizes were reduced considerably. This effect is obvious when μ_p values for 20S, 15S, and 12S membranes are compared with those of the corresponding UD membranes. For membrane 10S, the pore size reduction was modest from 10.4 nm to 7.1 nm. The σ_p values, on the other hand, increased after coating of SPPO.

Michaels found that σ_p values of different membranes, both biological and synthetic, were very close to each other (from 1.2 to 1.7) (13). On this basis, it was concluded that virtually all membrane filters, irrespective of their origin, were quite similar in their microstructure. However, membranes prepared in this study showed a wide range of σ_p values (1.7 to 3.3). A σ_p value as high as 7.4 was previously reported for a clay (montmorillonite) membrane (*17*).

The diameters of 50 pores measured from AFM images were plotted against the median ranks in a log-normal probability plot, which yielded a straight line with correlation coefficients of $r^2 > 0.97$. The mean pore size and the standard deviation are listed in Table I. The 10UD membrane with a MWCO of 94,000 g/mol showed a mean pore size of 38 nm. Dietz *et al.* reported a mean pore size of 25 nm for a polyethersulfone membrane having a MWCO of 100,000g/mol (*18*). Mean pore sizes measured by the AFM techniques were about 3.5 times larger than those calculated from solute transport data. Bessières *et al.* also observed that AFM images gave 2 to 4 times larger diameters than those obtained from solute (PEG) transport data (*16*). It was suggested that the pore size obtained from solute separation data corresponded to a minimal size of the pore constriction experienced by the solute while passing through the pore. On the other hand, the pore size measured by AFM corresponds to the funnel-shaped pore entrance and has a maximum opening at the entrance. The pore sizes obtained for 20U, 20UD, 20S, 15S, and 12S membranes were too large to be assigned for actual pore sizes. Pores were indistinct and agglomeration of a few small pores could easily be misinterpreted as one large pore, resulting in overestimation of the pore sizes. It was also difficult to distinguish between the pores and the depressions in the membrane surface.

The mean roughness parameter values are reported in Table II. Generally, the surface parameters of UD membranes increased from 20UD to 10UD, as the MWCO of the membrane increased, whereas the opposite tendency was observed for the S membranes. This result indicates that the surfaces of the S membranes were much smoother than the corresponding UD membranes.

Table II. Various roughness parameters measured from AFM images (500 nm x 500 nm) for different membranes.

Membrane	Roughness parameters (nm)		
	R_a	R_q	R_z
20UD	0.68* + 0.12**	0.86 ± 0.15	5.07 ± 1.32
20S	0.18 ± 0.04	0.23 ± 0.05	1.09 ± 0.16
15UD	0.61 ± 0.07	0.78 ± 0.08	4.03 ± 0.61
15S	0.14 ± 0.07	0.22 ± 0.04	1.06 ± 0.12
12UD	1.02 ± 0.12	1.31 ± 0.17	6.64 ± 0.61
12S	0.06 ± 0.01	0.09 ± 0.02	0.72 ± 0.10
10UD	2.02 ± 0.12	2.40± 0.46	14.02 ± 2.77
10S	0.07 ± 0.01	0.09 ± 0.01	0.59 ± 0.07

* mean value; ** standard deviation

Effect of Solvent in the Coating Solution on Membrane Performance.
SPPOH with an IEC value of 1.93 meq/g polymer was dissolved in different
solvents to prepare a 1 wt. % solution, and the solution was coated on the top-side
of a PES ultrafiltration membrane (supplied by Osmonics, HW 17). The coating was
repeated three times. The membranes were maintained in the hydrogen form
(SPPOH) and were stored in distilled water until testing. The solvents used were
chloroform/methanol mixtures with chloroform contents of 0, 18, 42, and 66 wt.%,
respectively. Intrinsic viscosity of SPPOH in different solvent mixtures was
measured at 25°C. Reverse osmosis experiments were performed using four
composite membranes and three electrolyte solutions.

The results for solute separation and those of flux are shown in Figures 2 and 3,
respectively. The flux of each membrane increased, whereas the solute rejection
decreased with an increase in the cationic radius from Li^+ to K^+. The solvent mixture
used to prepare coating solutions showed a marked effect on the membrane
performance. The solute separation increased slightly with an increase in the
chloroform content in the solvent mixture, with an exception for the membrane made
from 66 wt% chloroform in the coating solution. Flux, on the other hand, showed a
steep decline with an increase of chloroform content. Four PES substrate
membranes, which were the same as the one used for the preparation of the
composite membranes, were contacted with methanol and chloroform/methanol
mixtures of 18, 42, and 66 wt.% chloroform, respectively, and were dried after
draining the solvent from the membrane surface. The solvent exposure and the
drying were carried out similar to that for SPPOH-PES composite membrane
preparation. Then, pure water flux was determined for each membrane after one
hour of compaction at 2069 kPa gauge (300 psig). Pure water flux measurements at
1379 kPa gauge (200 psig) were 68, 62, 66, and 45 L/m^{2}h (1 L/m^{2}h = 0.59 GFD) for
the first, second, third, and fourth membrane respectively. Although the last
membrane showed a lower flux than other membranes, it was about six times higher
than the flux after SPPO coating. These results suggest that the effect of the
composition of the solvent mixture on the membrane properties was primarily due
to a change in the polymer morphology in the coating layer.

The intrinsic viscosities of the SPPOH polymer with an IEC value of 1.93 meq/g
in different solvent mixtures are given in Table III. The data show that the intrinsic
viscosity decreased with an increase in chloroform content. Thus, as the chloroform
content in the solvent mixture is increased, the intra-macromolecular SPPOH
interactions become more favorable than the SPPOH-solvent interactions. The
stronger polymer-polymer interactions lead to tighter SPPOH coils. The
macromolecular orientation in the coating solution is preserved in the coating layer
even after solvent evaporation. Thus, the macromolecular structure of the membrane
becomes tighter when the coating is made from a solution of a higher chloroform
content.

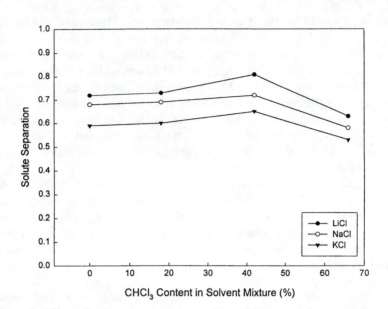

Figure 2. Effect of solvent system in coating solutions on the separation of different electrolytes. Operating conditions: pressure: 1379 kPa gauge (200 psig); temperature: 25°C.

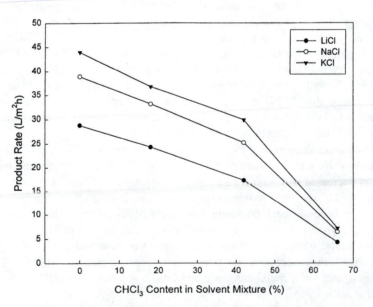

Figure 3. Effect of solvent system in coating solutions on the flux (given as product rate) of different electrolyte solutions. Operating conditions: pressure: 1379 kPa gauge (200 psig); temperature: 25°C.

Table III. Intrinsic viscosities of SPPOH polymer having an IEC value of 1.93meq/g in different chloroform-methanol solutions.

Solvent composition (wt.%)	Intrinsic viscosity at 25°C (dL/g)
CH$_3$OH	2.0
CHCl$_3$ (18)/CH$_3$OH (82)	1.5
CHCl$_3$ (42)/CH$_3$OH (58)	1.3
CHCl$_3$ (66)/CH$_3$OH (34)	0.9

The three-dimensional topographical picture obtained by AFM showed that the membrane surface was not smooth but consisted of many nodules. These nodules are considered to be aggregates of polymer aggregates as defined by Kesting (*19*). The composite membrane micrographs were further subjected to section analysis, using Nanoscope III software, to estimate the size range of the nodules on the surface of the four composite membranes. The sizes of nodules (diameter) for the composite membrane surfaces are shown in Table IV.

Table IV. Estimated nodule diameters of the SPPOH coating layer as a function of chloroform content in the coating solution.

Chloroform content in coating solution (wt.%)	Nodule diameter (nm)
Substrate membrane	71-153
0	85-125
18	54-70
42	37-51
66	20-32

Based on the results presented in Table IV, the SPPOH polymer appears to have a tighter macromolecular structure in a coated layer formed by using a higher chloroform content in the solvent system, resulting in a smaller nodule size.

Effect of Ion-Exchange from SPPOH Hydrogen Form to Cation Form on Membrane Performance. A solution of 1 wt.% SPPOH in methanol (PPO of [η] = 0.46 in chloroform at 25°C, IEC.= 2.0 meq/g) was coated on top of a commercial polyethersulfone substrate membrane (HW 17, supplied by Osmonics). Reverse osmosis experiments were carried out at 1000 kPa (gauge) in the presence of different electrolyte solutes. The results are summarized in Table V.

The flux in the presence of the electrolyte solute is higher than the pure water flux with the exception of LiCl, for which the mixture flux was nearly equal to the pure water flux. These results can be interpreted by the proton (H$^+$)-alkali metal cation (Me$^+$) exchange written as:

$$-SO_3H + Me^+hy \rightarrow -SO_3^- Me^+ + H^+hy$$

Table V. Reverse osmosis performance of SPPOH composite membranes for alkali chloride solutes.[a]

Solute	Pure water flux (L/m^{2}h)	Mixture water flux (L/m^{2}h)	Solute separation (%)
LiCl	63	61	87.8
NaCl	63	77	86.5
KCl	63	88	83.5
CsCl	63	91	77.0

[a] Feed concentration: 0.01 mol/L salt/water; feed pressure: 1000 kPa (gauge); feed temperature: 25°C. 1 L/m^2·h = 0.59 GFD (gallons/ft^2/day).

The energy involved in the above exchange reaction is the binding energy of SO_3^- Me^+ minus the hydration energy of Me^+. The latter energy difference is lowered with an increase of the radius of the cation when the size of the anion is as large as that of SO_3^-. Thus, the reaction is shifted to the right side as the cationic radius increases from Li^+ to Cs^+. Table VI shows the equilibrium constants for the exchange reaction for polystyrene sulfonate (20):

$$K_0 = [(-SO_3^-Me^+)(H^+)]/[(-SO_3H)(Me^+)]$$

Table VI. Ion-exchange equilibrium constant K_0 of sulfonated polystyrene for different alkali metal cations (20).

Cation	Log K_0
Li^+	-0.1
Na^+	0.17
K^+	0.37
Cs^+	0.38

The equilibrium constant K_0 increases from Li^+ to Cs^+. The same tendency is believed to be valid also for PPO. When the above exchange reaction takes place, SO_3H becomes $SO_3^-Me^+$ in the SPPO polymer. Comparing $-SO_3H$ and $-SO_3^-Me^+$, the former contains only one covalently bonded species, whereas the latter contains two ionic species. Therefore, the number of species increases in the SPPO polymer as the reaction proceeds from left to right, resulting in more water in the polymer phase due to the osmotic effect. Combining the ion exchange and osmotic effect, it is understandable that the uptake of water in the polymer membrane increases from Li^+ to Cs^+. As a result, the water flux of the membrane increases.

The pure water flux and solute separation data also depend on the metal cation, Me^+, when $-SO_3H$ is completely ion-exchanged to $-SO_3^-Me^+$. To assure the complete exchange of $-SO_3H$ to $-SO_3^-Me^+$, the membrane was immersed into 0.5 N alkali hydroxide and washed with distilled water before use in reverse osmosis experiments. Table VII shows the results of the reverse osmosis experiments. The pure water flux and mixture water flux decreased, whereas the separation increased with an increase in the cationic radius from Li^+ to K^+.

Table VII. The effect of counter-ions in SPPOH-Me^+ composite membranes on separation and flux in reverse osmosis of sodium chloride solution.[a]

Counter-ion	Pure water flux (L/m^2h)	Mixture water flux (L/m^2h)	Solute separation (%)
Li^+	90	91	64
Na^+	83	90	67
K^+	69	76	74

[a] Feed concentration: 0.04 mol/L NaCl/water; feed pressure: 1000 kPa (gauge); feed temperature: 25°C.

The number of charged species are independent of the Me^+ type because $-SO_3H$ is completely exchanged to $SO_3^-Me^+$. Me^+ may still be partially hydrated when it is near SO_3^-, and the degree of hydration may decrease as the ionic radius increases from Li^+ to K^+. Thus, less water is in the membrane phase and the flux decreases from Li^+ to K^+. The solute separation, on the other hand, will increase from Li^+ to K^+. The increase in solute separation with an increase in the ionic radius of the counter-ion was also observed when the solutes were nonelectrolyte organic molecules. The reverse osmosis results of the various SPPO-Me^+ membranes for glucose, sucrose, and raffinose are shown in Table VIII.

Table VIII. Effect of counter-ion of SPPO-Me^+ membranes on the separation of carbohydrates.[a]

| Counter-ion | Solute separation (%) | | |
	glucose	sucrose	raffinose
Li^+	65	70	80
Na^+	66	77	86
K^+	71	83	92

[a] Feed concentration: 100 ppm; feed pressure: 1000 kPa (gauge); feed temperature: 25°C.

Conclusions

Pore size distributions obtained from solute transport data and by atomic force microscopy (AFM) fit log-normal distributions reasonably well. The mean pore sizes obtained by AFM are larger than those obtained by using solute transport data. The coating solution solvent affects the performance of the resulting thin-film SPPOH composite membranes significantly. Solvents with stronger dissolving power result in coatings with higher fluxes without sacrificing the solute rejection. Ion-exchange of SPPOH membranes to the metal cation form affects both solute rejection and flux significantly.

Literature Cited

1. LaConti, A.B. in *Reverse Osmosis and Synthetic Membranes*, S. Sourirajan, Ed., National Research Council of Canada, Ottawa, (1977) pp. 211-229.
2. Plummer, C.W.; Kimura, G; LaConti, A.B. Development of sulfonated polyphenylene oxide membrane for reverse osmosis, Research and Development Progress Report # 551, Office of Saline Water, United States Department of Interior, (1970).
3. Huang, R.Y.M.; Kim, J.J. *J. Appl. Polym. Sci.* **1984**, *29*, 4017.
4. Huang, R.Y.M.; Kim, J.J. *J. Appl. Polym. Sci.* **1984**, *29*, 4029.
5. Agarwal, A.K. *Ph.D. Thesis*, University of Waterloo, 1991.
6. Kubota, N.; Yanase, S. Jpn. Kokai Tokkyo Koho JP 63, 229, 109.
7. Chowdhury, G.; Matsuura, T.; Sourirajan, S. *J. Appl. Polym. Sci.* **1994**, *51*, 1071.
8. Hamza, A.; Chowdhury, G.; Matsuura, T.; Sourirajan, S. *J. Appl. Polym. Sci.* **1995**, *58*, 620.
9. Hamza, A.; Chowdhury, G.; Matsuura, T.; Sourirajan, S. *J. Membrane Sci.* **1997**, *129*, 55.
10. Singh, S.; Khulbe, K.C.; Matsuura, T.; Ramamurthy, P. Membrane characterization by solute transport and atomic force microscope, submitted to J. Membrane Sci.
11. Tsang, K.-Y.C. B.A. Sc. Thesis, Chemical Engineering, University of Ottawa, 1995.
12. Lafréniere, L.Y.; Talbot, F.D.F.; Matsuura, T.; Sourirajan, S. *Ind. Eng. Chem. Res.* **1987**, *26*, 2385.
13. Michaels, A.S. *Sep. Sci. And Technol.* **1980**, *15*, 1305.
14. Yeoum, K.H.; Kim, W.S. *J. Chem. Eng. Japan* **1981**, *24*, 1.
15. Lipson, C.; Sheth, N.J. *Statistical Design and Analysis of Engineering Experiments*, McGraw-Hill, New York, (1973).
16. Bessières, A.; Meireles, M.; Coratger, R.; Beauvillain, J.; Sanchez, V. *J. Membrane Sci.* **1996**, *109*, 271.
17. Ishiguro, M.; Matsuura, T.; Detellier, C. *Sep. Sci. Technol.* **1996**, *31*, 545.
18. Dietz, P.; Hansma, K.; Inacker, O.; Lehmann, H.D.; Hermann, K.-H. *J. Membrane Sci.* **1992**, *65*, 101.
19. Kesting, *Synthetic Polymeric Membranes: A Structural Perspective*, 2nd Edn., Wiley, New York, 1985.
20. Marcus, Y.; Howery, D.W. *Ion Exchange Equilibrium Constants*, Butterworths, London, 1975.

Chapter 10

Functionalized Polysulfones: Methods for Chemical Modification and Membrane Applications

Michael D. Guiver[1], Gilles P. Robertson[1], Masakazu Yoshikawa[2], and Chung M. Tam[1]

[1]Institute for Chemical Process and Environmental Technology, National Research Council, Ottawa, Ontario K1A 0R6, Canada
[2]Department of Polymer Science and Engineering, Kyoto Institute of Technology, Matsugasaki, Sakyo-Ku, Kyoto 606, Japan

Lithiation chemistry is an extremely effective method for modifying polysulfones. Polysulfones (PSf) are activated rapidly and quantitatively with a lithiating reagent. Lithiated intermediates are converted to a wide array of functional group types by addition of suitable electrophiles. The scope and versatility of this chemistry is outlined in this paper. Functional groups in polysulfones tailor membrane properties, while the inherent stability of the polymer backbone is retained. In ultrafiltration (UF), hydrophilic groups can enhance water flux, reduce pore size, and have the potential for decreased fouling. Similarly, hydrophilic groups enable chlorine resistant PSf to be used in nanofiltration (NF) and reverse osmosis (RO). Gas separation membranes are highly material dependent, and significant changes in solubility selectivity, fractional free volume, and chain rigidity are possible through appropriate choice and site of functional group. Reactive groups provide attachment sites and crosslinking capability to membranes.

Functionalized polymers are of general interest in many technology applications (*1*). Customized polymers for reagents, catalysts, chromatography media, controlled release formulations, conductive materials, photoresists, and liquid crystals are just a few examples. Our research is concerned with the functionalization of aromatic polysulfones for tailoring properties in membrane applications. Polysulfone has overall thermal and chemical stability combined with good mechanical and membrane making qualities. It is a stable platform for functional group attachment and a good candidate material for membranes having tailored properties. However, there are few options for efficient modification chemistry because of the polymer's inherent chemical stability.

NRCC No. 37647

137

Polymer Lithiation.

Several polysulfone modification chemistries have been reported by Daly *et al.* (*2*) (e.g. nitration-amination and chloromethylation) and by Warshawsky *et al.* (*3*) (e.g. aminomethylation and chloromethylation). In the case of chloromethylation, the derivative is isolated before further reaction to other derivatives. Our research is focused on a functionalization method whereby a reactive intermediate is made and immediately reacted further without isolation, giving a variety of functional group polysulfones. This modification procedure is based on directed lithiation, taking advantage of the strong ortho-directing power of the sulfone group in the polysulfone backbone. The sulfone group is one of the best ortho-directing groups in lithiation chemistry due to both the strong electron withdrawing effect, and to the ability of the lone pairs of electrons on the sulfone oxygen atoms to complex with the lithiating agent. Butyllithium replaces the ortho-sulfone hydrogen atoms with lithium atoms, so that the aromatic carbon atoms to which lithium is attached nominally has negative charge. The activated site of lithiation is exclusively ortho to the sulfone linkage group (*4*).

Direct Lithiation. Figure 1 shows the conversion of Udel polysulfone (PSf) to lithiated polysulfone reactive intermediate, having a degree of substitution (DS) of 2.0 lithium atoms per repeat unit. Lithiation of Udel in the DS range 0 - 2.0 is achieved rapidly and near quantitatively simply by controlling the amount of butyllithium. It meets many criteria for useful modification procedures: a) practical for scale-up, using readily available reagents; b) rapid and near quantitative conversions; c) site specificity; d) absence of chain degradation or crosslinking, and e) versatility whereby lithiation can lead to a wide variety of functional groups. Further substitution of lithium up to a DS of nearly 3.0 is achieved using excess butyllithium, as shown in Figure 1. Preparation of lithiated intermediates with DS > 2.0 requires longer reaction times and the lithiated polymer precipitates from solution, unlike DS ≤ 2 which remains homogeneous in solution.

Polymers Amenable to Lithiation. A number of different polysulfones are amenable to the direct lithiation reaction, among those are Udel polysulfone, Radel-R polyphenylsulfone, and bis-6F-polysulfone, whose structures are shown in Figure 2. The first two are commercial thermoplastics from Amoco Performance Products. Udel lithiation is very convenient because high concentrations of polymer solution in THF can be reacted. Radel-R also undergoes near quantitative lithiation, but has limited solubility in THF. Bis-6F-polysulfone requires a measured excess of butyllithium to achieve a desired DS. This is apparently due to the presence of fluorine atoms which complex with butyllithium, thereby retaining some of the reagent passively until the reaction is quenched with electrophile. Many other aromatic sulfone containing polymers can undoubtedly be lithiated providing a) they are soluble in suitable solvents, b) there are no conflicting functional groups (e.g. ketones) present in the polymer structure, and c) the balance of electronic effects in the chain is favorable. This latter effect, combined with poor solubility, is

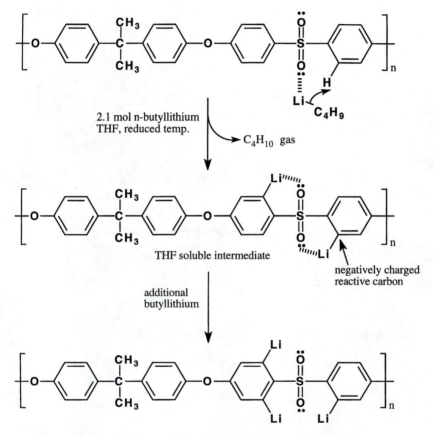

Figure 1.   Ortho-sulfone directed lithiation of polysulfone.

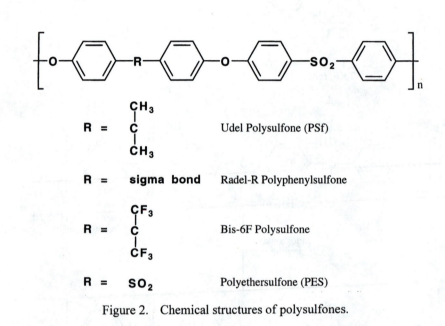

Figure 2. Chemical structures of polysulfones.

the most likely reason that attempts to lithiate polyethersulfone under homogeneous conditions have not succeeded. Polyethersulfone contains ether and sulfone linkages alternating between aromatic rings. The ether linkage itself is a weak directing group, as poly(2,6-dimethylphenyl)ether was lithiated in the reported work of Chalk and Hay (5). However, the alternating electron donating - electron withdrawing linkages present in polyethersulfone result in an overall unfavorable balance in the polymer chain.

Bromination - Lithiation. Lithiated polysulfones may also be prepared by metal-halogen exchange. Both Udel and Radel-R may be brominated to DS = 2.0 at the ortho-ether site (6). At reduced temperatures, lithium-bromine rather than lithium-hydrogen exchange occurs almost exclusively. While bromination encompasses an extra step, it enables functionalization to occur specifically around the ether linkage, as shown in Figure 3. Once both bromine sites are exchanged with lithium, excess butyllithium lithiates the ortho-sulfone site to a DS of nearly 3.0. Lithiating polysulfone with bromine substitution less than 2.0 results in incomplete bromine-lithium exchange because of a competing ortho-sulfone reaction. This is because of electronic effects; lithiation competes at ortho-sulfone segments when they are adjacent to bisphenol segments having no bromine atom. Thus, fully dibrominated polymer tends to inhibit ortho-sulfone lithiation.

Polymer Functionalization

Following lithiation to the desired DS (controlled by the quantity of lithiating agent added), an electrophile is added which reacts with the intermediate, forming a functional group. Negatively charged carbon atoms on lithiated polysulfone react with electrophilic carbon or heteroatom centers, either by addition reaction (e.g. CO_2) or by replacement of a leaving group (e.g. Cl in $ClSiMe_3$). A wide variety of functional groups can be attached to the polymer by using simple electrophiles (7) as shown in Table I.

More complex functional groups can also be attached by judicious choice of electrophile. The conditions for electrophile addition are highly dependent on the reactivity of the particular class of electrophile. Many electrophiles react easily and require minimum precautions during the addition step to prevent unwanted side reactions such as crosslinking. Typically, lithiation is conducted conveniently between -30°C to -78°C, then the temperature is adjusted if necessary, and excess electrophile is added to the intermediate. Some electrophiles have the potential for crosslinking to occur. This must be avoided or minimized by rapid addition of excess electrophile at an optimum temperature and conditions whereby the desired primary reaction occurs rapidly. Depending on the electrophile, factors to consider are: rate of addition, temperature, crosslinking potential, viscosity, and efficient mixing of the product. Conversion of the lithiated polymer to a functional derivative was usually close to quantitative, except in cases where side reactions of

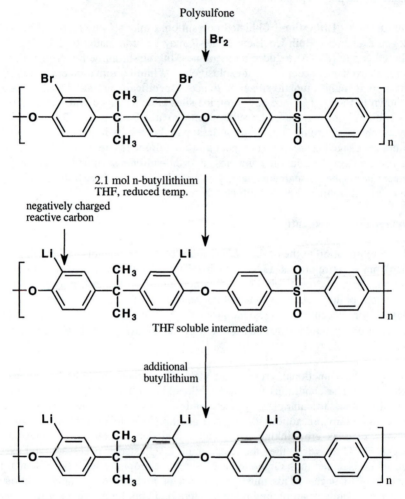

Figure 3.   Ortho-ether lithiation by halogen-metal exchange.

electrophiles could occur, such as in the case of alcohol derivatives with certain carbonyl containing electrophiles.

Table I.   A selection of functional group derivatives formed by reaction of lithiated polysulfones with various simple electrophilic reagents.

Electrophilic reagent	Polymer derivative	Comments
CO_2	-COOLi, -COOH	hydrophilic, reactive
SO_2	$-SO_2Li$, $-SO_2H$	sulfonic acid by oxidation
$CH_2=O$	$-CH_2OH$	hydrophilic, reactive
$CH(CH_3)=O$	$-CH(CH_3)OH$	1° & 2° alcohols
$CH(Ph)=O$	$-CH(Ph)OH$	
$C(CH_3)_2=O$	$-C(CH_3)_2OH$	
$C(Ph)_2=O$	$-C(Ph)_2OH$	3° alcohols
tosyl azide	$-N=N=N$	reactive
azide reduction	$-NH_2$	reactive
$ClSiR^1R^2R^3$	$-SiR^1R^2R^3$	hydrophobic, bulky group
$ClSn(CH_3)_3$	$-Sn(CH_3)_3$	bulky group
CH_3S-SCH_3	$-SCH_3$	
$ClP(Ph)_2$	$-P(Ph)_2$	bulky group
RBr	-R	hydrophobic
$BrCH_2CH=CH_2$	$-CH_2CH=CH_2$	reactive
$PhC\equiv N$	$-C(Ph)=NH$	reactive
I_2	-I	reactive
$PhN=C=O$	$-C=ONHPh$	hydrophilic

Carboxylated Polysulfone. CPSf is formed readily by reaction of carbon dioxide with lithiated polysulfone (*8*). The carboxylation reaction has a potential for forming ketone crosslinks by reaction of two lithiated chain sites with one molecule of carbon dioxide. An additional practical difficulty is that the product is a solid gel that must be mixed rapidly and efficiently with dry ice. We have effectively produced un-crosslinked material in 250 g lab scale reactors by using wide-neck flasks fitted with resin kettle flanges, and large excesses of freshly made dry ice. This material has also been produced in 4 kg batches with a specially designed reactor. RO, NF, UF, and pervaporation membranes have been made from carboxylated polysulfone.

Sulfinic Acid Derivative. This is produced by the reaction of liquefied sulfur dioxide gas, and dropping the liquefied SO_2 into a lithiated polysulfone solution with rapid stirring (*7*). The sulfinic acid group has been relatively little studied and can undergo some interesting chemical reactions which we are currently exploiting in the preparation of membranes. It has also been reported by Kerres *et al.* (*9*) that polysulfone sulfinic acid is readily oxidized to sulfonic acid using hypochlorite or permanganate, and has application as an ion exchange membrane. Previous processes sulfonated the bisphenol segment, whereas the lithiation route sulfonates the phenylsulfone segment. Higher degrees of sulfonation and absence of chain degradation are advantages over some other methods.

Alcohol Derivatives. Primary, secondary, and tertiary benzyl alcohols are prepared by the addition of the appropriate aldehyde or ketone to lithiated polysulfone (*10*). Ketones with protons alpha to the carbonyl group (e.g. acetone, acetaldehyde) do not form product quantitatively because of the side reaction involving carbanion removal of these acidic protons. In the case of benzophenone or benzaldehyde, both of which have no acidic protons , the formation of the carbinols are quantitative in spite of sterically hindered structures. In practical terms, the preparation of the -CH$_2$OH derivative is more difficult because formaldehyde gas, generated *in situ* from paraformaldehyde, has low mixing efficiency with the viscous lithiated polymer solution. Efforts to produce this product directly from para-formaldehyde and trioxane did not yield desired product.

Amine Derivatives. Aminated aromatic polymers have generally been prepared previously by a nitration - reduction sequence, where chain degradation may occur during nitration. The amine group, having two acidic protons, cannot be added to the polymer directly by lithiation because of incompatibility with the carbanionic lithiated polysulfone. We were successful in using an azide transfer reagent (tosyl azide) to quantitatively introduce azide groups, which are masked amines, into the polymer (*11*). A simple mild reduction of the azide group with borohydride converted it quantitatively to amine. As far as we are aware, this is the first time this chemistry has been exploited as a method of introducing amine groups efficiently into aromatic polymers. A one-pot preparation of aminated polysulfones *via* azides was developed (*12*). The polymeric azides can also be easily isolated, and cast into films. Heating or exposing polysulfone azides to UV results in loss of nitrogen and generation of a reactive nitrene. Crosslinkable films or membranes were prepared by this method, the probable crosslink being a stable nitrogen bridge.

Azide Cycloadditions. The reactivity of the polysulfone azides in 2 + 3 cycloadditions also provides access to a series of functional derivatives (*13*). Cycloaddition reactions of azides with substituted acetylenes give 1,2,3-triazoles, while with substituted olefins give 1,2,3-triazolines, as shown in Figure 4. Examples of polysulfone 1,2,3-triazole derivatives prepared by this method include carboxylic ester, hydroxyl, and trimethylsilyl. Norbornene, which is a strained olefin, readily gives a relatively stable bulky norbornyltriazoline pendant group.

Silyl Derivatives. Silyl electrophiles (e.g. chlorotrimethylsilane) are rather slow to react and require temperatures of between -5°C and -25°C depending on the alkyl or aryl substituent (*7*). The order of reactivity is ClSiMe$_2$Ph > ClSiMe$_3$ > ClSiMePh$_2$. We are currently investigating gas permeation properties of a series of silyl substituted polysulfones.

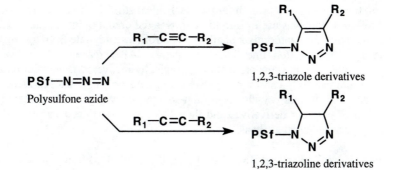

Figure 4.   Triazole and triazoline derivatives by 2 + 3 cycloaddition reactions of polysulfone azides.

Tricyclic Polymers. Interesting tricyclic polymer structures have been prepared by utilizing the dilithiation geometry around the ortho-sulfone site. This ring-closure method is not as practical for large scale preparation because of the exacting conditions required to favor intra-chain reaction, and minimize inter-chain (crosslinking) reactions. However, it offers an accessible route to novel polymers. The general reaction scheme is shown in Figure 5.

Experimental Methods for Lithiation

General Methods. Udel P3500 polysulfone, Radel R5000 polyphenylsulfone (Amoco Performance Products, Inc.) and 6F-polysulfone (prepared by synthesis) were dried at 120°C before modification. n-Butyllithium (10M in hexane) solution and other chemicals were obtained from Aldrich. Lithiation reactions were performed under an inert atmosphere of dry argon in oven dried glassware. Polymer solutions were prepared using anhydrous reagent grade tetrahydrofuran (THF), or freshly distilled over lithium aluminum hydride for small scale reactions. Adequate and controllable mechanical mixing is very important for handling the viscous solutions that occur. A high torque mechanical stirrer, with a stainless steel shaft and teflon paddle was used. Butyllithium was fed onto a point approximately two-thirds out from the center mixer shaft, at a controlled steady rate from a syringe pump via a flexible Teflon line into the reactor. After reacting the lithiated polymers with electrophiles, the products were typically recovered from solution by precipitation into carefully chosen non-solvents (e.g. alcohol) in a Waring blender, then washed several times with fresh non-solvent or water. The structural determination of polymer derivatives and their DS were determined by 400 MHz NMR, FTIR, and elemental analysis.

Polysulfone (Udel) Lithiation, DS = 2.0. Polysulfone solutions in tetrahydrofuran (THF) were reacted with an approximately 5% excess (of desired DS) of n-butyllithium at reduced temperatures (-30°C to -78°C). Above -5°C, crosslinking of the lithiated polymer occurred. During the first few drops of butyllithium addition, a colour change occurred. Lithiated polysulfone forms a homogeneous red-brown solution, and becomes increasingly viscous at DS ≤ 2. Typical polymer concentrations are approximately 10 wt% for DS ≤ 1, and 5 wt% for DS ≤ 2, however this is somewhat dependent on the electrophile and the ultimate state or viscosity of the product following addition of the electrophile. As an example, a solution of polysulfone (44.2 g, 100 mmol) in THF (800 mL) was cooled to -78°C by immersion in a dry ice/alcohol bath. n-Butyllithium (21.5 mL, 215 mmol) was injected drop-wise at a rate of 30 mL/h by a syringe pump. Appropriate quantities of butyllithium were used to obtain lithiated polymers having a required DS lower than 2.0, since its reaction with polysulfone is near quantitative. The mixture was stirred for 30 minutes following n-butyllithium addition, then reacted with an electrophile. The structure of lithiated polysulfone having a DS of 2.0 is shown in Figure 1.

Dilithiated (ortho-sulfone) Polysulfone

$$\downarrow \text{Cl}_2\text{Si}(\text{CH}_3)_2$$

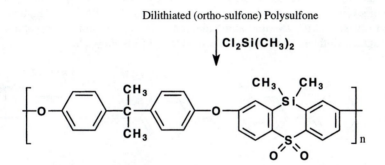

Figure 5.   Tricyclic polysulfone silane derived from ring closure about ortho-sulfone lithiated intermediates.

Polysulfone (Udel) Lithiation, DS = 3.0. A polysulfone solution (2.21 g, 5 mmol) in THF (123 mL) was cooled to -10°C. n-Butyllithium was added in 2 steps: the first portion (1.00 mL, 10 mmol) was injected dropwise over a period of 10 minutes while an excess second portion (1.50 mL, 15 mmol) was added more rapidly (< 5 min). After complete addition of butyllithium, the reaction solution turned into a paste as the trilithiated polymer came out of solution. The mixture was stirred at -10 °C for another 1.5 hr. For addition of most electrophiles, the temperature was reduced before addition. For example, to methylate the polymer, excess methyl iodide was poured promptly into the flask and the paste redissolved to a clear yellow solution of trimethylated polysulfone during a period of one hour. The structure of lithiated polysulfone having a DS of 3.0 is shown in Figure 1.

Polyphenylsulfone (Radel-R) Lithiation, DS = 2.0. Polyphenylsulfone (42.6 g, 107 mmol) was dissolved in cooled THF (-5°C, 1.2 L). The polymer has anomalous solubility properties, being more soluble at reduced temperature than at room temperature. The gray solution was cooled to -45°C and n-butyllithium (23.4 mL, 23 mmol) was injected. A bright lemon yellow colour developed that changed to a more viscous orange solution after a short time. Following lithiation, the solution was stirred for 25 min before addition of the electrophile at an appropriate temperature. Dilithiated polyphenylsulfone (DS = 2.0) is shown below.

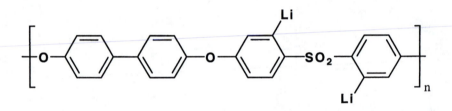

Bis-6F Polysulfone Lithiation, DS = 2.0. A solution of Bis-6F polysulfone (12.50 g, 23 mmol) in THF (1.25 L) was cooled to -25°C. n-Butyllithium (6.70 mL, 0.067 mol) was added slowly, and a viscous yellow solution formed. As the addition progressed, a less viscous red-brown solution formed After complete addition, the mixture was stirred for an additional 5 minutes. Note that 2.95 mol equivalents of butyllithium were required to obtain DS 2.0. The structure of lithiated Bis-6F polysulfone having a DS of 2.0 is shown below.

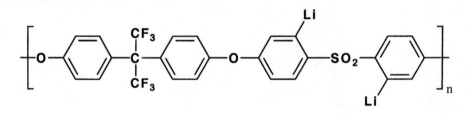

Dibrominated Polysulfone Lithiation, DS = 2.0. A solution of dibrominated polysulfone (60.0 g, 100 mmol) in THF (1.1 L) was cooled to -78°C using a dry-ice/alcohol bath (6). The polymer was lithiated with n-butyllithium (21.5 mL, 215 mmol) to give a soluble red lithiated intermediate (DS ~ 2.0) of the structure shown in Figure 3.

Dibrominated Polysulfone Lithiation, DS = 3.0. A solution of dibrominated polysulfone (4.5 g, 7.5 mmol) in THF (150 mL) was cooled to -78°C and lithiated with n-butyllithium (2.3 mL, 23.3 mmol). The first 15 mmol n-butyllithium were added drop-wise as before to prepare dilithiated polysulfone. The polymer was further lithiated at -78°C by adding the remainder of the metalating agent more rapidly and stirring for 30 min. The resulting trilithiated polymer was a purple precipitate having a DS of approximately 2.85. Trilithiated polymer is shown in Figure 3.

Experimental Methods for Introducing Functional Groups

The structures of modified polysulfones are indicated by the general structure shown below.

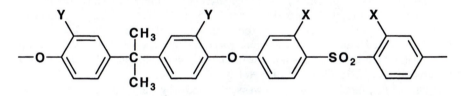

Polysulfone Carboxylic Acid Derivative, DS = 0.85. A 5 L 3-neck flask having a central resin kettle flange of internal diameter 10 cm was used for the reaction. A mechanically stirred solution of polysulfone (177 g, 0.40 mol) in THF (2 L) was lithiated to DS = 0.85 by addition of n-butyllithium (34 mL, 0.34 mol, 0.85 mol equivalents). A 0.5 kg block of dry ice was freshly prepared from liquid CO_2 and thoroughly mixed into the solution with a large rigid spatula (8). Additional blocks were mixed in rapidly, taking care to leave no unreacted solution. The mixture was allowed to warm up overnight under inert atmosphere, then THF was drained off, and the solid was mixed in a Waring blender with ethanol. The product was filtered, acidified with dilute HCl, and finally washed to give carboxylated polysulfone with DS = 0.85 as shown below.

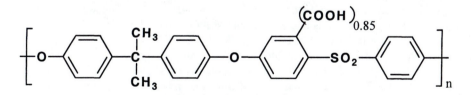

Polysulfone Sulfinic Acid Derivative. Sulfur dioxide gas was passed through a stainless steel condenser cooled with a dry ice bath. The liquefied gas was dripped into a mechanically stirred solution of polysulfone (110 g, 0.25 mol) in THF (2000 mL) that was lithiated (DS = 2) at -40°C (7). A heavy white precipitate formed immediately, and after a few minutes the reaction mixture changed to a less viscous heterogeneous suspension. Stirring was continued while the mixture was allowed to warm gradually to room temperature. The product was protected from light and was recovered from isopropanol. The resulting lithium salt of the sulfinated polysulfone was acidified by stirring in dilute HCl. In the sulfonic acid derivative, X is SO_2H and DS = 2.0.

Polysulfone Carbinol Derivatives. Dimethylcarbinol derivative (*10*) was prepared by adding cooled acetone (-78°C, 400 mL) to dilithiated polysulfone (from 110 g, 0.25 mol) dissolved in 1400 mL THF. A thick clear orange gel resulted, which changed to solution, then reverted to gel during a period totaling two hours. After warming to room temperature, the solution was precipitated into ethanol and the polymer was recovered by filtration. The polymer was immediately washed by stirring with fresh ethanol to remove yellow side-products, then washed with hot water. X is $CH_3–C(OH)–CH_3$, DS = 1.4.

Methylcarbinol derivative was prepared by adding 75 mL of cooled acetaldehyde (-30°C) to dilithiated polymer (0.25 mol). The resulting white gel was stirred at -70°C for two hours, then at room temperature for two hours. The mixture was precipitated into water-isopropanol (4:1) and the product was washed with isopropanol then water. If ethanol was used as a precipitant, the polymer remained in solution. X is $CH_3–C(OH)–H$, DS = 1.6.

Diphenylcarbinol derivative was prepared by adding a solution of benzophenone (182 g.) in THF (250 mL) cooled at -78°C into stirred dilithiated polysulfone (from 110 g, 0.25 mol) dissolved in 1200 mL THF. A thick green-yellow gel resulted, which was stirring for three hours. The polymer solution was precipitated into a large volume of water. The recovered polymer was sequentially washed with cold water, water-isopropanol (4:1), hot 2-propanol and finally boiling water. X is $C_6H_5–C(OH)–C_6H_5$, DS = 2.0.

Phenylcarbinol derivative was prepared by adding 280 g of cooled benzaldehyde into stirred dilithiated polysulfone (from 110 g, 0.25 mol) dissolved in 1400 mL THF. The resulting yellow gel was stirred for three hours at -20 °C and then allowed to warm up to room temperature. The mixture was precipitated into water-isopropanol (4:1) and the product was washed with ethanol. X is $C_6H_5–C(OH)–H$, DS = 2.0.

Polysulfone (ortho-ether) Azide, DS = 2.0. A solution of tosyl azide (59.1 g, 300 mmol) in THF (60 mL) was cooled and poured into a stirred solution of ortho-ether dilithiated polysulfone (100 mmol) in THF (1.1 L) prepared at -78°C (*11*). A precipitate formed immediately, which changed to an homogeneous suspension as it was stirred for 15 minutes at -78°C. The temperature was allowed to rise gradually to -50°C over a 90 minute period, then stirring was continued for 15 minutes. The resulting yellow solution was transferred to a beaker and water (~1 L) was added immediately by pouring in slowly until polysulfone azide precipitated from solution. The polymer was washed in the normal manner, but without heating. The polysulfone is photosensitive and thermally sensitive. Y is –N=N=N, DS = 2.0.

Polysulfone (ortho-ether) Diamine, DS = 2.0.   Sodium borohydride powder (18.9 g, 0.50 mol) was added to a yellow solution of polysulfone (ortho-ether) azide (DS = 2.0, 52.4 g, 100 mmol) in THF (900 mL) and absolute ethanol (90 mL) cooled at -60°C. The solution was stirred at room temperature overnight. Y is NH_2, and DS = 2.0.

'One-pot' Preparation of Polysulfone (ortho-sulfone) Diamine. A cooled solution of tosyl azide (14.8 g, 75 mmol)) in THF (15 mL) was poured into a mechanically stirred solution of dilithiated polysulfone (25 mmol) in THF (200 mL) prepared at -65°C (*12*). The stirred precipitate was gradually warmed to -50°C over 15 min. The resulting clear solution (-50°C) was rapidly poured into a vigorously stirred ethanol-water mixture (250 mL, 10:1) in a 1 L round bottom flask. Polysulfone azide precipitated from the resulting yellow solution within ~2 min. The mixture was left stirring until the temperature reached 10°C, then the polymer suspension was cooled to 0°C and $NaBH_4$ (14.2 g, 375 mmol) was added gradually. Following $NaBH_4$ addition, the ice bath was removed and the mixture was stirred overnight. The polymer was filtered and washed with ethanol, then water. X is NH_2, and DS = 2.0.

2 + 3 Cycloaddition Reactions of Azides. Polysulfone (ortho-ether) diazide (5.25g, 10 mmol) was dissolved in distilled DMF (52.5 mL) and 2-butyne-1,4 diol ($HOCH_2$-C≡C-CH_2OH, 3.45 g, 40 mmol) in DMF (3.5 mL) was added (*13*). The mixture was heated at 110°C for 4 days and recovered by precipitation from isopropanol.

$$HOCH_2 \qquad CH_2OH$$

Y is $-N \diagdown_{N} ^{N}$ DS = 2.0

Polyphenylsulfone (Radel-R) trimethylsilyl derivative. Chlorotrimethylsilane (43 mL, 340 mmol) was cooled to -25°C and poured promptly into a solution of lithiated polyphenylsulfone (DS = 2.0, 107 mmol) in THF (1.2 L) at -25°C (*7*). The solution immediately became less viscous and darkened to a red-burgundy colour. After 90 min of stirring at -25°C, the solution changed to a clear yellow colour. The solution was stirred for an additional 40 min and then the polymer was precipitated in 80% aqueous ethanol. The structure of the product is shown below.

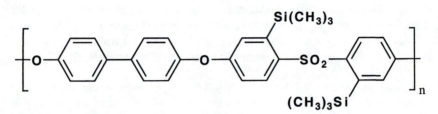

Membrane Applications

Ultrafiltration (UF). PSf is well known as an excellent polymer for fabricating UF membranes having a wide range of pore sizes, and it is resistant to aggressive cleaning cycles. Hydrophilizing PSf with COOH or OH groups can enhance membrane performance by reducing fouling, increasing water flux and narrowing or changing the pore size and pore size distribution.

UF membranes made from carboxylated polymer (CPSf, DS 0.85) have been cast at different concentrations from a variety of solvent systems with no additives, and compared to those made from Udel PSf. Polymer dopes were cast onto polyester non-woven backing material, then gelled into ice-cold water. The pore radii were estimated from non-linear least squares regression of a sieving curve using retention data of various molecular weight polyethylene glycol (PEG) solutes *(14)*. Figure 6 compares the change in pore radius as a function of the polymer concentration for different polymer solvent systems. The physical integrity of the formed membrane determines the lower polymer concentration limit of 17.5 wt%. The higher concentration limits are dependent on the practical solubility limit of polymer in solution or on the point at which permeation is negligible.

As expected, pore sizes decrease with increasing casting solution concentration, due to the density of polymer coils forming the membrane. For the same polymer casting dope concentrations, the pore sizes of PSf membranes are larger than those of CPSf. Above 25 - 27.5 wt% polymer casting solution, the hydrophobic dense skin layer of PSf membranes does not allow any water flux. However, the hydrophilic dense skin layer of CPSf membranes allow water permeation for casting solution concentrations up to 35 wt.%. Membrane formation behavior and pore size range is quite different from PSf, suggesting that carboxyl groups significantly alter the kinetics of the phase transition. Certainly, the polymer precipitation points of CPSf in different solvent-water systems are quite different from PSf *(15)*. From these data, tetramethyl urea (TMU) and N-methylpyrrolidinone (NMP) are stronger solvents for CPSf than dimethylacetamide (DMAc) and dimethylformamide (DMF), and the stronger solvents generally give the smallest pore sizes *(16)*. However, this effect could be associated with the loading capacity of the polymer-solvent system rather than the effect of polymer-solvent-nonsolvent system.

Pore radius provides only one parameter, whereas membrane performance requires a flux associated with given pore sizes. Figure 7 provides a comparative overview of performance by plotting the pressure-normalized flux to the pore radius for CPSf and PSf membranes (laboratory cast and a selection of commercially

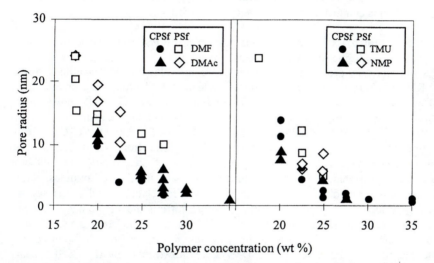

Figure 6. Effect of polymer concentration on pore size in casting solutions with different solvents.

available). PSf membranes incorporating the hydrophilic additive polyvinylpyrrolidinone (PVP) have generally better performance than PSf without additive, suggesting the importance of hydrophilicity in the polymer casting system. Asymmetric CPSf membranes, having typical rejection between 700 - 25,000 Daltons, are generally at the higher performance limit of those shown in the graph. CPSf appears especially promising as a material for membranes in the 700 - 2000 Dalton range.

Surface Modified Membranes. The lithiation reaction has also been applied to the heterogeneous surface modification of a variety of polysulfone membranes (*17,18*). Optimized conditions for the lithiation step were determined, but studies on surface morphology were not conducted. One advantage of heterogeneous modification is that a wider selection of electrophiles can be reacted, since crosslinking at the membrane surface is not problematical and may even be beneficial. Surface modification is conducted under non-aqueous conditions.

Reverse Osmosis (RO). An obvious advantage of PSf RO membranes is their resistance to chlorine-induced degradation reactions. Functional groups having a high water affinity are needed to impart the required level of hydrophilicity to PSf for RO membranes. Sulfonate, carboxyl, and hydroxyl groups are among those with this potential, and sulfonated RO membranes are already well known in the form of thin film composites (TFCs). We have demonstrated that carboxylated PSf membranes have NaCl rejection from 50% (NF) to greater than 99% (RO) (*19,20*). Unsupported membranes were cast onto glass plates from CPSf solutions in solvents containing LiCl, calculated by a molar ratio of LiCl to polymer. The cast films were subjected to elevated temperatures for a number of minutes before gelation into ice cold water. The membranes were exchanged with water several times to effect the complete removal of solvent from the membrane. Casting solutions containing inorganic salts (e.g. LiCl) resulted in higher flux membranes. Some examples of NaCl separation from membranes made from different casting formulations using 20 wt% CPSf in DMAc are shown in Table II. CPSf appears suited to NF membranes since reasonably high fluxes in the separation range of 46 - 65% NaCl are obtainable. The membrane having 45% NaCl separation had 400 MW polyethylene glycol (PEG 400) and raffinose rejections of approximately 90%, with fluxes of 37 gal/ft^2·d.

Our work has been confined to asymmetric membranes, but others have investigated CPSf for fabricating TFC NF membranes (*21*). Table III shows some NF data from TFCs derived from CPSf.

Gas Separation (GS). Permselectivity in GS membranes made from high T_g thermoplastic polymers is governed predominately by diffusivity, and to a lesser extent solubility. Increased chain rigidity and fractional free volume give rise to increased selectivity and increased permeability, respectively (*22*). Functional

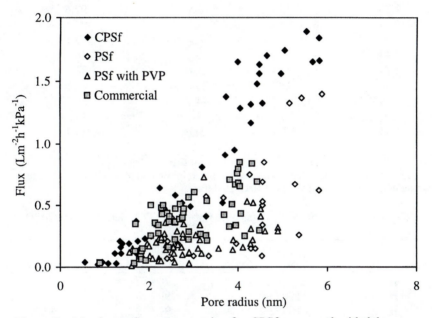

Figure 7.   Membrane flux vs. pore size for CPSf compared with laboratory cast and commercial PSf membranes.

Table II. Water flux and salt rejection of asymmetric CPSf membranes.

LiCl : Polymer	Casting conditions	Flux gal/ft^2.d	Rejection %
1.20 (PSf)	10 m, 95°C	0	0
0	10 m, 95°C	5.7	0
1.35	8 m, 95°C	0.5	92.6
2.13	8 m, 95°C	5.4	92.8
2.13	10 m, 95°C	2.9	97.1
2.44	8 m, 95°C	13.6	86.1
2.74	8 m, 95°C	16.0	88.7
3.04	8 m, 95°C	23.6	84.6
3.04	8 m, 95°C	44.1	69.3
0.6	5 m, 70°C	73.3	45.2
1.82	6 m, 95°C	3.0	100.0
1.82	10 m, 95°C	1.9	95.4

Cast from 20 wt% CPSf (DS 0.62) in DMAc (except first was PSf).
The cast solution was subjected to elevated temperature for a number of minutes before gelation in water. Solute: 3500 ppm NaCl at 500 psi.

Table III. NF rejection of solutes with flat-sheet TFC's derived from CPSf.

TFC preparation	Rejection Raffinose/Dextrose%	Rejection MgSO$_4$/NaCl %	Flux gal/ft^2·d
IP	100 / –	78 / 77	25
IP	100 / 90	99 / 70	4.0
IP	100 / 90	99 / 73	4.0
IP	100 / 87	97 / 90	8.3
SC	100 / 78	88 / 71	11.6
SC	100 / 78	– / 76	12.0
SC	100 / 92	– / 76	11.2
SC	100 / 80	84 / 75	12.4
SC	100 / 82	91 / 74	10.9
SC	100 / 86	90 / 79	11.2

IP: interfacially polymerized; SC: solvent coated; feed solution 2000 ppm MgSO$_4$; 500 ppm NaCl; operating pressure 200-225 psi.

groups that have particular steric shapes and that are sited around flexible chain linkages have a potential to bring about a simultaneous increase in selectivity and permeability by increasing chain rigidity and decreased chain packing. Our ongoing work with silyl derivatives in this direction has shown that they increase permeability considerably, with minimal reduction in selectivity (*23*). The lithiation route is a convenient way to produce an homologous series of functional groups having incremental changes in structure, and is therefore a potentially useful tool in studying structure-property relationships in gas separation membranes. Another approach in GS is to incorporate functional groups (e.g. basic) having a chemical interaction with one of the feed gases (e.g. acidic -CO_2), thereby increasing the permeability and selectivity. In the case of aminated polysulfone, high interchain hydrogen-bonding effects give rise to high chain stiffness. High O_2/N_2 ideal gas selectivities shown in Figure 8 (*24*) are evidence of this effect. Commercially viable GS membranes made from high value tailored thermoplastics may be potentially fabricated by coating a thin layer of functional polymer onto a supporting hollow fiber.

Other Membranes.

Functional polymers have usefulness in membranes areas outside those mentioned. The functionality may be active in itself, or provide reactive sites for grafting moieties for many purposes. Some examples include: reduced fouling, increased biocompatibility, crosslinking, affinity membranes, metal ion complexation membranes, chiral separations, and catalysis.

Affinity Membranes. Membranes for selective binding and fractionation of amino acids have been prepared by Staude *et al.* (*25*). A polymer with spacer arms terminated with epoxy groups was prepared by reaction of lithiated polysulfone, and cast into membranes. Affinity membranes having iminodiacetic acid (IDA) terminated side groups were prepared from the epoxide as shown in Figure 9, and complexed with Cu^{2+}. Residual coordination sites on Cu^{2+} are known to bind with imidazole or thiol side chains on amino acids, peptide sequences and proteins. The membranes exhibited selective binding of amino acids histidine, tryptophan, cysteine and tyrosine. Comparative elution volumes (mL) of the breakthrough curves for alanine (weak affinity) and histidine (strong affinity) were 4.3 and 70.4 respectively.

Ion Containing Membranes. A new method of sulfonating polysulfones was reported by Kerres *et al.* (*9*) based on oxidation of the sulfinic derivative with permanganate or with hypochlorite. It is a convenient alternative to existing procedures and offers good control of DS without chain degradation. Sulfonation is at the ortho-sulfone site rather that on the bisphenol ring. Applications for sulfonated polysulfones (SPSf) are in electro-membrane processes such as electrodialysis and polymer electrolyte fuel cells. The SPSf's were reported (*9*) to have low specific resistances and current efficiencies similar to Nafion. As

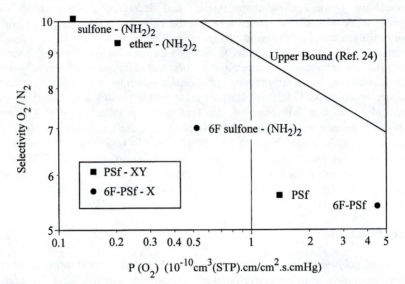

Figure 8. Gas permeability and ideal O_2/N_2 selectivity for aminated polysulfones.

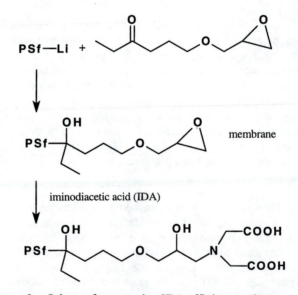

Figure 9. Scheme for preparing IDA affinity membrane.

examples, SPSf (115 μm) with IEC of 1.49 meq/g had specific resistances of 120 Ωcm for Na^+ and 18 Ωcm for H^+. SPSf (140 μm) with IEC of 1.37 was 176 Ωcm for Na^+. In comparison, Nafion 117 with IEC of 1.1 meq/g had specific resistances of 211 Ωcm for Na^+ and 13 Ωcm for H^+.

Metal Ion Chelation. Oxine (8-hydroxyquinoline) and other metal chelating groups have been attached heterogeneously to aminated polysulfone, in the form of either a fine powder or a membrane (*26*). Amine groups were first diazotized, then contacted with the metal chelating group to bind it covalently to the polymer or membrane surface, as shown in Figure 10. This application was for the detection of trace metal ions in water, whereby specific ions were chelated on a membrane, then eluted in a more concentrated solution for detection by atomic absorption. At pH 7.8, complete uptake of Co, Cd, Mn, and Ni ions was observed, and all ions had high elution values with acid. At pH 5.4, there was a high uptake of Cu and Pb ions with corresponding high elution.

Conclusions

Lithiation chemistry provides readily accessible routes to chemically modifying polysulfones with a wide variety of functional group types. Additional advantages of this modification technique include preparative convenience, rapid and high yield reactions, control over DS and site specificity, and the absence of polymer degradation. Examples of the introduction of different types of functionality are given, which include those that impart hydrophilicity, hydrophobicity, ionic character, and reactivity. An overview of some possible membrane applications of these polymers whereby the polymer properties are favorably influenced by introducing various types of functional groups is included for liquid separations, gas separations, and reactive membranes.

Literature Cited

1. Akelah, A.; Moet, A. *Functionalized Polymers and their Applications*, Chapman and Hall: London, 1990.
2. Daly, W. H.; Lee, S.; Rungaroonthaikul, C. In *Chemical Reactions in Polymers*, Kinstle, J. F.; Benham, J. L., Eds.; ACS Symp. Ser. 364: ACS, Washington DC, 1988, pp 4–23.
3. Kahana, N.; Arad-Yellin, R.; Deshe, A.; Warshawsky, A. *J. Polym. Sci., Polym. Chem.* **1990**, *28*, 3303, and references within.
4. Guiver, M. D.; ApSimon, J. W.; Kutowy, O. *J. Polym. Sci., Polym. Lett. Ed.* **1988**, *26*, 123.
5. Chalk, A. J.; Hay, A. S. *J. Polym. Sci. A–1*, **1969**, *7*, 691.
6. Guiver, M. D.; Kutowy, O.; ApSimon, J. W. *Polymer* **1989**, *30*, 1137.
7. Guiver, M. D.; ApSimon, J. W.; Kutowy, O. *U.S. Patent* 4,797,457. 1989; *U.S. Patent* 4,833,219. 1989.

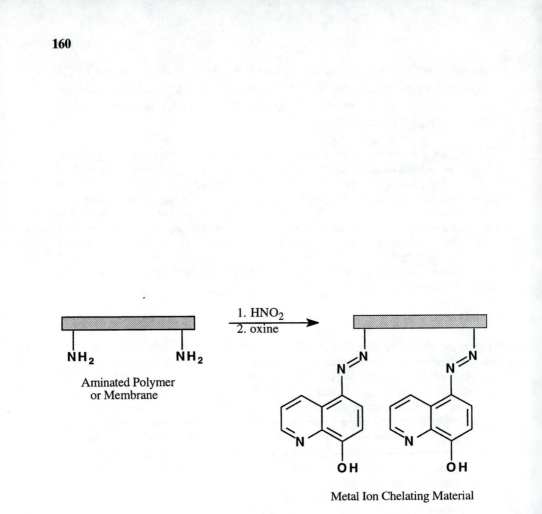

Figure 10. Formation of a metal ion chelating oxine derivative.

8. Guiver, M. D.; Croteau, S.; Hazlett, J. D.; Kutowy, O. *Br. Polym. J.* **1990**, *23*, 29.

9. Kerres, J.; Cui, W.; Neubrand, W.; Springer, S.; Reichle, S.; Striegel, B.; Eigerberger, G.; Schnurnberger, W.; Bevers, D.; Wagner, N.; Bolwin, K. In *Proceedings of Euromembrane'95*; Bowen, W. R.; Field, R. W.; Howell, J. A., Eds.; University of Bath, 1996, Vol. 1; pp 284–289.

10. Brownstein, S. K.; Guiver, M. D. *Macromolecules* **1992**, *25*, 5181.

11. Guiver, M. D.; Robertson, G. P. *Macromolecules* **1995**, *28*, 294.

12. Guiver, M. D.; Robertson, G. P.; Foley, S. *Macromolecules* **1995**, *28*,7612.

13. Guiver, M. D.; Robertson, G. P. *U.S. Patent* 5,475,065, 1995.

14. Dal-Cin, M. M.; Tam, C. M.; Guiver, M. D.; Tweddle, T. A. *J. Appl. Polym. Sci.* **1994**, *54*, 783.

15. Lau, W. W. Y.; Guiver, M. D.; Matsuura, T. *J. Appl. Polym. Sci.* **1991**, *42*, 3215.

16. Guiver, M. D.; Tam, C. M.; Dal-Cin, M. M.; Tweddle, T. A.; Kumar, A. In *Proceedings of Euromembrane'95*; Bowen, W. R.; Field, R. W.; Howell, J. A., Eds.; University of Bath, 1996, Vol. 1; pp 345–348.

17. Guiver, M. D.; Black, P.; Tam, C. M.; Deslandes, Y. *J. Appl. Polym. Sci.* **1993**, *48*, 1597.

18. Breitbach, L.; Hinke, E.; Staude, E. *Angew. Makromol. Chem.* **1991**, *184*, 183.

19. Guiver, M. D.; Tremblay, A. Y.; Tam, C. M. *U.S. Patent* 4,894,159, 1990.

20. Guiver, M. D.; Tremblay, A. Y.; Tam, C. M. *In Advances in Reverse Osmosis and Ultrafiltration*; Matsuura, T.; Sourirajan, S., Eds.; National Research Council, Ottawa, 1989, pp 53–70.

21. U.S. Dept. of the Interior, Bureau of Reclamation, Contract No. 1425-4-CR-81-19870 Final Report, June 1995; Water Treatment Technology Report No. 9.

22. *Polymeric Gas Separation Membranes*; Paul, D. R.; Yampol'skii, Y. P., Eds.; CRC Press, Boca Raton, 1994.

23. Park, H. C.; Won, J.; Kang, Y. S.; Kim, I. W.; Jho, J. Y.; Guiver, M. D.; Robertson, G. P. In *Proceedings of the 5th Pacific Polymer Conference*, Kyongju, Korea, October 26–30, 1997.

24. Robeson, L. M.; *J. Membr. Sci.* **1991**, *62*, 165.

25. Rodemann, K.; Staude, E. *J. Membr. Sci.* **1994**, *88*, 271.

26. Kan, M.; Guiver, M. D.; Robertson, G. P.; Willie, S. N.; Sturgeon, R. E. *React. Functional Polym.* **1996**, *31*, 207.

Chapter 11

Polyphosphazene-Based Cation-Exchange Membranes: Polymer Manipulation and Membrane Fabrication

Qunhui Guo[1], Hao Tang[2], Peter N. Pintauro[2,3] and Sally O'Connor[1]

[1]Department of Chemistry, Xavier University, New Orleans, LA 70125
[2]Department of Chemical Engineering, Tulane University, New Orleans, LA 70118

Poly[bis(3-methylphenoxy)phosphazene] was sulfonated in solution with SO_3 and solution-cast into ion-exchange membranes from N,N-dimethylacetamide. Water insoluble membranes were prepared with an ion-exchange capacity (IEC) as high as 2.1 mmol/g. For water insoluble polymers with an IEC < 1.92 mmol/g, there was no evidence of polymer degradation during sulfonation. The glass transition temperature of the sulfonated polymer increased from -28°C (for the base polymer) to -10°C for an IEC of 2.1 mmol/g. Equilibrium water swelling of a phosphazene membrane with an IEC of 0.95 mmol/g was 24% greater than that of a DuPont Nafion 117 cation-exchange membrane. The proton conductivity of a water-equilibrated 0.95 mmol/g IEC phosphazene membrane in the H^+ form ranged from 0.012 S/cm at 25°C to 0.058 S/cm at 60°C. The water diffusion coefficient in a 0.95 mmol/g IEC membrane, at saturated vapor conditions, ranged from 8.0×10^{-8} cm^2/s at 25°C to 4.1×10^{-7} cm^2/s at 60°C.

Polyphosphazenes are a potentially useful class of base-polymers for ion-exchange membranes because of their reported thermal and chemical stability and the ease of chemically altering the polymer by adding various sidechains onto the -P=N-backbone. Sulfonated polyphosphazene cation-exchange membranes, for example, may be an attractive alternative to perfluorosulfonic acid and polystyrene sulfonate membranes. The difficulty associated with producing such membranes lies in preparing the sulfonated polyphosphazene and, more importantly, in balancing the resulting hydrophilicity of the polymer to prevent dissolution in aqueous solutions.

In practice, there are three synthesis routes leading to water insoluble sulfonic acid membranes from polyphosphazene polymers. The first method involves crosslinking polyphosphazene membranes followed by heterogeneous sulfonation

[3]Corresponding author.

(*1*). In a second technique, homogeneous or heterogeneous polymer sulfonation is carried out, followed by film casting and crosslinking. These two methods require that the polyphosphazene contains sidegroups that can be used for crosslinking or in the second case, sulfonic groups can serve the purpose (*2*). The third route to a sulfonated polyphosphazene ion-exchange membrane is through appropriate balancing the hydrophilicity/hydrophobicity and, if possible, crystallinity of the polymer. In this case, a hydrophobic and/or semicrystalline polyphosphazene polymer is sulfonated to such an extent that it only swells and does not dissolve in aqueous media. This imposes some limitations on the magnitude of the ion-exchange capacity of the final membrane but, at the same time, offers significant processing advantages.

There are several reports in the literature on the sulfonation of phosphazene polymers although none of these studies was directed at fabricating an ion-exchange membrane nor was the structure of the sulfonated polyphosphazenes examined in detail. The sulfonation of aryloxy- and (arylamino)phosphazenes via reaction with sulfuric acid was studied by Allcock *et al.* (*3*) while the reaction of (aryloxy)polyphosphazenes with sulfur trioxide was studied by Montoneri and co-workers (*4,5*). During the initial stages of SO_3 addition, Montoneri *et al.* found no C-sulfonation, but rather the formation of a $\equiv N \rightarrow SO_3$ complex. When the SO_3/repeating monomer molar ratio was > 1, C-sulfonation was observed. A comprehensive review of the literature on sulfonated phosphazene polymers can be found elsewhere (*6*).

In a previous study by Wycisk and Pintauro (*6*), two semi-crystalline phosphazene polymers, poly[(3-methylphenoxy)(phenoxy)phosphazene] and poly[(4-methylphenoxy)(phenoxy) phosphazene], with an alkylphenoxy/phenoxy molar ratio of 1.0, were sulfonated successfully to varying degrees with SO_3. Water-insoluble cation-exchange membranes with good mechanical properties were fabricated with an ion-exchange capacity as high as 2.0 mmol/g. Polyphosphazenes with an ion-exchange capacity in the range of 2-3 mmol/g were also synthesized, but these polymers were found to be water soluble. Preliminary ^{13}C NMR and ^1H NMR analyses indicated that the methylphenoxy side groups were sulfonated preferentially. When poly[(4-ethylphenoxy)(phenoxy) phosphazene] was sulfonated with SO_3, the sulfonation rate was very slow (due to both steric interference by the ethyl substituents on the ethylphenoxy side groups and the poor reactivity of the phenoxy groups) and significant polymer degradation was observed.

In a more recent paper, Graves and Pintauro (*7*) examined the UV photocrosslinking of alkylphenoxy/phenoxy-substituted polyphosphazenes. Solution-cast films contained poly[(methylphenoxy) (phenoxy)phosphazene], poly[(ethylphenoxy)(phenoxy)phosphazene], or poly[(isopropylphenoxy) (phenoxy)phosphazene] (where the alkyl substituent was in either the *meta* or *para* position) and benzophenone photo-initiator (at a concentration of 1-25 mol%). Crosslinking was carried out at either 25°C or 70°C. The methylphenoxy/phenoxy phosphazenes were the best materials for crosslinking, indicating that steric effects of the alkyl group were playing a role during crosslink formation. For UV light-exposed poly[(3-methylphenoxy)(phenoxy)phosphazene] films, the glass transition temperature increased by approximately 25°C (from -15°C to 10°C) and the film swelling (in DMAc) decreased from infinity (complete solubilization) to 25% as the benzophenone concentration was increased from 0 to 25 mol%.

It is obvious from previous work that methyl substituents on the phenoxy side chains of the phosphazene polymer were activating the ring for electrophilic attack by SO_3 and were the optimum alkyl group for UV photo-crosslinking (6,7). In the present paper, we report on the sulfonation of poly[bis(3-methyl phenoxy) phosphazene], a semi-crystalline polymer with only methylphenoxy side groups. The chemical structure of the repeating polymer unit is shown in Figure 1. The method of sulfonation and the physical and ion-exchange properties of the resulting non-crosslinked sulfonated polymer membranes are described below.

Experimental

Poly[bis(3-methylphenoxy)phosphazene], purchased from Technically Inc., Andover, MA, was used as the base polymer without further purification. The molecular weight of this polyphosphazene, as determined by gel permeation chromatography (Waters Styragel HT 6E column in THF with polystyrene calibration standards), was 2.0×10^6 g/mol.

To sulfonate the polyphosphazene, a known weight of polymer (1.0 g) was first dissolved in 40 ml of 1,2-dichloroethane (DCE) and stirred for 24 h at 50°C. A given amount of SO_3 in 10 ml of DCE was then added dropwise to the polymer solution in a dry nitrogen atmosphere. The resulting precipitate was stirred for 3 h at 0°C followed by the addition of 50 ml of a NaOH solution (water/methanol solvent) to terminate the reaction. After evaporation of solvent at 70°C for 24 h, the polymer was pre-conditioned by soaking sequentially in distilled water, 0.1 M NaOH, distilled water, 0.1 M HCl, and distilled water (each soaking step was carried out for 48 hours). The polymer product was then dried thoroughly and dissolved in N, N-dimethylacetamide (DMAc) at a concentration of 5 wt%. Membranes were cast from this solution on a polypropylene plate and then dried at 70°C for 3 days (until there was no change in the film weight). The thickness of the dry membranes was between 200 and 600 μm.

The ion-exchange capacity (IEC, with unit of mmol/g of dry polymer) of sulfonated polyphosphazene membranes was determined by measuring the concentration of H^+ that exchanged with Na^+ when membrane samples were equilibrated with a NaCl solution. A known weight of dry polymer (0.2-0.4 g) in the acid form (after pre-conditioning) was placed into 100 ml of a 2.0 M NaCl solution and shaken occasionally for 48 h. Three 25 ml samples were then removed and the amount of H^+ released by the polymer was determined by titration with 0.01 M NaOH.

Equilibrium water swelling of sulfonated membrane samples at room temperature was measured in terms of the % increase in dry membrane weight, according to the following formula,

$$\text{Swelling} = (m_{wet} - m_{dry})/m_{dry} \times 100 \; (\%) \tag{1}$$

where m_{wet} and m_{dry} are the weights of the water swollen and dry membrane (in its H^+ form), respectively.

The diffusion coefficient of water in sulfonated polyphosphazene films at saturated vapor conditions was determined by performing vapor-phase sorption/desorption experiments with a McBain sorption apparatus (8,9). A small membrane specimen of known dry weight was suspended on a quartz spring (enclosed in a thermostated glass chamber) and allowed to equilibrate with a water vapor atmosphere of activity 0.98 (an activity < 1.0 was needed to avoid water condensation). After the membrane was fully equilibrated and absorbed no more water, the vapor pressure of water was lowered by evacuating the chamber and the decrease in membrane weight was monitored as a function of time. A typical water desorption curve for the polyphosphazene membranes is shown in Figure 2, where the weight loss is plotted as a function of the square-root of time. In this figure M_0 is the equilibrium, zero time, weight of water in the membrane sample and M_t is the water weight remaining in the membrane at a given time after desorption. The water diffusion coefficient was computed from the initial slope of the desorption curve and the following equation (10),

$$D = \pi \, \delta^2 \, (\text{slope})^2/16 \tag{2}$$

where δ is the thickness of the dry membrane. This method has been used previously to determine the diffusion coefficient of penetrants in polymers (11,12) and is accurate when the membrane is sufficiently thick so that errors due to polymer shrinkage at the membrane surface (where the water swelling is small) are minimized. Also, the membrane thickness must be much smaller than the sample width and length to ensure one-dimensional diffusion. In the present study the sulfonated polyphosphazene membranes were 600 μm in thickness, with a width and length of 0.5 cm.

DSC measurements were carried out using a TA Instruments DSC 2970 apparatus at a heating rate of 10°C/min, under a dry nitrogen atmosphere. Calibration was performed using indium. ^{13}C solid-state NMR spectra were recorded on a Bruker ASX 400 MHz spectrometer. A sample spinning rate of 10 KHz and proton decoupling at a level of 45 KHz were employed under magic angle spinning (MAS) conditions using dipolar decoupling with cross-polarization. Tetramethylsilane was used as a reference for ^{13}C.

The electrical conductivity of protons in water equilibrated membranes in the H^+ form was determined using an AC impedance method. Membrane samples were first soaked in deionized and distilled water for 24 hours. The conductivity was measured using a pair of pressure-attached, high surface area platinum electrodes, as described elsewhere (13). The mounted sample was immersed in deionized and distilled water at a given temperature and measurements were made from 1 Hz to 10^5 Hz using a Paar Model 5210 amplifier and a Paar Model 273

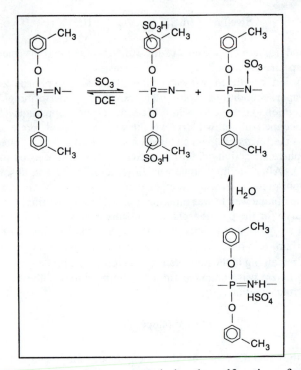

Figure 1. Chemical reaction sequence during the sulfonation of poly[bis(3-methyl phenoxy)phosphazene] with SO_3.

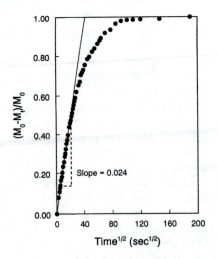

Figure 2. Membrane water weight loss vs square-root of time during a McBain balance vapor desorption experiment using a sulfonated polyphosphazene membrane.

potentiostat/galvanostat. Both real and imaginary components of the impedance were measured and the real Z-axis intercept was closely approximated. The cell constant was calculated from the spacing of the electrodes, the thickness of the membrane, and the area of the platinum electrodes.

Results and Discussion

Membrane Ion-Exchange Capacity (IEC) and Water Swelling.   The measured ion-exchange capacity of sulfonated poly[bis(3-methylphenoxy)phosphazene] polymer samples and the equilibrium swelling of the samples in water at 25°C are listed in Table I for different SO_3/POP molar ratios (where POP denotes a polyphosphazene monomer unit). When the SO_3/POP ratio was $\leq$ 0.64, the IEC of the membranes was detectable, but very low (0.01-0.06 mmol/g). A gradual increase in IEC was observed as the SO_3/POP ratio was increased from 0.64 to 1.92.

The overall reaction of SO_3 with poly[bis(3-methylphenoxy)phosphazene] is shown in Figure 1.   During polymer preconditioning with NaOH and water after sulfonation, most of the $\equiv$N$\rightarrow$SO$_3$ complex was hydrolyzed and the resulting $\equiv$N$^+$H$\cdots$HSO$_4^-$ product was flushed from the sample. The difference between the total sulfur content and the amounts of SO_3^- (equal to the measured ion-exchange capacity) and resulting $\equiv$N$^+$H$\cdots$HSO$_4^-$ gave the concentration of $\equiv$N$\rightarrow$SO$_3$ complex. The $\equiv$N$\rightarrow$SO$_3$ and $\equiv$N$^+$H$\cdots$HSO$_4^-$ species make the polymer more hydrophilic which affects polymer swelling in water, but their presence had essentially no bearing on the measurement of the polymer's ion-exchange capacity (the concentration of $\equiv$N$^+$H$\cdots$HSO$_4^-$ was so low that its decomposition and the elution of H_2SO_4 affected the IEC by no more than 7%).

Table I. Ion-exchange capacity and swelling of sulfonated
poly[bis(3-methylphenoxy)phosphazene].

SO$_3$/POP molar ratio	Ion-exchange capacity (mmol/g)	Membrane swelling in water[a] (wt.% dry membrane)
0.32	< 0.004	< 1.0
0.64	0.04	5.2
0.80	0.80	33
0.96	0.95	42
1.28	1.7	233
1.60	2.1	900
1.92	2.3	∞ (water soluble)

a) measured at 25°C.

Only a fraction of the total added SO_3 produced sulfonate fixed-charge sites on the polymer. The reaction yield for SO_3^- site creation varied from 1.6% at a SO_3/POP ratio of 0.64 to 44% at a SO_3/POP ratio of 1.6. These results are in agreement with the work of Montoneri *et al.* (*3,4*), who found that backbone nitrogen in polyphosphazenes were attacked first by SO_3 and are also consistent with our previous sulfonation studies with poly[(3-methylphenoxy)(phenoxy)phosphazene] (6), where no IEC was detected for a SO_3/POP ratio ≤ 0.8.

As expected, the equilibrium swelling of sulfonated polyphosphazene membranes in water was found to increase with increasing SO_3/POP molar ratio, that is, membrane ion-exchange capacity. The base polymer prior to exposure to SO_3 was highly hydrophobic and did not sorb any water. Once the polyphosphazene contacted SO_3 (even at a low SO_3/POP molar ratio), water uptake was observed, due to the presence of $\equiv N \rightarrow SO_3$ and $\equiv N^+H \cdots HSO_4^-$ species and a small number of sulfonic acid sites. As expected, the extent of polymer swelling increased with increasing IEC. Membranes prepared from the sulfonated polymers with a SO_3/POP ratio ≥ 1.92 (an IEC ≥ 2.1 mmol/g) were found to be water soluble. For comparison, a sulfonated poly[bis(3-methylphenoxy)phosphazene] membrane with an ion-exchange capacity of 0.95 mmol/g swelled 42% at room temperature, whereas a Nafion 117 perfluorosulfonic acid membrane (with an IEC of 0.909 mmol/g) swelled 34% (14).

IR and NMR Spectra. IR spectra of sulfonated bis(3-methylphenoxy) phosphazene polymer samples were collected for wave numbers in the 600-1,650 cm^{-1} range. There was no obvious change in the P-O-methylphenoxy stretching band (at 1,140 cm^{-1}) and the P-N band (at 1,243 cm^{-1}) for polymers sulfonated with a SO_3/POP ratio ≤ 1.28. A new peak corresponding to S=O (at 1,085 cm^{-1}) appeared and grew with increasing SO_3/POP. At a SO_3/POP ratio of 1.92, the P=N band decreased slightly, which was attributed to some degradation of the polyphosphazene main chain.

To determine the location of SO_3 attack on the polyphosphazene's methylphenoxy side groups, ^{13}C solid-state NMR spectra of the base and sulfonated polymers were collected. Six resonance peaks in the 100-180 ppm range were assigned to the six aromatic carbon nuclei on the methylphenoxy side-groups (aromatic carbon signals from sulfonated and non-sulfonated side-chains were indistinguishable). For a poly[bis(3-methylphenoxy)phosphazene] sample sulfonated at a SO_3/POP ratio of 0.76, the C4 (*para* position) signal (at δ =125.9 ppm) decreased significantly and shifted downfield by 14-15 ppm. This observation, along with subsequent calculations, is consistent with *para*-position sulfonation of the polyphosphazene's methylphenoxy side-groups. When poly[bis(3-methylphenoxy)phosphazene] was sulfonated at SO_3/POP molar ratios > 1.0, the intensities of the four resonance signals for aromatic carbons at the C5, C4, C2, and C6 positions decreased significantly, indicating that all phenoxy carbons except C1 and C3 were undergoing SO_3 attack.

DSC Measurements. The glass transition temperatures (Tg) of the base-polymer and sulfonated poly[bis(3-methylphenoxy)phosphazene] samples, as determined from DSC curves, are plotted in Figure 3 as a function of polymer ion-exchange capacity. No change in Tg was observed when the polymer was exposed to a low concentration of SO_3 (i.e., when there was essentially no aromatic C-sulfonation).

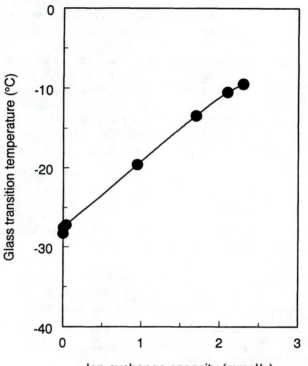

Figure 3.   Variation in glass transition temperature of sulfonated poly[bis(3-methyl phenoxy)phosphazene] as a function of polymer ion-exchange capacity.

The polymer's Tg increased with increasing ion-exchange capacity when the SO_3/POP molar ratio was > 0.64. The observed increase in Tg was associated with electrostatic interactions among fixed-charge groups in the ionomeric portions of the polymer and the general incompatibility of charged domains and nonpolar polymer backbone, which restricted polymer chain mobility and produced a rise in Tg (15,16).

Proton Conductivity. The proton conductivity of sulfonated polyphosphazene membranes in the H^+ form and equilibrated in water were determined at various temperatures ranging from 25°C - 60°C. The results for polyphosphazene membranes with an IEC of 0.8 and 1.0 mmol/g are plotted vs. the reciprocal temperature in Figure 4. The proton conductivity of a water-equilibrated Nafion 117 cation-exchange membrane with an IEC of 0.909 mmol/g (17) is also shown in this figure. As expected, the polyphosphazene membrane conductivity increased with increasing IEC and with increasing temperature. The proton conductivity of a 1.0 mmol/g IEC sulfonated phosphazene membrane was high, at approximately 60% of that for Nafion 117 (the lower conductivity was associated with the greater water swelling of the polyphosphazene membrane, coupled with a less well-defined "cluster network" micro-structure). The membrane conductivity data for Nafion 117 and the two polyphosphazene membranes in Figure 4 were fitted to the following equation,

$$\kappa = \kappa_0 e^{-\Delta E/RT} \tag{3}$$

where the constants κ_0 (a pre-exponential factor) and ΔE (the activation energy from proton conductivity) are constants listed in Table II.

Table II. Constants in the conductivity expression (Equation 3) for sulfonated poly [bis(3-methylphenoxy)phosphazene] and Nafion 117 membranes (membrane in the H^+ form and equilibrated in water).

Membrane type	κ_0 (S/cm)	ΔE (kJ/mol)
Sulfonated polyphosphazene (0.80mmol/g)	7.6×10^3	38.9
Sulfonated polyphosphazene (1.0mmol/g)	1.07×10^5	35.4
Nafion 117 (0.909 mmol/g)	20.0	14.2

The higher activation energy for the two polyphosphazene membranes, as compared to Nafion, was attributed to differences in membrane morphology, i.e., a difference in the distribution/spacing of sulfonate sites and/or variations in the size and spacing of ion clusters.

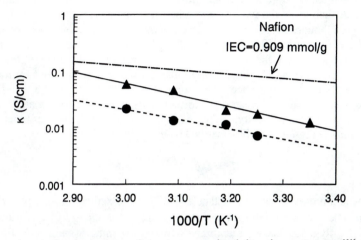

Figure 4. Arrhenius plot of proton conductivity in water-equilibrated poly[bis(3-methylphenoxy)phosphazene] membranes as a function of the reciprocal temperature. ● phosphazene membrane with an IEC of 0.80 mmol/g; ▲ phosphazene membrane with an IEC of 1.0 mmol/g.

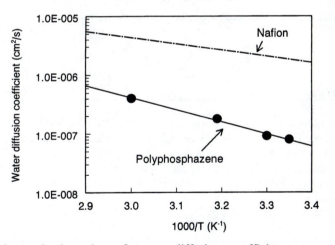

Figure 5. Arrhenius plot of water diffusion coefficient vs. reciprocal temperature for a 1.0 mmol/g IEC poly[bis(3-methylphenoxy)phosphazene] membrane and for Nafion 117.

Water Diffusion Coefficient. The water diffusion coefficient in a Nafion 117 cation-exchange membrane (*17*) and in a sulfonated polyphosphazene membrane with an ion-exchange capacity of 1.0 mmol/g is shown in Figure 5 as a function of the reciprocal temperature for temperatures between 25°C and 60°C. Surprisingly, the water diffusivity in the polyphosphazene film is considerably lower than that in Nafion, even though the polyphosphazene membrane swelled more in water. The low water diffusion coefficient for polyphosphazene was attributed to stronger water interactions with the P-N polymer backbone (as compared to the PTFE backbone of Nafion), a SO_3^- fixed-charge distribution that is different from that in Nafion, and/or the presence of some crystallinity in the sulfonated polyphosphazene, which increased the tortuosity of membrane pores and lowered the observed water diffusivity. The Nafion 117 and polyphosphazene water diffusion data in Figure 5 can be described by the following equation

$$D = D_0 e^{-\Delta E_d / RT} \qquad (4)$$

where D_0 is a pre-exponential constant and ΔE_d is the activation energy for water diffusion. The pre-exponential factor and the activation energy of diffusion for a sulfonated polyphosphazene membrane (IEC = 1.0 mmol/g) and a Nafion 117 membrane are shown in Table III.

Table III. Constants in the water diffusivity expression (Equation 4) for sulfonated poly[bis(3-methylphenoxy)phosphazene] and Nafion 117 membranes (H^+ form).

Membrane type	D_0 (cm^2/s)	ΔE_d (kJ/mol)
Sulfonated polyphosphazene (IEC = 1.0 mmol/g)	0.71	39.7
Nafion 117 (IEC = 0.909 mmol/g)	6.0 x 10^{-3}	20.4

Acknowledgments. This work was supported by the Office of Naval Research, the National Science Foundation (Grant Nos. ARI-9512258 and CTS-9632079), and the U.S. Department of Energy (Grant No. DE-FG01-93EW532023, "Hazardous Materials in Aquatic Environments of the Mississippi River Basin") through the Tulane/Xavier Universities' Center for Bioenvironmental Research.

Literature Cited

1.   Wycisk, R.; Pintauro, P. N.; Wang, W.; O'Connor, S. Proc. North American Membrane Society 7th Annual Meeting, Portland, OR, 1995.
2.   Nolte, R.; Ledjeff, K.; Bauer, M.; Mulhaupt, R. *J. Membrane Sci.* **1993**, *83*, 211.
3.   Allcock, H. R.; Fitzpatrick, R. J. *Chem. Mater.* **1991**, *3*, 1120.

4. Montoneri, E.; Gleria, M.; Ricca, G.; Pappalardo, G. C. *J. Macromol. Sci., Chem.* **1989,** *A26* (4), 645.

5. Montoneri, E.; Gleria, M.; Ricca, G.; Pappalardo, G. C. *J. Macromol. Chem.* **1989,** *190*, 191.

6. Wycisk, R.; Pintauro, P. N. *J. Membrane Sci.* **1996,** *119*, 155.

7. Graves, R.; Pintauro, P. N. *J. Appl. Polym. Sci.* **1998,** *68*, 827.

8. Felder, R. M.; Huvard, G. S. In *Methods of Experimental Physics*; Fava, R. A., Ed.; Academic: New York, 1980, Vol 16, Chapter 17, Part C, p 315.

9. Park, G. S. In *Synthetic Membranes: Science, Engineering and Applications*; Bungay, P. M.; Lonsdale, H. K.; de Pinho, M. N., Eds.; D. Reidel: Boston, 1986; pp 57-107.

10. Crank, J., *The Mathematics of Diffusion*, Oxford, London, 1956, p 228.

11. Odani, H.; Uchikura, M.; Taira, K.; Kurata, M.; *J. Macromol. Sci. Phys.* **1980,** *B17* (2), 327.

12. Ponangi, R.; Pintauro, P. N. *Ind. Eng. Chem. Res.* **1996,** *35*, 2756.

13. Zawodzinski, T. A.; Neeman, M.; Sillerud, L. O.;Gottesfeld, S., *J. Phys. Chem.* **1991,** *95*, 6040.

14. Proceedings of the First International Symposium on Proton Conducting Membrane Fuel Cells I, Gottesfeld, S.; Halpert, G.; Landgrebe, A., Eds.; The Electrochemical Society, Inc., Pennington, NJ, 1995, p 193.

15. Eisenberg, A. *Macromolcules* **1970,** *3*, 147.

16. Eisenberg, A.; Hird, B.; Moore, R. A. *Macromolcules* **1990,** *23*, 4098.

17. Halim, J.; Büchi, F. N.; Haas, O.; Stamm, M.; Sherer, G. G. *Electrochim. Acta* **1994,** *39*, 1303.

18. Yeo, S. C.; Eisenberg, A., *J. Appl. Polym. Sci.* **1977,** *21*, 875.

Chapter 12

Development of Radiation-Grafted Membranes for Fuel Cell Applications Based on Poly(ethylene-*alt*-tetrafluoroethylene)

H. P. Brack, F. N. Büchi, J. Huslage, M. Rota, and G. G. Scherer

Paul Scherrer Institut, Elektrochemie, CH–5232 Villigen PSI, Switzerland

We have adapted a previously developed method for preparing radiation-grafted membranes for fuel cell applications to poly(ethylene-*alt*-tetrafluoroethylene), or ETFE, base polymer films. This method has been used earlier with poly(tetrafluoroethylene-*co*-hexafluoropropylene), or FEP, base polymer films and consists of: (i) polymer film irradiation, (ii) subsequent grafting with styrene and crosslinkers, and (iii) sulfonation of the polystyrene-based graft component. The ETFE-based grafted films and membranes in the present work have mechanical properties that are superior to those of their FEP counterparts. This improvement may result from the (i) higher molecular weight of the ETFE materials, (ii) better compatibility of ETFE with the graft component, and (iii) reduced extent of radiation-induced chain scission occurring in the case of ETFE due to (a) its lower inherent susceptibility to chain scission and (b) the lower pre-irradiation doses that it requires. The performance of these membranes in fuel cell tests is briefly discussed.

In a fuel cell, a fuel, e.g., hydrogen, and an oxidant, e.g., oxygen from air, are electrochemically converted to water, thereby generating electricity and heat.

$$2H_2 + O_2 \Rightarrow 2H_2O + \text{electricity} + \text{heat}$$

In a low temperature ($\leq 100°C$) polymer electrolyte fuel cell (PEFC) a proton-conducting polymer membrane acts as both the (i) solid electrolyte to conduct protons from the anode to the cathode and (ii) separator for the gas phases (H_2 and

O$_2$) and prevent their direct and uncontrolled reaction. Figure 1 shows a cross-sectional view of a PEFC. In the PEFC, hydrogen is oxidized at the anode (H$_2$→2H$^+$+2e$^-$), and the protons are conducted through the membrane to the cathode where they react with oxygen (1/2O$_2$+2H$^+$+2e$^-$→H$_2$O).

Desirable properties of PEFC membranes have been discussed elsewhere (*1*), and they include: (i) high acidity and low ionic resistance to maximize proton conduction and thus minimize ohmic losses in the PEFC, (ii) low gas crossover in the hydrated state, and (iii) thermal, hydrolytic, oxidative, and reductive stability.

Radiation Grafting Method

Radiation grafting can be used to combine the chemical stability of fluoropolymer base films with the ion-exchange or proton-conducting properties of grafted polymeric branches having acid groups. This synthetic method is shown schematically in Figure 2 for a poly(ethylene-*alt*-tetrafluoroethylene), or ETFE, polymer with styrene as the graft component.

Much of the early important work on the radiation grafting of polymers was performed and reviewed by Chapiró (*2*). The properties of radiation-grafted films and membranes depend on such parameters as: (i) the nature of the base polymer film, for example, perfluorinated or non-perfluorinated; (ii) irradiation type, dose and dose rate; and (iii) the nature of the graft component, for example, the extent of crosslinking, type of crosslinker, and the grafting level.

Radiation grafted membranes have found applications as battery separators, in electrodialysis, and in water treatment. Fluorinated base polymer films are of interest in PEFC applications because of the rigorous stability requirements. Our early work (*3*) focused on poly(tetrafluoroethylene-*co*-hexafluoropropylene), or FEP, base films because FEP has been reported (*4*) to have better radiation resistance than PTFE in terms of the radiation-induced degradation of mechanical properties (tensile strength and flexibility). The usefulness of radiation-grafted membranes based on FEP as the electrolyte in fuel cells has been demonstrated (*5*). The development of similar membranes based on poly(ethylene-*alt*-tetrafluoroethylene), or ETFE, is reported in the present work. A continuing theme in both our earlier work with FEP and the present work with ETFE has been the development of progressively thinner membranes based on thinner base polymer films (25-50 µm) to minimize ohmic losses in the fuel cell resulting from the membrane area resistance (mΩcm^2).

ETFE Base Polymer. ETFE offers several potential advantages in comparison to FEP as a base polymer material: (i) ETFE is readily available in higher molecular weights. Desirable mechanical properties such as flexibility and breaking strength generally improve with increasing molecular weight; (ii) because ETFE is not perfluorinated, the monomer solubility and thus monomer diffusion into the

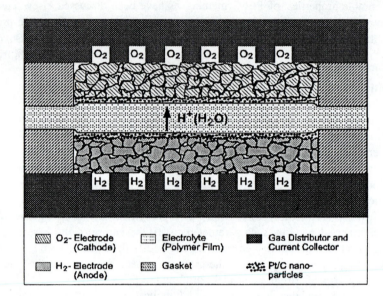

Figure 1. Cross-sectional view of polymer electrolyte fuel cell (not to scale).

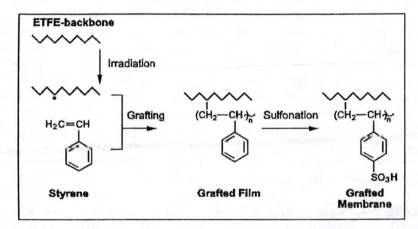

Figure 2. Pre-irradiation grafting of styrene into an ETFE film.

polymer during the grafting process is expected to be enhanced for ETFE base polymer films. For example, ETFE swells slightly in benzene and styrene and FEP does not measurably swell in either; (iii) more radicals are formed per unit irradiation dose (6) in ETFE, thus lowering the irradiation dose required to achieve a particular level of grafting; (iv) the extent of radiation-induced C-C bond scission reactions, which would lead to chain scission and a decrease in molecular weight, is expected to be smaller for ETFE (7). In this chapter, we compare the polymer chemistry of ETFE and FEP base polymer films occurring in our irradiation and grafting steps and the properties of the resulting membranes.

Experimental Methods

Grafted Film and Membrane Synthesis. Non-porous, dense Nowoflon ET-6235 (ETFE) films from Nowofol GmbH of Siegsdorf, Germany (Mw = 680,000 g/mol, 25 to 100 μm thickness) and FEP type A films from DuPont of Circleville, Ohio, USA (Mw = 325,000 g/mol, 25 to 75 μm thickness) were used as base polymer films. Our membrane preparation method (3) consists of: (i) polymer film pre-irradiation by gamma irradiation in air (γ/air) at a dose rate of 6.0 kGy/h or by electron irradiation under a nitrogen atmosphere (e⁻/N₂) typically using an accelerating voltage of 2.2 MV, a beam current of 5 to 20 mA, and a dose rate of 15.0 kGy/s; (ii) subsequent grafting with a solution of 3.5 M styrene (S, Fluka purum grade), 0.33 M divinylbenzene (DVB, Fluka technical grade, 70-85% active), and 0.19 M triallylcyanurate (TAC, purum grade) in benzene for 4.5 h at 60°C under nitrogen atmosphere; and (iii) sulfonation of the polystyrene-based graft component in a solution of 30 vol% chlorosulfonic acid in 1,1,2,2-tetrachloroethane at 66°C for 4 hours. The sulfonated membranes are next treated (a) with 0.15 M NaOH (aq) to hydrolyze any acid chloride groups, (b) with approximately 10 vol% HCl (aq) to regenerate the sulfonic acid form, and (c) in approximately 90°C de-ionized water to swell the membrane.

Film and Membrane Characterization. Irradiated 100 μm ETFE and 75 μm FEP films were equilibrated for one week at 25°C in air prior to their characterization by melt flow index (MFI) measurements or differential scanning calorimetry (DSC). The MFI values were measured according to DIN 53'735 using a mass of 1 kg at 300°C and 372°C, respectively. Their relative heats of fusion, H_f, were determined by DSC under N_2 atmosphere at a scan rate of 10 and 20°C min⁻¹ and using constant integration limits of 160-280°C and 140-300°C for ETFE and FEP films, respectively. The grafting yield was determined according to the following equation: mol% = $N_g/(N_g+N_{bp})$ x 100%, where N_g is the number of moles of monomer incorporated in the grafted chains and N_{bp} is the number of moles of monomer incorporated in the ungrafted base polymer film. Stress-strain measurements of the radiation-grafted films and sulfonated and swollen membranes (90°C in deionized water) were performed according to DIN 53'455 using a jaw speed of 10 mm/min. The *ex-situ* resistance of the swollen membranes was determined by impedance spectroscopy in a two electrode arrangement (*8*).

Fuel Cell Testing. Membranes were tested in fuel cells using our previously reported method (*5*). Test parameters include: (a) a nominal cell temperature of 60°C, (b) humidified hydrogen and oxygen gases at 1 atm, (c) Nafion-impregnated electrodes having a platinum loading of 0.8 mg/cm^2, and (d) active areas of 28 cm^2. The membrane stability in the PEFCs was evaluated by monitoring the *in-situ* membrane area resistances measured by a current pulse method (*9*). The membrane performance over their lifetime was evaluated based on the polarization characteristics of the PEFCs. In this measurement the cell is polarized away from the open circuit voltage (OCV) to lower voltages by progressively decreasing the external resistance and thereby drawing more current.

Results and Discussion

Irradiated Base Polymer Films. Base polymers can be activated for grafting by gamma or electron beam irradiation. Gamma irradiation was advantageous to us earlier (*3*) because we have an in-house gamma irradiation source. Unfortunately our gamma source is limited to the irradiation of smaller samples (less than ~ 12x12 cm^2), and gamma irradiation has disadvantages for the scale up to pilot or production level because of the long irradiation times required (hours) and the safety risks involved in using radioactive sources. Electron beam irradiation is more attractive in these respects. The irradiation atmosphere and temperature play important roles in determining the effects of the radiation processing on the base polymer (*7,10*). Irradiation under inert atmosphere leads to less oxidative degradation (*7*) of the polymer than irradiation in air, but irradiation under inert atmosphere is more expensive because of the additional handling necessary. Irradiation at elevated temperatures tends to increase the extent of radiation-induced crosslinking relative to chain scission (*2,7*) because of the increased radical mobility and thus rates of combination reactions, especially at temperatures above the glass or melting transitions. To provide a basis for evaluating these sample irradiation options, we investigated the effects of radiation type and atmosphere on (i) molecular weight and crystallinity changes occurring in irradiated base polymer films, (ii) their grafting behavior, (iii) mechanical properties of the grafted films and membranes, and (iv) membrane properties relevant to PEFC applications.

The results of MFI and DSC measurements on irradiated 100 µm ETFE and 75 µm FEP films are shown in Table I. A more detailed dose dependence of the MFI values is given elsewhere (*1,10*). These MFI results indicate that chain scission reactions can be minimized or even eliminated by using ETFE as the base polymer film and by carrying out the pre-irradiation under inert atmosphere. In contrast, although chain scission reactions in FEP can be minimized by performing the irradiation under inert atmosphere, chain scission reactions dominate under both oxygen-containing (air) and inert atmosphere.

Table I. Irradiation of 100 μm ETFE and 75 μm FEP films.

Film	Irradiation type	Dose (kGy)	MFI (g/10min)	Relative H_f* (J/g)
ETFE	unirradiated	-	1.1	67.2
ETFE	γ/air	40	9.4	72.4
ETFE	e^-/O_2	40	3.3	77.7
ETFE	e^-/N_2	40	0.8	72.9
FEP	unirradiated	-	1.3	24.4
FEP	γ/air	60	12.9	28.8
FEP	e^-/O_2	60	8.9	29.9
FEP	e^-/N_2	60	5.5	27.2

* H_f: Heat of fusion.

Our DSC measurements indicate that the crystallinity increases upon irradiation. Changes in crystallinity, both increases and decreases, have been observed upon irradiation of many polymer systems (2). Increases in crystallinity are attributed to the crystallization of mobile short chain segments formed by chain scission reactions. Heating of the sample by the incident radiation is also expected to play a role in the crystallization. It is important for us to understand these changes in crystallinity in our pre-irradiation process because grafting is believed to occur primarily (2,11) in the more permeable, higher free volume amorphous domains and not in the crystalline domains of the base polymer film.

The difference in the behavior of the two polymers upon irradiation can be best understood based on the difference in typical relative bond strengths in fluoropolymers (12): C-F > C-C > C-H, so that C-C bond scission is expected to dominate in the case of FEP and C-H bond scission in ETFE. In this case, radical combination reactions would lead to crosslinking only in the case of the secondary carbon-centered radicals resulting from C-H bond scission in ETFE films irradiated under inert atmosphere. The combination reaction of primary radicals resulting from C-C scission would not lead to crosslinking. In the presence of oxygen, peroxy radicals are rapidly formed (7), and the reaction of peroxy radicals do not, in general, lead to the formation of thermally stable crosslinks (7).

Grafting Yields. An important concept in the radiation-grafting method is the grafting front mechanism (1,2,11). If the base polymer film does not swell in the grafting monomer, grafting begins at the surface. This grafted surface region swells in the monomer, however, and monomer can diffuse further into the film interior and react. In this manner the grafting front on each side of the film can move further into the film interior with time until they meet, thus providing bulk and not just surface modification. The e^-/N_2 pre-irradiation doses necessary to obtain graft levels useful for membranes for fuel cell applications are shown in Figure 3. The graft levels shown are high enough so that the resulting membranes are grafted through the film thickness and thus exhibit bulk proton conductivity, as will be shown later based on resistance measurements. Previous investigations (*13*) of membranes based on 50 μm FEP films grafted with sulfonated polystyrene, FEP-*g*-SPS, have shown that the graft component is distributed fairly homogeneously through the thickness at graft levels of > 13 wt.%. It should be noted that the same graft level can be obtained by a variety of means, for example, by varying the irradiation dose, radiation source, base polymer film type and thickness, grafting time, and the type and concentration of the grafting monomer(s). The influence of these parameters, as well as radiation-induced changes in crystallinity and molar mass, on the graft distribution through the film thickness and in the presumably microphase-separated structure (*11*) is not yet known in detail.

Figure 3 shows that ETFE base polymer films require lower irradiation doses than FEP to achieve a comparable grafting level on a molar basis. Comparing the grafting results of Figure 3 with the MFI measurement results in Table I and those reported earlier (*10*) indicates that membranes can be prepared from ETFE films using lower irradiation doses, and thus with less undesirable chain scission, than in the case of FEP films. As discussed previously (*1*), the greater graft levels and grafting rates obtained with ETFE base films may result from higher monomer diffusion rates into the base polymer film, better compatibility of the graft and base polymer components, and a greater extent of radical formation per unit irradiation dose in these films (*6*). The results in Figure 3 show that higher irradiation doses are required to obtain the same grafting level for thinner films. This problem has been attributed (*1*) to the greater extent of orientation, and thus lower permeability, of the thinner extruded fluoropolymer films.

Mechanical Properties. The difference in their chain scission behavior gives ETFE an advantage over FEP as a base polymer film because such desirable mechanical properties as strength and flexibility tend to improve with increasing molecular weight. Improved mechanical properties of the membrane can be quite important in the fuel cell application because of the presence of mechanical and swelling stresses (*14*). Indeed, ETFE-based grafted films and membranes have superior mechanical properties, as shown in Figure 4 for some γ/air radiation-grafted and crosslinked (DVB and TAC) films and membranes based on 100 μm ETFE and 75 μm FEP.

The ETFE-based grafted films and membranes are generally more robust than the FEP-based ones. This is likely to be a result of the greater extent of chain scission during the pre-irradiation processing of FEP. The mechanical properties of both ETFE and FEP-based grafted membranes improve if the pre-irradiation

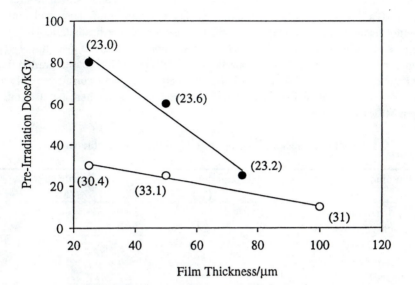

Figure 3. e⁻/N₂ pre-irradiation dose as a function of (○) ETFE and (●) FEP film thickness. The graft levels on a mole% basis are given in parenthesis.

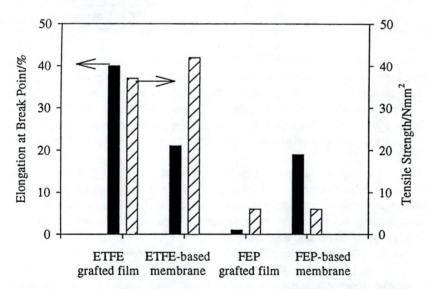

Figure 4. Mechanical properties of crosslinked γ/air radiation-grafted ETFE- and FEP-based grafted films and membranes.

processing is performed using electron beam irradiation under inert atmosphere. For example, as shown in Table II the elongation at break is 23% and the tensile strength is 16 MPa for a FEP-based membrane having the same graft level and grafting solution as the FEP-based membrane prepared using e^-/N_2 pre-irradiation. The break point and tensile strength for the FEP-based membrane are given in Figure 4. The improvement in the mechanical properties of the ETFE and FEP-based grafted films and membranes prepared by irradiation under inert atmosphere correlates well with the reduced extent of chain scission occurring under inert atmosphere (as determined from MFI measurements).

Table II. Mechanical properties of additional membranes*

Membrane	Elongation at break (%)	Tensile strength (MPa)
e^-/N_2 irradiated FEP-based membrane	23	16
e^-/N_2 irradiated FEP-based membrane (air dried)	7	16
Uncrosslinked γ/air pre-irradiated ETFE-based membrane	85	10
Nafion 117 membrane	>100	Not determined

* All membranes characterized in water-swollen state, except as noted.

The mechanical properties of the FEP-based membrane are superior to those of the corresponding grafted film, perhaps due to the plasticizing effect of water in the swollen membrane. The water content of the membrane has an important influence on the mechanical properties of both ETFE and FEP-based membranes. For example, the percent elongation at break of the above mentioned FEP-based membrane (e^-/N_2 pre-irradiation) decreases from 23 to 7% as the membrane dries (Table II).

The differences in the mechanical properties of the ETFE-based films and membranes are not so readily explained. Both Young's modulus (not shown) and elongation at break decrease significantly when the grafted film is sulfonated and swollen. An important point, however, is that the chemistry and properties such as acid resistance of ETFE and FEP are quite different (*15*). For example, subjecting unirradiated ETFE base polymer films to our sulfonation process results in a water swelling of 3%, IEC of 0.2 mEqg^{-1}, and the development of infrared bands due to sulfonic acid groups. The same treatment does not produce any of these changes in the FEP base films. Some sulfonation of the base polymer and perhaps chain scission occurs during the preparation of the ETFE-based membranes, and this may have an influence on the membrane mechanical properties and microstructure.

The extent of crosslinking also has a significant effect on the mechanical properties of the membrane. The elongation at break point was 85% and the tensile strength was 10 MPa for an uncrosslinked ETFE membrane having a graft level corresponding to that of the crosslinked ETFE membrane. The mechanical properties of the crosslinked ETFE are shown in Figure 4. The uncrosslinked membrane is weaker, as expected, and it can be more elongated than the DVB/TAC-crosslinked membrane (Figure 4) because DVB acts as a stiff crosslinker (16). The relationship between the extent of crosslinking and mechanical properties may be quite complex in our system because crosslinking can result from the pre-irradiation processing, grafted added crosslinkers like DVB and TAC, and grafted polystyrene segments connecting two fluoropolymer chains.

In comparison, the swollen Nafion 117 membrane (a perfluorinated membrane from DuPont that is not covalently crosslinked) does not break even when subjected to elongations of up to 100% under our test conditions. The poorer mechanical properties of our radiation grafted membranes probably results in large part from the polystyrene-based graft component, which is expected to be stiff and brittle due to the aromatic groups. Crosslinking using DVB makes the membrane still less elastic and can even result in membranes that shatter because they are too inelastic (16) to absorb the local swelling stresses of several thousand kilopascals that can develop in the membranes during drying and wetting. It is of interest to further improve the mechanical properties of our membranes because they become quite important in scaling up the PEFC to cells with larger active areas (higher power) due to the larger defect free membrane areas required and the greater and less homogeneous stresses encountered.

Membrane Properties. The importance of the proton conducting and gas separation functions of the membrane in the PEFC has been stated in the introduction. Some important membrane properties relevant to these two functions are given in Table III for our radiation-grafted membranes in comparison to the Nafion membranes.

The ETFE- and FEP-based membranes have IECs that are much higher than those of the Nafions, and these higher IECs tend to improve proton conductivity and reduce the membrane resistance. It is necessary, however, to crosslink our radiation-grafted membranes to extend their lifetime (5) in the rigorous fuel cell application. In addition, crosslinking can control swelling and improve the gas separation properties of radiation-grafted membranes. An undesirable consequence of crosslinking, however, can be an increase in the membrane resistance. This problem has been overcome using a grafting solution containing both crosslinkers, TAC and DVB (5). The resulting membranes are much more stable and perform well relative to uncrosslinked ones, and the membrane resistance is not too greatly increased in comparison to that of uncrosslinked membranes (5).

Table III. Properties of radiation grafted ETFE- and FEP-based membranes and Nafion membranes.

Property	ETFE-based membrane		FEP-based membrane		Nafion 117	Nafion 112
Base polymer film Thickness (μm)	100	25	75	25	N.A	N.A.
Water swollen membrane Thickness (μm)	152	39	125	41	200	60
Ion exchange capacity [IEC](mEq/g)	2.4	2.4	1.8	2.0	0.91	0.91
Swelling in water (wt%)	44	27	33	22	37	37
In-situ area resistance at 60°C (mΩcm^2)	135	66	120	70	190	70
In-situ specific resistivity at 60°C (Ωcm)	9	17	9.6	17	9.5	11

N.A.: not applicable.

The membrane area resistances of the radiation-grafted membranes given in Table III are not proportional to the membrane thickness, and the specific resistance of the thinner membranes is higher. This behavior has been attributed to inhomogeneities in the graft component distribution through the membrane thickness (1), and we are examining ways of overcoming this challenge.

The *ex-situ* area resistances as a function of graft level of the membranes based on 100 μm ETFE films are shown in more detail in Figure 5. The area resistances of these water-swollen membranes decrease with increasing graft level. This decrease in area resistance is most pronounced at graft levels between 50 and 60 wt% although the water content in terms of water molecules per acid group, H_2O/RSO_3H, is fairly constant over this grafting range. Such behavior was observed in our earlier work (*11,14*) with FEP base polymers, and the grafting level at which a sharp decrease in resistance occurs has been referred to as the *critical degree of grafting*. It occurs at a grafting level above that at which the membrane is first grafted through the film thickness. The critical grafting level shown in Figure 5 of the crosslinked membranes based on 100 μm ETFE is considerably higher, both on a weight and mole% basis, than that reported (*11,14*) for the uncrosslinked and crosslinked membranes based on 75 μm FEP. The influence of base polymer film type and thickness, radiation dose, graft component, and membrane microstructure in determining this critical degree of grafting is not yet known in detail.

Fuel Cell Tests. ETFE-and FEP-based membranes have been tested in fuel cells as described previously (5). *In-situ* resistances were measured to learn about membrane stability during long-term tests. An increase in the membrane resistance indicates that proton-conducting groups have been lost due to degradation reactions. Infrared spectroscopic analysis of FEP-based membranes after fuel cell testing and HPLC analysis of the product water when the membrane resistance increases indicated (17) that the non-fluorinated graft component is the weak point in our present radiation-grafted membranes.

As shown in the examples in Figure 6, many of our radiation-grafted membranes based on 75 μm FEP and 100 μm ETFE base polymer films have demonstrated lifetimes on the order of 1,000 hours or more before significant membrane degradation occurs (5). More recent tests (18) have indicated that the lifetimes of our membranes can be extended to 3,000 - 4,000 hours by restricting the PEFC to operating at higher cell voltages (greater than 600 mV) characteristic of electrical efficiencies greater than 50% (based on the lower heating value of H_2).

The lifetimes of our present radiation-grafted membranes is limited by the susceptibility of the α-hydrogen of the polystyrene graft component (5,17) to abstraction reactions and the subsequent oxidative degradation of the graft component. Although our membranes already have demonstrated (14) much better stability in the PEFC application compared to that of current commercial radiation-grafted membranes, future efforts will be directed to improving the lifetimes by using monomers for grafting which do not have an α-hydrogen. For example, the stability of radiation-grafted membranes in electrochemical applications has been improved using α,β,β-trifluorostyrene (19) or α-methylstyrene (20) as the grafting monomer.

We reported (10,21) that the polarization properties of PEFCs using our radiation-grafted membranes compare favorably with those obtained using commercial and experimental Nafion membranes of approximately similar thicknesses. For example, maximum power densities of 450, 390, and 270 mW/cm^2 were achieved (10,21) under the same test conditions with membranes based on 75 μm FEP and 100 μm ETFE base polymer films and Nafion 117, respectively. Maximum power densities of 510, 490, and 440 have been achieved (18) under the same test conditions with thinner membranes based on 25 μm FEP and 25 μm ETFE base polymer films and Nafion 112, respectively.

Our results indicate that the base polymer film type (ETFE or FEP), radiation source, and irradiation atmosphere only affect membrane mechanical properties and do not appear to directly affect the chemical stability in the PEFC application. We have optimized these parameters to improve the mechanical properties of our membranes. This result is quite important, however, due to the high mechanical and swelling stresses on the membrane (14) in the application, especially in cells of larger active areas.

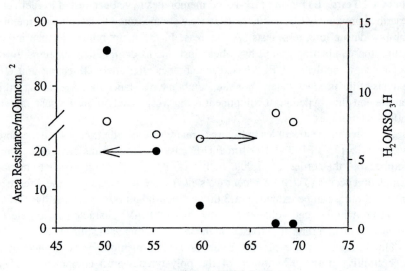

Figure 5. Area resistances (●) and water content (○) of swollen membranes based on γ/air pre-irradiated 100 μm thick ETFE films grafted with S/DVB/TAC.

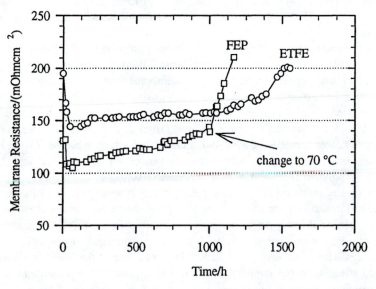

Figure 6. *In-situ* membrane resistances of (○) 100 μm ETFE- and (□) 75 μm FEP-based membranes tested at a nominal temperature of 60°C (periodically up to 75°C).

Acknowledgments

Financial support from the Swiss Federal Office of Energy (BFE) is gratefully acknowledged. R. Hodann of Nowofol GmbH, D-83310 Siegsdorf and Dr. N.D. McKee of DuPont Fluoroproducts, Wilmington, DE, USA are thanked for technical product information and the gift of some ETFE and FEP materials. Technical assistance by F. Geiger of the Paul Scherrer Institute (PSI), MFI measurements by Professor H.G. Bührer, A. Tarancón, M. Wyler, and M. Benz of Technikum Winterthur Ingenieurschule (TWI), CH-8401 Winterthur; stress-strain measurements by Dr. C. Löwe and C. Walder of the Swiss Federal Laboratories for Materials Testing and Research (EMPA-Dübendorf), CH-8600 Dübendorf; ESR measurements by Professor E. Roduner and G. Hübner of the University of Stuttgart, Germany; and sample irradiation by M. Steinemann of PSI and C. Günthard of Studer AG, CH-4658 Däniken are all gratefully acknowledged.

Literature Cited

1. Brack, H.P., Scherer, G.G. *Macromol. Symp.* **1997**, *126*, 25.
2. Chapiró, A. In *Radiation Chemistry of Polymeric Systems*; Mark, H.; Marvel, C.S.; Melville, H.W. Eds. High Polymers; Interscience: New York, 1962; Vol. XV.
3. Gupta, B., Büchi, F.N.; Scherer, G.G. *J. Polym. Sci.: Part A: Polym. Chem.* **1994**, *32*, 1931.
4. Sperati, C.A. In *Polymer Handbook*, 3rd ed.; Bandrup, J.; Immergut, E.H., Eds.; Wiley-Interscience: NY, 1989; V/35.
5. Büchi, F.N.; Gupta, B.; Haas, O.; Scherer, G.G. *J. Electrochem. Soc.* **1995**, *142*(9), 3044.
6. Hübner, G.; Brack, H.P.; Scherer, G.G.; Roduner, E. unpublished results.
7. Charlesby, A. In *Irradiation Effects on Polymers*; Clegg, D.W.; Collyer, A.A., Eds., Elsevier: Essex, England, 1991; chapter 2.
8. Halim, J.; Büchi, F.N.; Haas, O.; Stamm, M.; Scherer, G.G. *Electrochim. Acta* **1994**, *39*(8/9), 1303.
9. Büchi, F.N.; Marek, A.; Scherer, G.G. *J. Electrochem. Soc.* **1995**, *142*(6), 1895.
10. Brack, H.P.; Büchi, F.N.; Huslage, J.; Rota, M.; Scherer, G.G. *PMSE* **1997**, *77*, 368.
11. Gupta, B.; Scherer, G.G. *Chimia* **1994**, *48*, 127.
12. Banks, R.E. In *Fluoropolymers '92*; Banks, R.E.; Dunn, A.S.; Fielding, H.C.; Johns, K.; Peyrical, J.P.; Smith, E.H.; Willoughby, B.G. Eds.; Fluoropolymers Conference 1992; Rapra Tech.: Shrewsbury, UK, 1992; Vol. I., paper 1, p.1.
13. Gupta, B.; Büchi, F.N.; Staub, M.; Grman, D.; Scherer, G.G. *J. Polym. Sci.: Part A: Polym. Chem.* **1996**, *34*, 1873.

14. Scherer, G.G.; Brack, H.P.; Büchi, F.N.; Gupta, B.; Haas, O.; Rota, M.; *Hydrogen Energy Progress XI*, Proceedings of the 11th World Hydrogen Energy Conference, Stuttgart, Germany, June 1996; Vol. 2, p. 1727.

15. *Modern Fluoropolymers*; Schiers, J. Ed.; Wiley Series in Polymer Science; Wiley: Chichester, England, 1997.

16. Dardel, F.D.; Arden, T.V. In *Ullmann's Encyclopedia of Industrial Chemistry*, 5th ed.; Elvers, B.; Hawkins, S.; Ravenscroft, M.; Schulz, G., Eds.; VCH: Weinheim, 1989; Vol. A14.

17. Büchi, F.N.; Gupta, B.; Haas, O.; Scherer, G.G. *Electrochim. Acta* **1995**, *40*(3), 345.

18. Brack, H.P.; Büchi, F.N.; Huslage, J.; Scherer, G.G. *Extended Abstracts*, 194th Meeting of the Electrochemical Society, Boston, MA, 1998, abstract 1105.

19. D'Agostino, V.F.; Lee, J.Y.; Cook, E.H. *U.S. Patent* 4,012,303, 1977.

20. Assink, R.A.; Arnold, C.; Hollandsworth, R.P. *J. Membrane Sci.* **1991**, *56*, 143.

21. Rota, M.; Brack, H.P.; Büchi, F.N.; Gupta, B.; Haas, O.; Scherer, G.G. *Extended Abstracts*, 187th Meeting of the Electrochemical Society, Reno, NE, 1995, Vol. 95-1, p. 719.

Chapter 13

Surface Modification of Microporous Polypropylene Membranes by Polyelectrolyte Multilayers

T. Rieser, K. Lunkwitz, S. Berwald, J. Meier-Haack, M. Müller, F. Cassel, Z. Dioszeghy, and F. Simon

Institute of Polymer Research Dresden e. V., Hohe St. 6, 01069 Dresden, Germany

Plasma-induced grafting of ionic monomers onto hydrophobic polypropylene membranes led to permanently charged surfaces. In addition, polyelectrolyte multilayers were built up by alternating adsorption of oppositely charged polyelectrolytes onto the grafted membrane surface. The modified membranes were characterized by FTIR-ATR and XPS. Streaming potential measurements were used to determine the electrokinetic properties of the charged surfaces. Filtration experiments were performed with human serum albumin (HSA) solutions to study the fouling behavior of the membranes. Membrane surfaces and the protein were both negatively charged under the applied conditions. The irreversible adsorption of HSA decreased remarkably for membranes that were modified by polyelectrolyte multilayers, as compared to untreated and plasma-treated samples.

Membranes with high mechanical strength and long lifetime are currently gaining interest in different membrane applications. Polyolefin-based membranes satisfy these requirements very well in many cases. On the other hand, the hydrophobic olefinic surface has to be modified by chemical or physical treatments for applications that require functionalized or hydrophilic membrane surfaces, e. g. biosensors, pervaporation, microfiltration, and ultrafiltation (1). In addition, the hydrophobicity of polymer materials normally increases membrane fouling, one of the major factors limiting the use of membranes (2-3).

Surface modification of membranes, besides bulk modification of polymers, is a promising approach to provide membranes with tailor-made separation properties and a reduced tendency for fouling (4). Coating of porous membranes with a thin film may lead to a change of the pore size in a defined way and to an increase of the retention of solutes (5). Hydrophilization of polymer surfaces gives rise to a steric repulsion force between organic solute molecules and the membrane surface due to an extensive hydration of the protecting coating as it is very well known for poly(ethylene oxides) (6). In addition, surface modification of porous membranes is a versatile tool to prepare other membrane types, such as gas separation membranes or pervaporation membranes (7-8).

In this report the modification of hydrophobic polypropylene membranes by polyelectrolytes is described for the alteration of the interface interactions between the solid membrane and the feed components. Layer-by-layer deposition of polyelectroytes was carried out on microporous membranes based on a method developed by Decher et al. to prepare supported ultrathin multilayers on glass slides and silicon substrates (9-10). The simple adsorption procedure involves alternating dipping of a charged surface into an aqueous solution of an oppositely charged polyelectrolyte. After each adsorption step, the samples have to be rinsed extensively with water to remove the weakly bound polyelectrolytes. Recently, McCarthy et al. reported on the modification of gas separation membranes (11) and Kunitake et al. built up multi-enzyme films on an ultrafilter applying this useful preparative technique (12).

The goal of this work was to study filtration and fouling properties of microfiltation membranes that were modified by layer-by-layer deposition of polyelectrolytes. The microporous polypropylene membranes were first plasma-treated to create peroxide species on the chemically inert polyolefinic surface. Mitchell and Perkins reported that the formation of hydroperoxides during an oxidation procedure of polyolefins is the most likely reaction (13). Plasma-induced graft polymerization of hydrophilic/charged monomers led to a first polyelectrolyte layer that was covalently attached to the membrane surface. The charged substrates were used to construct two additional polyelectrolyte layers by consecutively alternating adsorption of anionic and cationic polyelectrolytes (Figure 1). The covalently fixed first polyelectrolyte layer guaranteed the stability of the multilayer assembly during the microfiltration.

Several advantages of membranes modified by polyelectrolyte multilayers in comparison to just grafted membranes have to be pointed out. For example, defects in the grafted polyelectrolyte layer may be compensated for in the case of polyelectrolyte multilayers. In addition, the layer assembly could act as an enhanced steric barrier for feed molecules which leads to an increased retention. Proteins with an isoelectric point in the acidic or basic range can be filtered with the same membrane after adjustment of the surface charge by a simple adsorption step. Finally, a low tendency to protein adsorption is expected from polyelectrolyte complex layers on substrates. Experimental results of Kabanov et al. show that linear polyions are able to replace proteins quantitatively in interpolyelectrolyte complexes in solution when the protein is bound unspecifically to the linear polyelectrolytes (*14*). Hence, applying polyelectrolyte complexes on polymer substrates could lead to interesting anti-fouling coatings.

The modified membranes were characterized by FTIR-ATR, XPS, and streaming potential measurements. Retention and fouling behavior were examined with human serum albumin as test substance. The membrane surface and the protein carried the same charge under the applied conditions, thus the electrostatic interactions were repulsive.

Experimental Section

Materials.   Acrylic acid (AC) was obtained from Aldrich (Steinheim, Germany) and distilled before use. Poly(diallyldimethyl ammoniumchloride) (PDADMAC), MW = 99,000 g/mol, and poly(acrylic acid) (PAC), MW = 90,000 g/mol, were purchased from the Frauenhofer Institute for Applied Polymer Research (Teltow, Germany) and Polyscience Europe GmbH (Eppelheim, Germany), respectively. Human serum albumin and Celgard 2400 microporous polypropylene membranes (PP membrane) were provided by the Hoechst AG (Wiesbaden, Germany). The porosity of the membranes was 28-40 % and the pore dimensions were 0.04 μm width x 0.12 μm length (specification of the manufacturer).

Modification Methods.   Plasma treatment was carried out with a Tepla 440G plasma reactor (Technics Plasma GmbH, Kirchheim, Germany) supplied with a microwave plasma source. The experimental conditions were: 70 W microwave power; 10 sccm CO_2 mass flow; 10 Pa gas pressure; 3 min treatment time. Some of the plasma-treated samples were reacted with SO_2 for 1 h to prove the hydroperoxide formation. The plasma-induced graft polymerization was carried out with the plasma-activated membranes in aqueous monomer solutions under nitrogen atmosphere at 70°C. All samples were placed in a beaker filled with distilled water and heated overnight, followed by filtration for 1 h with pure water to remove non-grafted homopolymer and unreacted monomers. The grafted amount of the polymer was determined by weighing. In addition some of these grafted membranes were used for construction of polyion multilayer assemblies. Polyelectrolytes were consecutively adsorbed from aqueous solutions containing 20 mM of the polyion (calculated as monomer). The pH of the cationic polyelectrolyte solution was adjusted to 8 with 1 m NaOH. The pH of the polyanion was 3.6 after dissolving of PAC in water and used at this value. The adsorption time was 30 min. After each deposition step the samples were rinsed and filtered with pure water to remove weakly bound polyions.

Characterization Methods.   The modified membranes were characterized by streaming potential measurements with an electrokinetic analyzer (EKA) from A. Paar KG (Graz, Austria). FTIR-ATR measurements of the grafted samples were performed with an IFS 66 spectrometer from Bruker (Germany). Plasma-treated polypropylene membranes were analyzed by X-ray photoelectron spectroscopy (XPS) (ESCAlab 220i, Fisons Instruments, England) equipped with an unmonochromized Mg K $\alpha_{1,2}$ excitation source.

The water flux of the polypropylene membranes was determined using dead-ended stirred filtration cells (model 8050, Amicon, Witten, Germany) with a volume of 50 ml and an active membrane area of 13.4 cm^2. Before each filtration the membranes were wetted with isopropanol. Water volumes of 25 ml were filtered through the membranes under a pressure of 300 kPa until a stable water flux was reached. The average value of the water flux for each membrane was calculated from the last two filtration measurements.

Human serum albumin adsorption was carried out as follows. The protein was dissolved in deionized water. The filtration cell was filled with 50 ml of 0.1 wt.% protein solution (pH = 6.4). Volumes of 2 x 12.5 ml were filtered through the membrane under a pressure of 300 kPa. The protein retention was calculated from the protein concentration in the retentate. The protein concentration was determined by UV/VIS spectroscopy using the typical protein adsorption at 287 nm. A calibration curve was drawn by plotting absorbance intensity towards known concentrations of HSA to determine the unknown concentrations of the protein in the feed and permeate. UV/VIS spectra were obtained with a UV/VIS spectrophotometer from Dr. Lange GmbH (Berlin, Germany). After filtration, the membranes were rinsed several times with water to remove reversibly adsorbed protein. The water permeation flux was determined after each rinsing step. When the flux remained constant it was assumed that all reversibly adsorbed protein was removed. The amount of irreversibly adsorbed human serum albumin was determined by the weight increase of the sample.

Results and Discussion

Plasma Treatment. Polypropylene is a very hydrophobic and chemically inert material. Treatment of the polymer surface with oxygen-containing plasmas introduces a variety of functional groups, including ether, carbonyl, ester, acid, and peroxide groups, etc. (*15*) as it is schematically shown in Figure 1. The incorporation of oxygen was proved by FTIR-ATR (Figure 2, curve b) and XPS measurements. The broad absorption band from 1,850 - 1,650 cm^{-1} is assigned to the stretching vibration of C=O double bonds present in carbonyl, acid, and ester groups but gave no direct prove for hydroperoxides created on the polypropylene surface. Therefore, plasma-treated samples were reacted with SO_2 gas, (Table I), which led to sulfate groups that were confirmed by XPS (*16,17*). The reaction is given as follows:

$$-C-OOH + SO_2 \rightarrow -C-O-SO_2OH$$

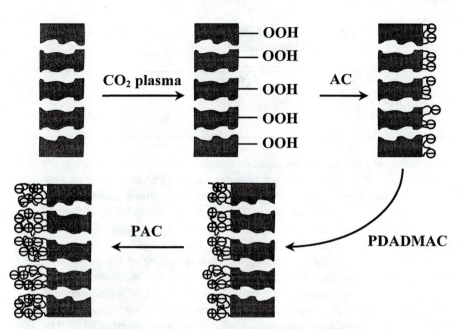

Figure 1. Simplified scheme of the functionalization of microporous polypropylene membranes by plasma treatment and build-up of multilayer assemblies by consecutive polyelectrolyte adsorption. Hydroperoxides (R-OOH) are reactive in the plasma-induced graft polymerization.

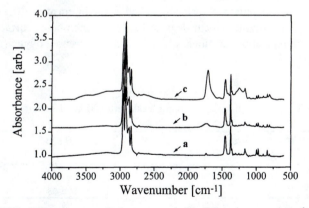

Figure 2. FTIR-ATR spectra of a) untreated, b) CO_2 plasma-treated (70 W, 3 min, 10 sccm, 10 Pa) and c) PAC-grafted PP membranes (5 wt.%, 70°C, 60 min).

The atomic ratio [element]:[carbon] of an untreated and plasma-actived sample after derivatisation with SO_2 is presented in Table I.

Table I. XPS measurements of plasma-activated polypropylene membranes after derivatisation with SO_2.

Treatment time	0 sec	180 sec
[O] : [C]	0.0119	0.2028
[N] : [C]	-	0.0032
[S] : [C]	-	0.0045

Another indirect prove for peroxide formation was the successful grafting of acrylic acid onto the plasma-treated polymer surface (Figure 2, curve c). Peroxides created at the polypropylene substrate initiated grafting reactions of vinyl monomers by thermal decomposition. The intensive absorbance band from 1,820 - 1,550 cm^{-1} was caused by the C=O double bond stretching vibration of the associated carboxylic acid groups.

During plasma treatment functionalization and etching of the polymer surface occurred simultaneously. In this work CO_2 plasma was used to modify polypropylene membranes, which normally causes milder attacks on polymer substrates (18) than an oxygen plasma. Generally, an elevated plasma power and long treatment times favor degradation of the polymer materials (19). The fine lamellar structure of the polypropylene membranes formed by stretching of isotacic polypropylene films (20) was particularly sensitive to plasma etching effects, therefore mild plasma treatment conditions were chosen. All experiments were carried out with 70 W discharge power, 10 sccm gas flow, and a pressure of 10 Pa.

According to the XPS results displayed in Table II, the incorporation of oxygen functionalities in the polymer surface reached a maximum after 30 sec treatment time. For longer plasma treatment degradation of the membrane surface dominated over the functionalization reactions, which caused a decreased oxygen content after 90 sec and 180 sec.

Table II.
Results of XPS after CO_2 plasma treatment (70 W, 10 sccm, 10 Pa).

Treatment time	0 sec	30 sec	90 sec	180 sec
[O] : [C]	traces	0.2944	0.2811	0.2761
[N] : [C]	-	traces	traces	traces

Nevertheless, 180 sec treatment time was chosen because a slight pore etching was desired to compensate for the pore size reduction by the subsequent grafting reaction. Such plasma-treated membranes showed an increase of the permeate water flux of about 10 to 15%.

Plasma-Induced Graft Polymerization. Graft polymerization of vinyl monomers onto plasma-activated samples was initiated by thermolysis of peroxides at 70°C. In many studies about plasma-oxidized polymer surfaces a continuous decomposition of peroxides was observed during storage of the modified samples at room temperature (*21*), which led to reduced graft polymer yields (g.y.). Hence, for reproducibility of the following grafting reactions with vinyl monomers, the crucial point was to start the reactions always after approximately the same storage time.

The graft polymer yield was determined by the percentage increase in weight of the samples according to equation 1, where W_g and W_0 represent the weights of the plasma-treated and the grafted membranes, respectively.

$$\text{Graft polymer yield (g.y.)} = [(W_g - W_0)/W_0] \times 100 \ (\%) \tag{1}$$

In principle a variety of ionic vinyl monomers are suited for grafting reactions and a subsequent multilayer build-up. In this study, acrylic acid was chosen to create a first polyelectrolyte layer on the membrane surface, because it was found that the reproducibility of the graft polymerizations was the best for acrylic acid in the range of low graft polymer yield. High polymer graft yields led to an undesired decrease of the membrane pore sizes and, consequently, to a dramatic water flux decline.

In Figure 3 the influence of the reaction time on the poly(acrylic acid) graft yield is shown for two different monomer concentrations. The reaction was carried out at constant temperature. An obvious increase of the amount of grafted polymer was obtained within the first 60 min of the graft polymerization, than the extent of further grafting seemed to be suppressed for longer reaction periods, which likely resulted from the diminishing amount of initiating peroxide species. The total amount of grafted poly(acrylic acid) and the reaction velocity at the beginning of the grafting procedure were higher for the 5 wt.% monomer concentration, compared to that of 3 wt.%. It can be assumed that the probability for an oxygen radical to start a chain growth reaction is much higher for enhanced monomer concentrations in the surrounding solution.

Polyelectrolyte Multilayer Build-up. Plasma-treated polypropylene membranes were grafted with different amounts of poly(acrylic acid) and consecutive polyelectrolyte adsorption was carried out with a 20 mM PDADMAC and a 20 mM PAC solution. The influence of the modification steps on the electrokinetic behavior of the membranes was studied by streaming potential measurements, which allowed to calculate the corresponding zeta potential according to a method described in detail elsewhere (22). The zeta potential characterizes the surface charge of a substrate at the shear plane of the solution/solid interface; however, the surface charge of a solid itself cannot be measured directly (23). Dissociation of ion groups and/or adsorption of electrolyte ions give rise to a surface charge at a polymer substrate. The detected negative zeta potential of the untreated hydrophobic polypropylene membrane is explained by the preferred adsorption of electrolyte anions (24) (Figure 4). Plasma treatment introduced carboxylic acid groups on the membrane surface, confirmed also by XPS, which dissociated in dependence of the pH and caused a corresponding shift of the isoelectric point (IEP) to the acidic pH range. As the grafting yield of PAC increased, a further shift of the IEP to lower pH was observed, reflecting the increasing density of carboxyl groups at the membrane surface.

Before the layer-by-layer deposition of polyelectrolytes was carried out on the grafted membranes, the pH of the cationic polyelectrolyte solution was adjusted to 8.0 with 1 m NaOH. At high pH the grafted PAC layer on the membrane surface provided a much higher concentration of dissociated carboxyl groups, which led to an increase of the negative zeta potential, as shown in Figure 4. In general, the increase of the negative surface charge on a substrate resulted in a rapid adsorption of an oppositely charged polycation and an increase in the adsorbed amount (25). After 30 min adsorption time a reverse of surface charge, due to the quaternized ammonium group of the PDADMAC component, was observed by a defined shift of the isoelectric point (IEP) from the acidic to the basic pH range and by a sign-reverse of the zeta potential from minus to plus for pH values lower than the IEP (Figure 5). After subsequent adsorption of poly(acrylic acid) the surface charge was reversed again, but the IEP of the grafted PAC layer could not be obtained, which can be explained by the influence of the oppositely charged underlying PDADMAC layer. The same effect is responsible for the measured IEP of 7.8 of the PDADMAC layer, whereas the value of the IEP is expected to be in the range of 8 and 10 for quaternized ammonium groups.

Filtration Experiments.

Water Permeation Measurements. The water permeation properties as a function of different PAC graft yields are displayed in Figure 6 as relative water flux reduction calculated according to equation 2. $J_{W,M}$ and $J_{W,0}$ are the initial water flux of the untreated and PAC-grafted membrane, respectively.

$$\text{Relative Water Flux Reduction (RWFR)} = [(1 - J_{W,M}/J_{W,0})] \times 100\ [\%] \qquad (2)$$

Increasing the graft polymer yields led to a dramatic loss of water flux, caused by a constriction and even partial coating of the membrane pores with a thin graft

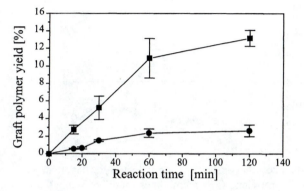

Figure 3. PAC graft yield of CO_2 plasma-treated PP membranes (70 W, 3 min, 10 sccm, 10 Pa) as a function of reaction time for different AC concentrations (5 wt.%, -■-; 3 wt.%, -●-) at 70°C.

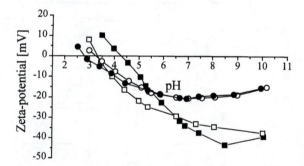

Figure 4. Influence of plasma treatment and PAC graft yield on zeta potentials of: untreated (-■-), CO_2 plasma-treated (70 W, 3 min, 10 sccm) (-□-), and grafted PP membranes (-O-, 3.3 % g.y.; -●-, 5.3 % g.y.); g.y. = graft yield.

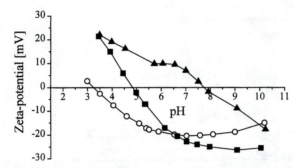

Figure 5. Influence of consecutive polyelectrolyte layer deposition on zeta potentials of polypropylene (PP) membranes modified with: 1 layer (3.3 % PAC graft yield, -O-), 2 layers (PAC-grafted/PDADMAC, (-▲-), and 3 layers (PAC-grafted/PDADMAC/ PAC, -■-).

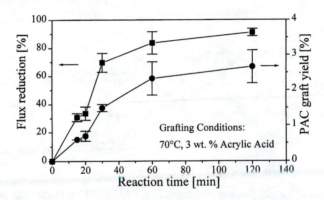

Figure 6. Graft yield (-●-) and flux reduction (-■-) of CO_2 plasma-treated PP membranes (70 W, 3 min, 10 sccm, 10 Pa) as a function of reaction time.

polymer film. Nevertheless, a systematic dependency of water flux and graft polymer yield was observed, which indicated that defined and reproducible permeation properties could be obtained by the plasma-induced graft polymerization step.

Both, plasma treatment and grafting of acrylic acid resulted in negatively charged substrates, which could be used for the build-up of polyelectrolyte multilayer assemblies. However, starting the polyelectrolyte layer construction by grafting resulted in two important advantages. Firstly, the stability of polyelectrolyte multilayer assemblies on just plasma-treated membranes did not appear to be stable enough during the filtration process. Secondly, a large number of polyelectrolyte layers had to be build up to observe any influence of the modification on the microfiltration properties. However, an effective and stable modification was obtained with one and two further polyelectrolyte layers on grafted membranes. The stability of the polyelectrolyte multilayer system on the grafted membrane surface was investigated by recording the water flux during 24 h after each complexation step.

It was expected, that adsorption of polycations onto PAC-grafted membranes would lead to a further blocking of membrane pores and, hence, to a decrease of the water flux. However, an enhancement of the water flux was detected instead of a decrease (Figure 7). The effect obviously occurred for samples covered with ca. 1.5 wt.% PAC and is currently not very well understood. Adsorption of PAC as the third polyelectrolyte layer resulted in water fluxes that were similar to those of membranes covered with a polycation. This effect was compensated for by a larger number of polyelectrolyte layers. Increasing the layer number resulted finally in a strong decrease of the water flux. The experiment was repeated with more than five samples to prove the reproducibility. However, construction of a fourth layer resulted in a dramatic flux decline (not shown in Figure 7). It is proposed that the adsorbed polyelectrolyte may affect the swelling behavior of the grafted membrane in water and, therefore, the pore diameter of the membranes. A similar result was obtained by Osada *et al.* for complexation of a poly(methacrylic acid)-grafted membrane with poly(ethylene glycol). Expanding of micropores was explained by polymer-polymer complexation (*26*).

Human Serum Albumin (HSA) Filtration.   One important purpose of this study was to investigate the fouling behavior of polypropylene membranes modified by polyelectrolyte multilayers. Therefore, HSA filtration experiments were performed and the susceptibility to fouling was investigated regarding the relative water flux reduction (RWFR) after HSA filtration and the irreversibly adsorbed amount of protein. The water flux was measured before ($J_{W,M}$) and after ($J_{W,P}$) HSA filtration. The RWFR values were calculated from equation 2. In addition, the samples were dried and weighted to determine the residual amount of protein adsorbed onto the polymer membrane. Irreversible protein adsorption was mainly responsible for a severe flux decline during bioseparations. Therefore, in general, cleaning circles have to be applied to regain the loss in flux.

The results for polypropylene membranes modified by three polyelectrolyte layers are presented in Figure 8. The water flux of these membranes in dependence of the PAC graft yield has already been presented in Figure 7. Filtration experiments were carried out after dissolving HSA in water. The pH of the solution was 6.4. Hence, the membrane surface and the protein (IEP_{HSA} = 4.8 (3)) were negatively charged and the surface/feed molecule interactions were repulsive.

Table III. Retention of polypropylene membranes modified by three
polyelectrolyte layers in dependence of the PAC
graft yield.

PAC graft yield	0[a]	0[b]	2.5	3.2	4.8
Retention (%)	76	69	94	94	95

[a]Untreated polypropylene membrane
[b]CO_2 plasma-treated membrane (70 W, 3 min, 10 sccm, 10 Pa)

The irreversibly adsorbed amount of protein was minimized remarkably as compared to the untreated and plasma-treated samples under the applied conditions, combined with an increased retention (94%) (Table III). The retention, R, was calculated from the protein concentration in the feed (C_f) and in the permeate (C_p) according to equation 3.

$$R = [(C_f - C_p)/C_f] \times 100 \ [\%] \tag{3}$$

The polyelectrolyte multilayer assemblies might act as a steric barrier for the protein molecules by preventing the penetration of HSA molecules into the pore channels and fouling of the internal membrane surface. No influence of the PAC graft yield on retention and adsorbed amount of protein was observed for the polyelectrolyte multilayer membranes. On the other hand, the RWFR increased with larger amounts of PAC graft yields. This result is explained as follows. In the case of membranes with increased graft yields the membrane pore size was reduced compared to the unmodified substrate or substrates with low polymer graft yields. In consequence of this effect, the mechanical blocking of pores with diminished pore diameters by protein aggregates may be enhanced. These results indicated that membrane fouling was caused mainly by protein adsorption onto the external membrane surface. The RWFR values may be minimized in the future by optimizing the polymer graft yield of the first polyelectrolyte layer.

During pressure-driven membrane processes, macro-solute adsorption caused by intermolecular interactions between membrane and feed molecules and filtration-induced particle deposition have to be considered (27). Hence, it was not possible to

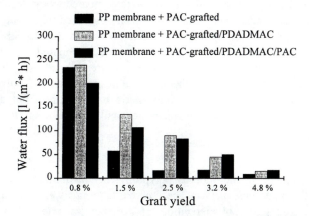

Figure 7. Influence of polyelectrolyte layers on water flux of modified PP membranes.

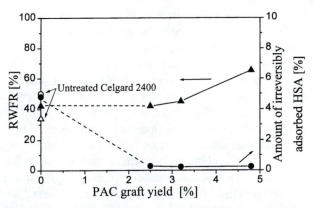

Figure 8. Influence of polyelectrolyte layer assemblies on membrane fouling. RWFR values for untreated (-Δ-), plasma-treated (--▲--), grafted polypropylene (PP) membranes (-▲-), and irreversibly adsorbed amounts of HSA on untreated (-O-), plasma-treated (--●--) and PAC-grafted PP membranes (-●-) were determined.

correlate anti-fouling properties of polyelectrolyte complex layers with physico-chemical interactions based on the results describe above. Therefore, polypropylene films were spin-coated from the membrane material to study HSA adsorption without interfering filtration-induced influences. *In situ* FTIR-ATR studies (*24,28*) were performed to investigate protein adsorption onto untreated and plasma-treated planar surfaces. These spectroscopic results were compared to the results obtained for polypropylene substrates that had been modified by polyelectrolyte multilayers. No protein adsorption was observed within the sensitivity of the method for the latter case over 2 days of investigation compared to untreated and plasma-treated films, for which HSA adsorption occurred immediately in both cases (*29*). These results confirmed that the anti-fouling properties of polyelectrolyte multilayer coatings resulted from not only steric barrier effects but also altered physico-chemical interactions between the protein and membrane surface.

Conclusions

CO_2 plasma treatment of microporous polypropylene membranes led to peroxide formation on the membrane surface. These peroxide groups were used as initiators for plasma-induced graft polymerization of poly(acrylic acid). Two additional polyelectrolyte layers were built up by subsequent layer-by-layer deposition of polyelectrolytes. The covalently attached first polyanion layer was necessary for the stability of the multilayer assemblies under the applied filtration conditions. A systematic dependence of the water permeate flux on the graft polymer yield was obtained. Hence, filtration properties could be tailor-made by plasma-induced graft polymerization. The reverse of surface charge after each polyelectrolyte deposition step was demonstrated by streaming potential measurements. An increase in the water flux was observed after complexation of the grafted polyanion layer with a polycation and is explained by alteration of the thickness of the grafted layer by polymer-polymer complexation. This effect was compensated for by increasing the number of the polyelectrolyte layers on the membrane surface.

The irreversibly adsorbed amount of HSA decreased remarkably for membranes modified by polyelectrolyte multilayer assemblies and the retention increased compared to the virgin and plasma-treated membranes. *In situ* FTIR-ATR spectroscopy of polypropylene films is a versatile tool to investigate physico-chemical interactions between macro-solutes and polymer surface. In the case of substrates modified by polyelectrolyte multilayers, no HSA adsorption was observed under static conditions within the sensitivity of the method. It can be concluded from these results that reduced membrane fouling was caused also by alteration of physico-chemical interactions between polymer substrate and macro-solutes due to the polyelectrolyte multilayer coating of the membrane surface.

In the future, microporous membranes modified by polyelectrolyte multilayer assemblies will be used to design membranes suited for the filtration of different protein types by fitting the pore size and membrane charge to the corresponding macro-ion.

Literature Cited

1. Garg, D.-H.; Lenk, W.; Berwald, S.; Lunkwitz, K.; Simon, S. *J. Appl. Polym. Sci.* **1996**, *60*, 2087.
2. Carme, G.; Davis, R.H. *J. Membr. Sci.* **1996**, *119*, 269.
3. Pincet, F.; Perez, E.; Belfort, G. *Langmuir* **1995**, *11*,1229.
4. Koehler, J.A.; Ulbricht, M.; Belfort, G. *Langmuir* **1997**, *13*, 4162.
5. Ulbricht, M.; Belfort, G. *J. Membr. Sci.* **1996**, *111*, 193.
6. Gölander, C.-G.; Herron, J.N.; Lim, K.; Claesson, P.; Stenius, P.; Andrade, J.D. In *Poly(Ethylene Glycol) Chemistry: Biotechnical and Biomedical Applications*; Harris, J.M. Ed.; Plenum Press: New York, USA, 1992, Chapter 15, pp 221-245.
7. Quinn, R.; Laciak, D.V. *J. Membr. Sci.* **1997**, *131*, 49.
8. Hirotsu, T.; Arita, A. *J. Appl. Polym. Sci.* **1991**, *42*, 3255.
9. Decher, G.; Hong, J.-D. *Ber. Bunsen-Ges. Phys. Chem.* **1991**, *95*, 1430.
10. Schmitt, J.; Decher, G. *Thin Solid Films* **1992**, *210/211*, 831.
11. Leväsalmi, J.-M.; McCarthy, T.J. *Macromolecules* **1997**, *30*, 1752.
12. Onda, M.; Lvov, Y.; Ariga, K.; Kunitake, T. *J. Ferment. & Bioengin.* **1996**, *82*, 502.
13. Mitchell, J.; Perkins, L.R. *Appl. Polymer Symposia* **1967**, *4*, 167.
14. Kabanov, V.A. In *Macromolecular Complexes in Chemistry and Biology*; Dubin, P.; Bock, J.; Davis, R.; Schulz, D.N.; Thies, C., Eds.; Springer-Verlag, Heidelberg, Germany, 1994, Chapter 10, pp 151-174.
15. Chan, C.-M.; Ko, T.-M.; Hiraoka, H. *Surface Science Reports* **1996**, *24*, 1.
16. Briggs, D.; Kendall, C.R. *Int. J. Adhesion and Adhesives* **1982**, 13.
17. Gerenser, L. J.; Elman, J.F.; Mason, M.G.; Pochan, J.M. *Polymer* **1985**, *26*, 1162.
18. Hettlich, H.-J.; Otterbach, F.; Mittermayer, C.; Kaufmann, R.; Klee, D. *Biomaterials* **1991**, *12*, 521.
19. Poncin-Epaillard, F.; Chevet, B.; Brosse, J.C. Eur. Polym. J. 1990, 26, pp 333.
20. Sarada, T.; Sawer, L.C. *J. Membr. Sci.* **1983**, *15*, 97.
21. Suzuki, M.; Kishida, A.; Iwata, H.; Ikada, Y. *Macromolecules* **1986**, *19*, 1804.
22. Schwarz, S.; Jacobasch, H.-J.; Wyszynski, D.; Staude, E. *Angew. Makrom. Chem.* **1994**, *221*, 165.
23. Müller, R.H. *Zetapotential und Partikelladung in der Laborpraxis*; Wissenschaftliche VerlagsgesellschaftmbH: Stuttgart, Germany, 1996, Vol 7, pp 1-254.
24. Müller, M.; Werner, C.; Grundke, K.; Eichhorn, K.J.; Jacobasch, H.-J. *Macromol. Symp.* **1996**, *103*, 55.

25. Chen, W.; McCarthy, T. J. Macromolecules 1997, 30, 78.
26. Osada, Y.; Honda, K. Ohta, M. *J. Membr. Sci.* **1986**, *27*, 327.
27. Zeman, L.J.; Zydney, A.L. Micofiltration and Ultrafiltration: Principles and Applications; Marcel Dekker, Inc.: New York, USA, 1996, Chapter 9, pp 397-442.
28. Müller, M.; R. Buchet, R.; Fringeli, U.P. *J. Phys. Chem.* **1996**, *100*, 10810.
29. Müller, M.; Rieser, T.; Lunkwitz, K.; Berwald, S.; Meier-Haack, J.; Jehnichen, D. *Macromol. Rapid Commun.* **1998**, *19*, 333.

Chapter 14

Postsynthesis Method for Development of Membranes Using Ion Beam Irradiation of Polyimide Thin Films

X. L. Xu[1,3], M. R. Coleman[1,3,4], U. Myler[2], and P. J. Simpson[2]

[1]Department of Chemical Engineering, University of Arkansas, Fayetteville, AR 72701
[2]Department of Physics, The University of Western Ontario, London, Ontario N6A 3K7, Canada

The impact of ion beam irradiation on the permeation properties and morphology of polyimides are reported. Polyimide-ceramic composite membranes that were irradiated with H^+ ions over a range of fluences exhibited large increases in permeance and selectivity for several industrially important gas pairs. For example, there was up to a seven-fold increase in the oxygen permeance and a three-fold increase in O_2/N_2 selectivity for the composite membranes following irradiation. Analysis of N^+ irradiated polyimides using atomic force microscopy (AFM) and positron annihilation spectroscopy with variable energy positron (PAS) indicated that there also was a significant evolution in the morphology and microstructure of the polymer surface with increasing ion fluence. A review of ion beam modification of polymer structure and properties is included in this paper.

Synthesis of materials having both high permeabilities and selectivities combined with mechanical stability is very important for the development of membrane-based gas separations systems. Two approaches have been thoroughly investigated for the development of high performance membrane materials (1-6): (i) synthesis of glassy polymers with structures tailored using established structure-property relationships, and (ii) modification of polymeric permselective layers using diverse post-synthesis techniques (e.g. pyrolysis and fluorination). Polymers which have

[3]Current address: Chemical and Environmental Engineering Department, University of Toledo, 2801 West Bancroft, Toledo, OH 43606.
[4]Corresponding author.

been characterized for membrane or barrier applications typically exhibit a trade-off of a decrease in permeability for an increase in selectivity (*1,7*). While selective alterations of transport properties through chemical modification using structure-property principles have successfully improved membrane properties, the benefits achievable using these principles appear to be approaching a new upper limit as indicated by the rather extensive trade-off curves for gas permeabilities and selectivities compiled by L. M. Robeson *et al.* (*8,9*). Carbon molecular sieving membranes which have been formed from pyrolysis of asymmetric polymeric membranes and which lie well above this upper bound of the trade-off curve have been characterized (*10-12*). The attractive transport properties of molecular sieving materials combined with the upper bound seen for many polymers have led to investigations of a number of post-formation modification techniques for developing membrane materials with combined high permeabilities and selectivities (*11-14*).

Ion beam irradiation has long been recognized as an effective method for the synthesis and modification of diverse materials including polymers (*15-20*). Previous studies (*21-26*) with a number of polymers indicate that there is a gradual modification in the structure, morphology, and microstructure of the polymer surface layer (depths from a few hundred Angstroms to a few micrometers depending on ion irradiation conditions) with ion irradiation fluence (number of ions/irradiated surface area). There is an evolution in the backbone structure from the virgin polymer to a graphite-like material following irradiation at increasing ion beam fluences. This evolution in backbone structure is accompanied by significant changes in the microporosity of the irradiated polymers (*26-29*). As would be expected, the modification in chemical structure and microporosity of polymers following ion irradiation can result in a dramatic evolution in the physical properties of irradiated polymers (e.g. electrical conductivity and transport properties). This wide range of changes in the structure and properties of polymers with ion irradiation fluence could provide a great opportunity to modify polymer properties in a controlled manner (e.g. permeability and selectivity) by selecting appropriate irradiation conditions.

A number of factors are important in determining the properties of polymers following ion irradiation including the ion type, incident ion energy, ion fluence, and chemical structure of the virgin polymer (*22*). The evolution of the chemical structure, microporosity, and permeation properties of polyimides with ion irradiation fluence is currently being investigated by our group. A review of the extensive literature on the impact of ion irradiation on the structure, morphology and properties of glassy polymers is presented in this paper. The effect of H^+ irradiation on the permeation properties of several gases in a fluorine containing polyimide as well as preliminary results of an investigation of the effect of N^+ irradiation on the microporosity of a structurally related polyimide are also discussed in this paper.

Background

Transport in Polymer Matrix.   Transport of a penetrant in a polymer matrix occurs through a solution-diffusion mechanism in which the penetrant dissolves into the polymer at the high pressure side, diffuses through the polymer because of the concentration gradient, and desorbs on the low pressure side (*30-32*).   The permeability can be written as a product of a thermodynamic parameter, S_A, called the solubility coefficient and a kinetic parameter, D_A, called the diffusion coefficient, as shown in equation 1. The ideal selectivity, $\alpha^*_{A/B}$, of the membrane for component A relative to component B is the ratio of the pure-gas permeabilities:

$$P_A = S_A \cdot D_A \tag{1}$$
$$\alpha^*_{A/B} = P_A/P_B = (S_A/S_B) \cdot (D_A/D_B) \tag{2}$$

The ideal selectivity can be factored into the product of a solubility selectivity, S_A/S_B, and diffusivity selectivity, D_A/D_B, as shown in equation 2. The solubility of a penetrant in a polymer matrix is determined by the: (i) inherent condensability of the penetrant, (ii) polymer-penetrant interactions, and (iii) amount and distribution of free volume in the polymer matrix. The diffusion coefficient is determined by the chain packing and chain segmental mobility of the polymer as well as the size and shape of the penetrant molecules.

The solubility of gases in glassy polymers can be discussed in terms of the "Dual Mode" sorption model (*33-34*) which states that there are two environments in the polymer matrix into which a penetrant can sorb.   The first sorption environment is penetrant scale packing defects or microvoids which are formed as the polymer is quenched through the glass transition temperature.   Solubility in these microvoids can be modeled using Langmuir type sorption in which the solubility sharply increases with pressure and reaches an asymptote as the microvoids become saturated.   The second sorption environment is the amorphous densely packed region which is similar to the sorption environment in liquids or rubbery polymers and is modeled using Henry's law at low or moderate pressures. The sorption of penetrants in the polymer matrix is, therefore, controlled by the overall free volume and distribution of free volume between the packing defects and the densely packed region.

A penetrant molecule can be visualized as residing in the polymer in a sorbed cage surrounded by polymer chains (*33-35*). For a penetrant to make a diffusive jump within the matrix, a transient gap between the polymer chains greater than the sieving diameter of the penetrant must be generated. Thermally activated motions of the polymer chain segments generate these transient gaps through which diffusive jumps of the penetrant occur. Modifications of the polymer structure which inhibit chain packing and/or increase chain segmental mobility can be used to shift the distribution of transient gaps in the matrix to larger sizes to increase the diffusivity of each of the penetrants. If these modifications simultaneously change the size distribution of the transient gaps to retard diffusion of one penetrant relative to another, the diffusivity selectivity will be increased. The average size and size distribution of transient gaps required for diffusion to occur are determined by the chain packing and chain segmental mobility of the polymer as well as the size and shape of the penetrant molecules.

Transport studies with diverse materials, ranging from glassy polymers to carbon molecular sieves, confirm that the development of membranes relies on the ability to manipulate the microstructure of a material to fulfill the following conditions: (i) high fractional free volume (FFV), (ii) narrow free volume distribution which can precisely sieve a given gas pair, and (iii) a rigid microstructure (*10,35*). While selective alterations of transport properties through chemical modifications using structure-property concepts have successfully improved polymer transport properties, conventional chemical synthesis can be expensive and time consuming. Therefore, post-synthesis treatments of polymers have been the focus of considerable attention for the development of improved membrane materials. Ion beam irradiation, which results in a significant modification in the microstructure of a polymer, is being investigated by our group as a post-synthesis technique to develop materials with attractive gas transport properties.

Ion Beam Modification of Polymer Structure and Properties. Energetic ions penetrating into a polymer matrix lose energy according to several different ion-polymer interaction mechanisms which depend primarily upon the ion type, energy of the incident ion, and the chemical structure and packing of the polymer (*22*). For the ion energy range which is typically used in industrial processes and which was used for this research, only the following energy transfer mechanisms must be considered:

(i) Inelastic or electronic energy loss, which includes the process of electronic excitation, ionization of the atoms in the polymer, and the exchange of electrons between the incident particle and the atoms in the irradiated polymer;

(ii) Elastic or nuclear energy loss, which primarily involves the atomic collision between the incident particle and the atoms in the irradiated polymer.

The depth profiles of energy loss for several irradiating ions at 180 keV in the polyimide 6FDA-6FpDA and for N^+ at 140 keV in the polyimide 6FDA-MDA, which were estimated using a well established Monte Carlo simulation method (*36*) (SRIM code), are shown in Figures 1 and 2. The effect of ion type on the depth profile of energy loss for H^+, N^+, and Ar^+ irradiation of 6FDA-6FpDA is shown in Figure 1. In the case of H^+ ion implantation, the electronic energy loss mechanism is dominant and nuclear energy loss is negligible. The mean projected range (Rp) of the H^+ ions in this polymer is more than 2 µm and the energy loss is relatively flat over a large portion of the depth range. For N^+ implantation of 6FDA-6FpDA,

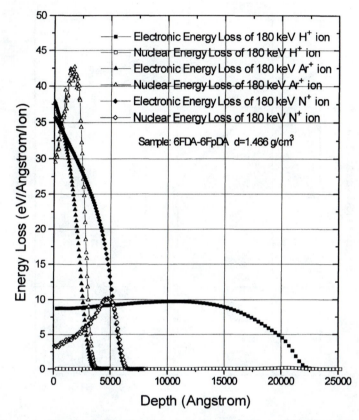

Figure 1. Depth profile of energy transfer mechanism of incident ions in 6FDA-6FpDA estimated using SRIM simulation. The depth profile of energy loss for H+ at 180 keV, N+ at 180 keV, and Ar+ at 180 keV in 6FDA-6FpDA are given.

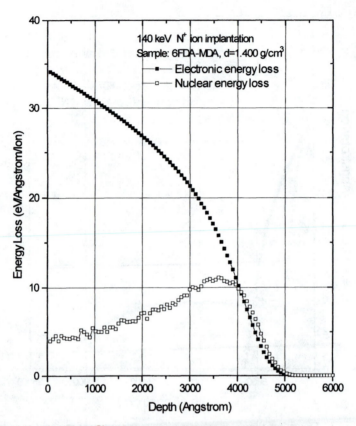

Figure 2. Depth profile of energy transfer mechanism of incident ions in 6FDA-MDA estimated using SRIM simulation. The depth profile of energy loss for N+ at 140 keV in 6FDA-MDA is given.

both energy loss mechanisms contribute to the total energy transfer from the incident ion to the irradiated polymer. However, the electronic energy loss is significantly greater than the nuclear energy loss for the majority of the ion penetration depth. The mean projected range for the N^+ ions in the 6FDA-6FpDA is less than 0.6 µm. Finally, the mean projected range of Ar^+ ions in the 6FDA-6FpDA is less than 0.3 µm and nuclear energy loss is much greater than electronic energy loss for a large portion of the mean projected range. As can be seen in Figure 1, the depth of modification is decreased but the total energy transfer from the incident ions to the polymer is increased dramatically for irradiating ions of increasing mass. Also the relative contribution of the nuclear energy loss becomes greater with increasing ion size. Similar results can be seen in Figure 2 for SRIM simulations of N^+ ion energy losses in 6FDA-MDA at 140 keV. The mean projected range of N^+ ions in 6FDA-MDA is 0.41 µm. The different energy loss mechanisms (i.e. electronic energy loss and nuclear energy loss) have been shown to contribute quite differently to modifications in polymer structure and properties.

Ion beam irradiation through energy transfer of the incident ion to the polymer can induce four different changes in the polymer backbone structure (*22*): (i) degradation of the polymer chain with formation of small volatile molecules, (ii) crosslinking between the polymer chains, (iii) formation of new chemical bonds such as double bonds, and (iv) oxidation or other chemical reactions in the presence of chemical atmosphere during the irradiation process. The evolution in backbone structure of several polymers following ion beam irradiation has been investigated using both Fourier transfer infrared spectroscopy (FTIR) and Raman spectroscopy (*22-25*). FTIR and Raman spectra of irradiated polyimides show a significant evolution in the polymer backbone structure with ion irradiation fluence. Typical infrared spectra for B^+ irradiated Kapton® polyimide films over a wide range of fluences also indicated that there were three regimes of types of ion beam modification of structure in these materials (*23*). In the low fluence range, the polymer maintained its basic structure with only a small decrease in characteristic absorption bands of the base polymer. In the middle dose range, the characteristic adsorption bands for functional groups were further decreased and new adsorption bands began to appear indicating a significant change in the polymer backbone structure. FTIR results for this system in the high ion dose range indicated a dramatic change in the chemical structure of the polymer with almost a complete loss of characteristic absorption bands for all functional groups. Raman spectra of high dose irradiated samples show the development of peaks which indicate the presence of polycrystalline graphite, amorphous carbon and mono-crystalline graphite structures (*21,25*). Therefore, high fluence implantation leads to the formation of a carbon-rich material at the surface layer of the polymer in which the microstructure is quite different from that of the precursor polymer. Poly(methyl methacrylate) and polystyrene exhibited similar trends in evolution of chemical structure with ion irradiation (*22*).

The evolution of the chemical structure of the polymers induced by ion irradiation also leads to modifications in many polymer properties (e. g. electronic conductivity and optical properties). For example, the polyimide Kapton® can be modified from a very good insulator (10^{-16}~10^{-17} Ω^{-1} cm^{-1}) to a poor conductor (10^{2}~10^{3} Ω^{-1} cm^{-1}) with increasing ion irradiation fluence (21,22,25). This large increase (20 orders of magnitude) in the electrical conductivity of the polyimide films following ion irradiation was attributed to the change in structure from the virgin polymer to a graphite-like material which was induced by ion irradiation.

Davenas and Xu (27,28) showed that ion beam irradiation can also significantly influence the diffusion and sorption of iodine in polyimide films. In this study, a series of ion-irradiated polyimides were held at a well defined distance from an iodine source in a glass reactor and were exposed to iodine vapor at a temperature of 45°C for 64 hours. Rutherford backscattering analysis (RBS) was used to determine the depth profile of iodine concentration in the irradiated polyimide films which were exposed to the iodine vapor. The RBS analysis was performed using 2 MeV α particles from a van de Graaf accelerator with a detection angle of 160°. For the non-irradiated control film the iodine penetrated to a depth of 2 μm and the iodine diffusion obeyed Fick's law. There was a significant evolution in the concentration profile of iodine in the modified polymer with increasing ion fluence. The concentration of iodine within the modified layer increased with ion irradiation fluence, passed through a maximum and then decreased at high fluences. In the low fluence range, there was an enrichment of iodine concentration in the ion beam modified surface layer and a significant reduction in iodine diffusion beyond the modified layer. In the high fluence range, no iodine was detected in the modified layer which acted as a barrier layer for iodine diffusion. Between these two regimes there was an intermediate fluence regime in which the quantity of iodine in the modified layer gradually decreased to zero with increasing ion fluence. Therefore, the results for the iodine diffusion studies of irradiated polyimides indicated that there were three regimes of ion beam modification in the diffusion properties and the microstructure environment of the polymer.

Because the transport properties of small molecules are determined by the polymer morphology (i.e. free volume and chain segmental mobility), ion irradiation would be expected to have a significant impact on the permeability and selectivity of polymers. The permeability of H_2 and CH_4 in ion irradiated 6FDA-based polyimide thick films were investigated to further probe the effect of irradiation on the microstructure of polymer surface layers (26,28). A series of films were irradiated with He$^+$ at 2 MeV in the low fluence range, whereas a second series of films were irradiated with N$^+$ at 170 keV in the high fluence range. Samples irradiated with He$^+$ in the low fluence range exhibited sharp increases in both CH_4 and H_2 permeabilities with a corresponding decrease in selectivity. For example, implantation of a polyimide film with He$^+$ at a dose of 3.6×10^{14} ion/cm^2 resulted in a 10-fold increase in H_2 permeability and a 50-fold increase in CH_4 permeability.

These permeation results were attributed to an increase in fractional free volume in the implanted layer in the low ion fluence range which would favor an increase in permeability for CH_4 over H_2.

The polyimide films which were irradiated with N^+ in the high fluence range exhibited very different properties than the He^+ irradiated samples (26,28). There was an increase in H_2 permeability with a corresponding increase in H_2/CH_4 selectivity for the N^+ irradiated samples. For example, implantation of a polyimide film with N^+ at a dose of 2×10^{15} ion/cm^2 resulted in a 70% reduction in the CH_4 permeability with a slight increase in H_2 permeability. This decrease in methane permeability combined with the increase in H_2 permeability resulted in a 47% increase in H_2/CH_4 selectivity. Even though the irradiated layer represented only 4% of the total film thickness, a dramatic impact on transport properties was seen following high fluence irradiation of the polymer. The results of gas permeation studies imply that high dose irradiation caused a significant modification in both the free volume and free volume distribution in the irradiated layer. A small micropore size combined with a very narrow size distribution would explain the simultaneous increases in selectivity and permeability and is consistent with results seen using other analytical techniques. As will be discussed later, the evolution in permeation properties and iodine diffusion in these polyimides agreed well with results of positron annihilation spectroscopy studies performed by our group of the evolution in the microstructure for a N^+ irradiated fluorine containing polyimide.

In summary, studies of ion irradiation effects on polymer structure and properties using a variety of analytical techniques indicate that ion bombardment can cause a large evolution in the chemical structure and micropore size and size distribution of polymers. There are also three fluence ranges of effect of ion irradiation on the structure and properties of polymers. The modification in the free volume and rigidity of the polymer matrix combined with the attractive H_2/CH_4 separation properties of the irradiated polyimides demonstrate the potential of this technique for development of high- performance gas separations membranes. In this paper, we report recent progress on the investigation of ion-beam irradiation of polyimides as a post-synthesis modification technique for development of advanced membrane materials.

Experimental

Materials Used in This Study. The two fluorine containing polyimides shown in Figure 3 were used for a preliminary study of the effect of ion irradiation on the microstructure and permeation properties of aromatic polymers. Polyimides were chosen for this study because of their excellent potential for applications as membranes for gas separations. The 6FDA-6FpDA was kindly supplied by Professor William Koros of the University of Texas at Austin. The 6FDA-MDA was synthesized via a general reaction of a dianhydride with a diamine with final imidization effected using the chemical method developed by Husk (*6*). The diamine monomer (MDA) was purchased from Aldrich Chemical Co. and was purified using crystallization from methanol. The dianhydride monomer (6FDA) was purchased from Hoechst-Celanese Corporation and purified using vacuum sublimation. All solvents were purchased from Aldrich Chemical Co. and were purified using vacuum distillation. For the 6FDA-MDA synthesis, equimolar quantities of the diamine and dianhydride monomers were charged with dimethylacetamide (DMAC) to a flask under an argon atmosphere. Following dissolution of the monomers, the reaction mixture was heated to 50°C for approximately two hours. The polyamic acid solution was then cooled to 45°C, and a solution of triethyl amine and acetic anhydride was added to the reaction mixture to promote chemical imidization. Following heating at 100°C for 1.5 hours, the polymer solution was poured into an excess of methanol to precipitate the polyimide.

Preparation of Polymer Films and Composite Membranes. Positron annihilation spectroscopy (PAS) and atomic force microscopy (AFM) studies of N$^+$ irradiated 6FDA-MDA were performed using free-standing thick films (25 to 50 μm) cast from solution using standard techniques (*4,6*). The 6FDA-MDA was dissolved in methylene chloride to form a solution which was between 5 and 10 wt% polymer. The solutions were filtered to remove any residual particles which could cause defects in the dense films. Free-standing films of 6FDA-MDA were cast into glass dishes in a glove bag to regulate the rate of solvent evaporation. After the films were dried, the dishes were submerged in water to assist in the removal of the film from the glass dish. The films were air dried for an additional 12 hours at room temperature followed by drying in a vacuum oven for several days at temperatures up to 250°C to remove any residual solvent.

The permeation measurements in H$^+$ irradiated 6FDA-6FpDA were made using polyimide-ceramic composite membranes with 2-μm-thick selective polymer layers deposited onto porous ceramic substrates as described in detail elsewhere (*37*). The use of composite membranes allowed determination of permeances of polymers in which the entire thickness had been modified by the ion beam. Anopore® porous ceramic membranes purchased from Whatman Inc. with an average pore size of 0.02 μm were used as the porous substrates for the composite membranes. Dilute solutions of 6FDA-6FpDA in methylene chloride (0.4 wt % to 0.5 wt %) were prepared in a controlled dry environment to avoid humidity which could cause the polymer to precipitate during the film formation process. A predefined quantity of solution determined based upon the desired thickness of the polymer layer was deposited on the ceramic support and covered to control the evaporation rate of the solvent. The composite membranes were allowed to dry in air for at least 12 hours and then dried in a vacuum oven at 100°C for more than one hour to remove any residual solvent. Two methods were used to estimate the thickness of the selective polymer layer of the composite membranes: i) weight difference method including an estimation of the surface area of polymer coverage and ii) cross-section imaging using scanning electronic microscopy (SEM). The results of SEM analysis (Figure 4) and weight difference method for estimating the thickness of the polymer layer were consistent.

Ion Irradiation Procedure. The free-standing 6FDA-MDA films for the PALS and AFM studies were irradiated with N^+ ions at 140 keV at ion fluences ranging from $2x10^{14}$ N^+/cm^2 to $5x10^{15}$ N^+/cm^2 which encompassed the entire ion beam modification range. The 6FDA-6FpDA composite membranes for the permeation studies were irradiated with H^+ ions at 180 keV at ion fluences ranging from $1x10^{14}$ H^+/cm^2 to $2x10^{15}$ H^+/cm^2. Ion implantations were performed using a IM-200 implantor at the Ion Beam Laboratory, Shanghai Institute of Metallurgy, Chinese Academy of Sciences. A small ion beam current density ($\leq$ 1 $\mu A/cm^2$) was used to avoid overheating of the samples. The mean projected range (Rp) of the irradiating ions and energy loss in the samples were estimated using SRIM Code, as shown in Figures 1 and 2. The Rp of the H^+ ion at 180 keV in 6FDA-6FpDA was 2.09 μm and the longitudinal straggling (ΔRp) was about 0.13 μm. As shown in Figure 1, the total thickness modified by ion irradiation Rm, which is defined as Rp + ΔRp, would be 2.2 μm for H^+ irradiation of 6FDA-6FpDA. Therefore, the thickness of the composite membranes was equal to or less than 2.2 μm to allow modification of the entire thickness of the selective layer of the polymer layer. The total thickness modified by 140 keV N^+ ion irradiation is equal to 0.48 μm which represents a very small portion of total thickness of the free standing film.

Gas Permeation Measurements. The pure-gas permeances of O_2, N_2, He, and CH_4 were measured at room temperature in each of the composite membranes using Millipore test cells. For each permeation measurement, the feed gas was applied to the upstream side of the membrane at a pressure of 50 psig and the permeate side was held at atmospheric pressure. Once steady-state was reached ($\sim$ 12 hours), the volumetric flow rate of the permeate stream was determined using a soap-bubble flowmeter. The permeance of each of the gases was measured for each membrane before and after exposure to the ion beam. This allowed a direct assessment of the effect of ion beam irradiation on the permeation properties of a specific membrane and accounted for any differences in the transport properties of the virgin membranes. The O_2/N_2 selectivity of the bulk material at 25°C was 5.0 (*38*). Composite membranes with O_2/N_2 selectivities greater than 2.7 were used in this study.

Atomic Force Microscopy. Atomic force microscopy was used to monitor the evolution in the surface morphology of a series of N^+ irradiated 6FDA-MDA films. The atomic force microscope used in this study was a Digital Instruments

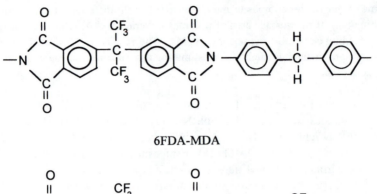

6FDA-MDA

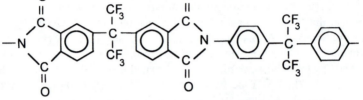

6FDA-6FpDA

Figure 3. Structure of the polyimides used in this study.

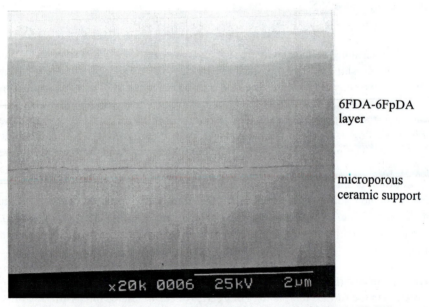

6FDA-6FpDA layer

microporous ceramic support

Figure 4. Image of scanning electron microscopy of 6FDA-6FpDA-ceramic composite membrane following ion irradiation. The thickness of the permselective polymer layer agreed with values estimated using weight difference method.

NanoScope® III at the High Density Electronic Center (HiDEC) at the University of Arkansas. All the measurements were performed in air using a contact mode. The AFM was equipped with a software system which allowed a variety of analyses of AFM data, including section analysis, bearing analysis, and roughness analysis.

Positron Annihilation Spectroscopy Measurements. Positron annihilation lifetime spectroscopy (PALS) has been widely used to investigate the free volume distribution of bulk materials and has also been used in studies of bulk polymers for membrane applications (*39,40*). It is not practical to use this "bulk" positron method in the study of ion-beam modified materials, because the layers of interest are typically only a few hundred nanometers to a few micrometers thick. However, the development of variable-energy positron beam facilities makes available sources of positrons with energies from ~0.1 to 60 keV which allows analysis of thin surface layers. For technical reasons, variable-energy positron experiments usually measure the energy spectrum of the annihilation gamma-rays instead of measuring the positron lifetime. The gamma-rays are Doppler-shifted from 511 keV due to the need to conserve the momentum of the annihilating electron. Usually a parameter, S, which is defined as counts in the central portion of the annihilation line divided by the total number of counts, is used as a measure of the Doppler-broadened 511 keV lineshape. Different types of defects have different S parameters because of the electronic environment at the site of positron annihilation. Because it can be correlated to defect concentration and defect type, the S parameter is often used to characterize free volume in materials.

Positron Annihilation Spectroscopy (PAS) using variable-energy positrons is an effective method for determining depth profile vacancy-based structures of solids and has been applied to the studies of ion beam induced defects in metals and semiconductors (*41,42*). The positron annihilation spectroscopy measurements of 6FDA-based polyimide samples were performed at the slow-positron-beam facility, Positron Beam Laboratory, at the University of Western Ontario, Canada. Monoenergetic positron beams with an energy between 0.5 keV and 22 keV were implanted in the samples. For each sample, two separate spectra, each of which consisted of three runs were collected. Inspection of the spectra showed that for each sample both spectra were identical within reasonable statistical fluctuations. Therefore, both spectra were averaged to form the final spectrum used in the preliminary evaluation of the effect of ion irradiation on polymer microstructure.

Results and Discussion

Modification of Polymer Structure and Morphology. Thick films of 6FDA-MDA were bombarded at ion fluences in the following ranges: (i) low fluence at 2×10^{14} ion/cm^2, (ii) intermediate fluence at 1×10^{15} ion/cm^2, and (iii) high fluence at

$5x10^{15}$ ion/cm^2. The base 6FDA-MDA film was transparent with a light yellow color. The color of the film changed with the implantation fluence from yellow to brown to black with a metallic luster as the irradiation dose was increased from $2x10^{14}$ ion/cm^2 to $5x10^{15}$ ion/cm^2. Atomic force microscopy was used to investigate the impact of ion irradiation on the surface morphology of these films. The results of this AFM analysis were described in detail in a previous paper (29). A two-level-structure, which was composed of bold chains that, in turn, consisted of a nodule-like structure, were observed in the AFM image of the base 6FDA-MDA film. Almost no differences between the AFM image of the reference film and the film implanted with N$^+$ in the low fluence range ($2x10^{14}$ ion/cm^2) was observed. An image composed of small cone-like structures of smaller dimension than the nodule-like structure of the reference polymer was observed for the sample implanted in the intermediate fluence range ($1x10^{15}$ ion/cm^2). At an irradiation fluence of $5x10^{15}$ ion/cm^2, the surface of the sample was very flat and a small nodule-like structure appeared instead of the small cones observed at the intermediate dose. Detailed roughness and bearing analyses of the AFM images indicated that free-standing polyimide films had deep surface valleys which extended to a depth of several micrometers. Ion-beam irradiation, even at a small dose, alters the microstructure of the surface layer and forms a modified layer which eliminates these initial deep valleys. This AFM analysis indicated that small fluence irradiation-induced microvoids in the surface layer of the polymer and high fluence irradiation resulted in a large number of small size microvoids at the surface.

Positron annihilation spectroscopy was used to investigate the evolution in microporosity in the surface layer of the N$^+$ ion irradiated films used in the AFM studies. A plot of the S-parameter for the virgin and the implanted samples of the 6FDA-pMDA over a range of fluences as a function of positron energy (correlated to depth) is shown in Figure 5. The depth of the damaged layer extended to a positron energy of approximately 5 keV which corresponds to a depth of more than 0.4 μm which is in agreement with the result of the theoretical calculation using the SRIM code (Rp = 0.41 μm). There is a significant variation in the depth profile of the S parameter (which correlates with concentration of packing defects) in the ion beam modified films as shown in Figure 5.

Defect modeling was performed in order to clarify the microstructural evolution induced by ion irradiation. The defect models for the positron annihilation spectra of 6FDA-MDA samples are given in Figure 6. A value of $4.33x10^9$ sec^{-1} for the free annihilation rate for a positron in the 6FDA-MDA and a value of $3x10^{14}$ sec^{-1} for the specific trapping rate for vacancies were used in the model. The virgin polymer had a very low concentration of defects. A two layer model for the base polymer, which consists of a layer from the surface to 1,800 Å with a defect concentration of $3.6x10^{-6}$ defects per atom and a layer from 1,800 Å to 7,600 Å with a defect concentration of $1.2x10^{-6}$ defects per atom, was obtained from the fit of the PAS results. The defect density increased significantly with dose for the low (Figure 6b)

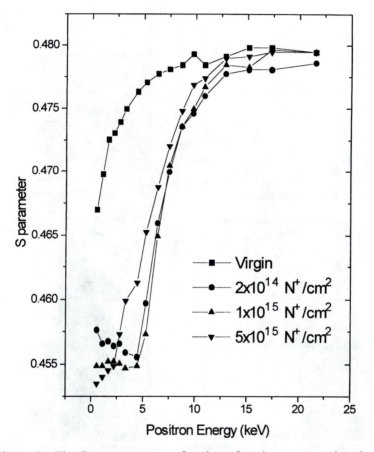

Figure 5. The S-parameters as a function of positron energy (can be interpreted as depth) of 6FDA-MDA samples including virgin, implanted at $2x10^{14}/cm^2$, $1x10^{15}/cm^2$, and $5x10^{15}/cm^2$ of 140 keV N^+ ions.

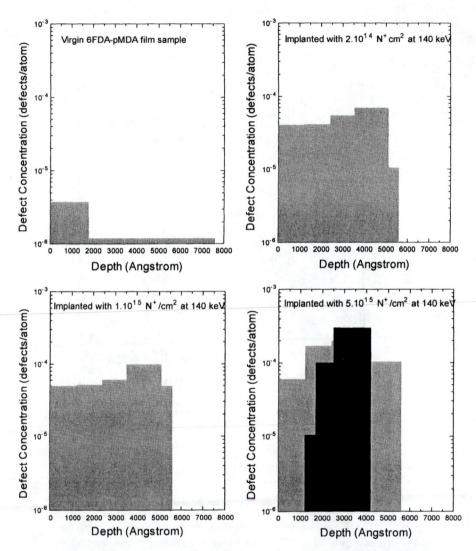

Figure 6. Defect model of positron annihilation spectra of 6FDA-pMDA samples. Virgin sample has a very low defect concentration. At small and medium dose irradiation the defect density increase with the fluence. At high dose irradiation, a different defect type (in dark) must be added to explain the experimental results.

and medium fluence (Figure 6c) irradiation ranges. The spectra for the low dose $(2x10^{14}$ ion/cm^2) and medium dose $(1x10^{15}$ ion/cm^2) samples can be fitted reasonably well with the defect distributions resembling the theoretical prediction. At high dose irradiation, all attempts to achieve a reasonable fit using only one type of defect failed to reproduce the measured data. An additional defect type (shown in black) was included in the model to quantify the experimental results. The appearance of the new defect type in the high dose irradiation range implies a significant evolution in the microporosity of the polymer and will be investigated further for ion irradiation of several polyimides over a wide range of ion fluence.

Modification of Gas Transport Properties. The effect of H^+ ion irradiation in the low fluence range on the permeation properties of several gases in 6FDA-6FpDA are shown in Figures 7 to 9. The results are reported in terms of a normalized permeance and selectivity for each of the gas pairs to allow a direct comparison of the modification in the transport properties with ion fluence and accounts for any differences in transport properties in the virgin membranes. The normalized permeance is defined as the ratio of the pure-component permeance following irradiation to the permeance in the virgin membrane. The normalized selectivity is defined as the ratio of the selectivity following ion irradiation to the selectivity in the virgin membrane. The transport properties of the 6FDA-6FpDA composite membranes prior to H^+ irradiation are given in Table I. The O_2/N_2 selectivities in each of these membranes are below the selectivity of the bulk material. Thus, the polymer layers of the unmodified membranes were not defect-free.

Table I. Gas permeation properties for polymer-ceramic composite membranes prior to H^+ ion irradiation.

Sample	Permeance[a] (GPU)			Selectivity[b]		
	He	O_2	N_2	He/N_2	O_2/N_2	N_2/CH_4
MR3-3	48.6	4.5	1.20	40	3.7	2.0
MR3-4	58.8	5.8	1.95	30	3.0	1.6
F25-5	55.2	6.3	1.85	30	3.4	1.9
F25-6	50.8	5.1	1.73	29	2.9	1.8
F25-4	24.8	2.1	0.77	33	2.7	-

(a) 1 GPU = 10^{-6} cm^3(STP)/cm^2•s•cmHg; feed pressure: 50 psig.
(b) Ratio of pure-gas permeances.

The gas permeation properties in each of the composite membranes following H^+ ion irradiation are given in Table II. For each of the gases studied, there was a significant evolution in the permeation properties following H^+ ion irradiation. While irradiation at a very low fluence $(1x10^{14}$ ion/cm^2) had little impact upon the

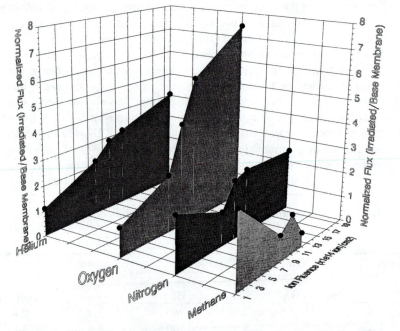

Figure 7. Evolution in relative permeance of He, O_2, N_2, CH_4 in 6FDA-6FpDA composite membranes following H^+ irradiation.

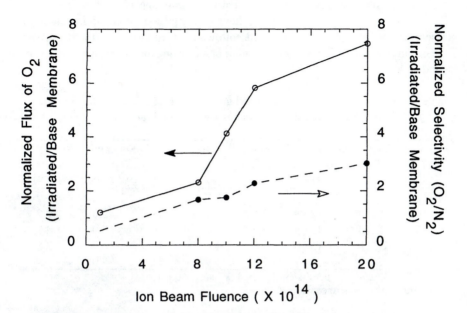

Figure 8. Effect of H$^+$ fluence on the relative permeance and selectivity of the O$_2$/N$_2$ gas pair in irradiated 6FDA-6FpDA.

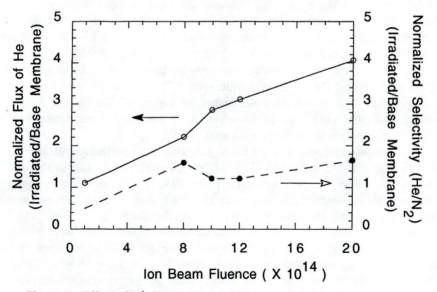

Figure 9. Effect of H$^+$ fluence on the relative permeance and selectivity of the He/N$_2$ gas pair in irradiated 6FDA-6FpDA.

He and O_2 permeances, there was an increase in the permeances of both N_2 and CH_4. However, at larger ion fluences, there was a general trend of increases in permeances for each of the gases tested following ion irradiation with the largest increases in permeance following ion irradiation occurring for the smaller gases (i.e. O_2 and He).

Table II. Ion implantation conditions and gas permeation properties for polymer-ceramic composite membranes after H^+ ion irradiation.

Sample	Ion fluence (H^+/cm^2)	Permeance[a] (GPU)			Selectivity[b]		
		He	O_2	N_2	He/N_2	O_2/N_2	N_2/CH_4
MR3-3	1×10^{14}	53.7	5.3	2.7	20	2.0	1.2
MR3-4	8×10^{14}	129.8	13.4	2.7	48	5.0	2.7
F25-5	1×10^{15}	157.6	26.0	4.4	36	6.0	2.9
F25-6	1.2×10^{15}	158.0	29.6	4.4	36	6.7	2.8
F25-4	2×10^{15}	100.7	15.3	1.9	53	8.0	-

(a) 1 GPU = 10^{-6} cm^3(STP)/cm^2·s·cmHg; feed pressure: 50 psig.
(b) Ratio of pure-gas permeances

The normalized O_2 permeance and O_2/N_2 selectivity as a function of H^+ ion fluence are shown in Figure 8. At a very low fluence (1×10^{14} ion/cm^2), irradiation gave rise to a slight increase in permeance for both of the gases but resulted in an overall decrease in the selectivity. Because the O_2 permeance increased more sharply than N_2 permeance with increasing ion implantation fluence, there was a simultaneous increase in the O_2 permeance and O_2/N_2 selectivity. For the fluence range from 1×10^{15} ion/cm^2 to 2×10^{15} ion/cm^2, the oxygen permeances following irradiation were from 4 to 7.5 times greater than the permeances in the unmodified membranes, whereas the O_2/N_2 selectivities were 1.75 to 3 times greater following irradiation than those in the virgin membranes. As can be seen in Table II, at ion fluence greater than 1×10^{15} ion/cm^2 the selectivity in the modified membranes were greater than the selectivity in the bulk material.

The permeation properties for the He/N_2 gas pair in the modified membranes are shown in Figure 9. Irradiation with H+ in the low fluence range resulted in a steady increase in He flux with implantation fluence. The He/N_2 selectivity initially increased with ion fluence, reached a maximum at a fluence of 8×10^{14} ion/cm^2 and then leveled off with the increasing fluence. For the fluence range from 1×10^{15} ion/cm^2 to 2×10^{15} ion/cm^2, the helium permeances following irradiation were from 2 to 4 times greater than the initial permeances, whereas the He/N_2 selectivities were up to 1.6 times greater following irradiation than those in the virgin membranes. The effect of ion irradiation in the low fluence range on the nitrogen and methane permeance was quite different from the impact of irradiation on He and O_2

transport. The normalized permeances of N_2 and CH_4 in the irradiated 6FDA-6FpDA membranes as a function of ion fluence are shown in Figure 7. While there was approximately a two-fold increase in the N_2 permeance following H^+ irradiation over the entire fluence range studied, irradiation at a fluence of 8×10^{14} ion/cm^2 resulted in only a 20% increase in the nitrogen permeance. Although very low dose irradiation resulted in a three-fold increase in methane flux, the general trend was for a decrease in the methane permeance. Indeed, at a fluence of 1.2×10^{15} ion/cm^2, irradiation resulted in a decrease in the methane permeance relative to the unmodified membrane. Current work includes extending the H^+ irradiation fluence range to encompass samples modified in both the middle and high fluence regimes.

The large variations in permeation properties with H^+ ion fluence support the PAS results which indicated that there is an evolution in micropore structure within the polymer layer following ion beam irradiation. As mentioned earlier, ion irradiation can also induce significant modifications in polymer backbone structure (i.e. breaking of chemical bonds and crosslinking) and microstructure (i.e. micropore size and distribution) (22). Positron annihilation spectroscopy of N^+ irradiated 6FDA-MDA indicated that there was an increase in the concentration of packing defects in the polyimide following irradiation at low ion fluence. This evolution in microstructure with ion beam fluence may explain the evolution in permeation properties for the irradiated membranes. The increase in the concentration of packing defects at low ion fluence would be expected to result in an increase in permeability for each of the gases studied. Simultaneously, crosslinking of polymer chains which occurs as a result of recombination of broken bonds and free radicals should result in the formation of a rigid polymer network. This increase in rigidity following crosslinking would be expected to result in an increase in selectivity with ion fluence. Therefore, these modifications in the microstructure of the irradiated polymer could explain the simultaneous increase in permeance and selectivity which were induced by H^+ ion irradiation. The evolution in backbone structure, microporosity, and surface structure is currently being investigated for irradiated composite membranes using a variety of analytical techniques.

Conclusions

Analysis of N^+ irradiated samples of 6FDA-MDA films using positron annihilation spectroscopy using variable energy positron and atomic force microscopy indicated that there was a significant evolution in the microstructure of the polymer surface layer over the entire ion fluence range. The result of the modification in polymer microstructure agreed well with previously published studies of iodine diffusion and H_2/CH_4 permeation in the ion-beam-modified Kapton polyimide. Irradiation of 6FDA-6FpDA polyimide-ceramic composite membranes with H^+ ions over a range of fluences resulted in a general increase in both permeance and selectivity of several gas pairs. These results demonstrate the potential of ion beam irradiation as a post-synthesis modification technique for developing materials with both high fluxes and selectivities. A comprehensive investigation of the impact of ion beam irradiation over a wide range of fluences on chemical structure, morphology, and transport properties of aromatic polymers is currently underway in our laboratory.

Acknowledgments

The authors would like thank Professor W. J. Koros of the University of Texas at Austin for supplying the 6FDA-6FpDA. The support of the National Science Foundation through the Presidential Faculty Fellows Program (CTS-9553267) in funding this project is gratefully acknowledged. The authors gratefully acknowledge Dr. John Shultz for his help with the Scanning Electron Microscope and the High Density Electronic Center at the University of Arkansas for the use of the equipment. The authors would also like thank Professor Xianghuai Liu and Mr. Zexin Lin of the Ion Beam Laboratory, Shanghai Institute of Metallurgy, Chinese Academy of Sciences for the ion implantation.

Literature Cited

1. Koros, W. J.; Fleming, G. K.; Jordan, S. M.; Kim, T. H.; Hoehn, H. H. *Prog. Polym. Sci.* **1988**, *4*, 339.
2. Hoehn, H. H. In *"Material Science of Synthetic Membranes"*, ACS Symp. Series No. 269, D.R. Lloyd (Ed.), American Chemical Society, Washington, D.C., 1985, Chapter 4.
3. Coleman, M. R.; Koros, W. J. *J. Membrane Sci.* **1990**, *50*, 285.
4. Coleman, M. R.; Koros, W. J. *J. Polym. Sci., Polym. Phys.* **1994**, *32*, 1915.
5. Jones, C. W.; Koros, W.J. *Carbon* **1994**, *32*, 1419.
6. Kim, T. H.; Koros, W. J.; Husk, G. R.; O'Brien, K. C. *J. Membrane Sci.* **1988**, *45*, 37.
7. Kesting, R. E.; Fritzsche, A. K. *Polymeric Gas Separation Membranes*, Wiley-Interscience Publication, New York 1993.
8. Robeson, L. M. *J. Membrane Sci.* **1991**, *62*, 165.
9. Robeson,. L. M.; Burgoyne; W.F.; Langsam, M.; Savoca, A. C.; Tien, C.F. *Polymer* **1994**, *35*, 4970.
10. Singh, A; Koros, W. J. *Ind. Eng. Chem. Res.* **1996**, *35*, 1231.
11. Geiszler, V.C.; Koros, W.J. *Ind. Eng. Chem. Res.* **1996**, *35*, 2999.
12. Wang, S; Zeng, M; Wang, Z. *Sep. Sci. and Tech.* **1996**, *31*, 2299.
13. Damie, A. S.; Gangwal, S. K.; Spivey, J. J.; Longanbach, J.; Venkataraman, V. K. *Key Engineering Materials* **1991**, *61-62*, 273.
14. Le Roux, J. D.; Paul, D.R.; Kampa, J.; Lagow, R.J. *J. Membrane Sci.* **1994**, *94*, 121.

15. Martin, P.; Daudin, B.; Dupuy, M.; Ermolieff, A.; Olivier, M.; Papon, A. M.; Rolland, G. *J. Appl. Phys.* **1990**, *67*, 2908.

16. Radermacher, K.; Mantl, S.; Kohlhof, K.; Jager, W. *J. Appl. Phys.* **1990**, *68* 3001.

17. Kuniyoshi, S: Kudo, K.; Tanaka, K. *Appl. Surf. Sci.* **1989**, *43*, 447.

18. Grummon, D.S.; Schalek, R.; Ozzello, A.; Kalantar, J.; Drzal, L.T. *Nucl. Instr. and Meths B* **1991**, *59*, 1271.

19. Dearnaley, G. *Nucl. Instr. and Meths B* **1990**, *50*, 358.

20. Buchal, C. Nucl. *Instr. and Meths B.* **1991**, *59*, 1142.

21. Davenas, J.; Xu, X. L.; Boiteux, G.; Sage, D. *Nucl. Instr. and Meths B* **1989**, *39*,754 .

22. Xu, X. L.. Dissertation of Doctorat d'état es Sciences at University of Lyon, France, 1987.

23. Xu, D.; Xu, X. L.; Du, G. D.; Wang, R.; Zou, S. C.; Liu, X. H. *Nucl. Instr. and Meths B* **1993**, *80/81*, 1063.

24. Xu, X. Xu, D.; Xu, H. L.; Wang, R.; Zou, S. C.; Du, G. D.; Xia, G. Q. *J. Vac. Sci. Technol. A Surface and Films* **1994**, *12*, 3200.

25. Xu, D.; Xu, X. L.; Zou, *S. C. Mater. Lett.* **1991**, *12*, 12.

26. Xu, X. L.; Dolveck, J. Y.; Boiteux, G.; Escoubes, M.; Monchanin, M.; Dupin, J. P; Davenas, J. *J. Appl. Polym. Sci.* **1995**, *55*, 99.

27. Davenas, J.; Xu, X. L., *Nucl. Instr. and Meths B*, **1992**, 71, 33.

28. Xu, X. L.; Dolveck, J. Y.; Boiteux, G.; Escoubes, M.; Monchanin, M.; Dupin, J.P.; Davenas, J. *Mat. Res. Soc. Symp. Proc.***1995**, *354*, 351.

29. Xu, X.; Coleman, M. R. *J. Appl. Polym. Sci.* **1997**, *66*, 459

30. Koros, W. J.; Chan, A. H.; Paul, D. R. *J. Membrane. Sci.* **1977**, *2*, 165.

31. Barrer, R. M.; Barrie, J. A.; Slater, J. *J. Polym. Sci.* **1958**, *27*, 177.

32. Koros, W. J. *J. Polym. Sci., Polym. Phys.* **1985**, *23*, 1611.

33. Barrer, R. M., *Diffusion in and Through Solids*, Cambridge, The University Press, New York, 1941.

34. Crank, J.; Park, G. S., In *"Diffusion in Polymers"*, Academic Press, New York, John Wiley & Sons (1971).

35. Koros, W.J.; Coleman, M.R.; Walker, D. R. B. *Annual Rev. Mater. Sci.* **1992**, *22*, 47.

36. Ziegler, J. F., Biersack, J. P.; Littmark, U. In *"The Stopping and Range of Ions in Solids"*, Ziegler, J.F. (Ed) , Vol. 1, Pergamon Press, New York, 1985

37. Rezac, M. E.; Koros, W. J. *J. Appl. Polym. Sci.* **1992**, *46*, 1927.

38. Costello, L.; Koros, W. J. *J. Polym. Sci.: Part B: Polym. Phys.* **1995**, *33*, 135.

39. Yu, Z.; Yahsi, U.; McGervey, J. D.; Jamieson, A. M.; Simha, R. *J. Polym. Sci.: Part B: Polym. Phys.* **1994**, *32*, 2637.

40. Li, X. S.; Boyce, M. C., *J. Polym. Sci.:Part B: Polym. Phys.* **1993**, *31*, 869.

41. Tandberg, E.; Schultz, P.J.; Aers, G.C.; Jacman, T.E. *Canadian J. Phys.* **1989**, *67*, 275.

42. Simpson, P.J.; Vos, M.; Michell, I.V.; Wu, C.; Schultz, P.J. *Phys. Rev. B* **1991**, *44*, 12180.

Chapter 15

Preparation and Properties of Surface-Modified Polyacrylonitrile Hollow Fibers

A. Higuchi[1], P. Wang[2], T.-M. Tak[3], T. Nohmi[4], and T. Hashimoto[1]

[1]Department of Industrial Chemistry, Seikei University,
Tokyo 180–8633, Japan
[2]Chia Yiu Enterprise Corporation, Tokyo 160, Japan
[3]Department of Natural Fiber Science, Seoul National University,
Suwon 440–774, South Korea
[4]Research Laboratory, Nohmi Bosai Ltd., Tokyo 160, Japan

A controlled reaction on the surface of polyacrylonitrile hollow fibers was developed to change the CN unit to a CNH_2 unit. Ion-exchange measurements and IR spectra for the chemically modified fibers suggested the existence of NH_2 units on the surface of the modified fibers. The modified fibers had a smaller molecular weight cut-off than the unmodified fibers. Serum albumin was immobilized on the surface of the modified and unmodified fibers using the succinimide reaction. Ultrafiltration experiments for optical resolution of racemic amino acids were also performed using the immobilized albumin membranes. The immobilized albumin membranes prepared from chemically modified hollow fibers demonstrated efficient optical resolution of racemic phenylalanine.

Surface reactions of active reagents on membranes are useful modifications to introduce various functional groups such as SO_3H, NH_2, COOH or OH (*1-18*). For example, Nabe *et al.* prepared five different chemically modified membranes and investigated protein fouling using these membranes (*1*). Membranes modified by direct sulfonation had the lowest surface energy and the shortest grafted chain length and exhibited the highest volumetric flux with BSA solution (*1*). In previous studies (*2-5*), we developed several surface reactions on polysulfone hollow fibers. A negatively charged group, $-CH_2CH_2CH_2SO_3^-$, and a hydroxide group, $-CH(CH_3)CH_2OH$, were introduced on the surface of the polysulfone hollow fibers via Friedel-Crafts reactions. It was demonstrated that the modified hollow fibers showed excellent properties for the anti-absorption of proteins and solutes compared to those of unmodified and conventionally sulfonated fibers (*2-5*).

Recently, special attention has focused on affinity membranes for the purification of proteins and optical isomers for biotechnology applications (*19-22*). Affinity membranes are generally made from porous membranes activated by the covalent attachment of affinity ligands to their interior pore surfaces. When membranes made from conventional polymers such as polyacrylonitrile or polysulfone are selected as the base materials, the membranes must be activated. One of the most promising methods for activation of the membranes is surface modification.

This paper describes the surface reaction of polyacrylonitrile (PAN) hollow fibers with hydrazine and Raney nickel by a one-step reaction that introduces an NH_2 unit on the surface. Bovine serum albumin (BSA) was subsequently immobilized on the surface of the modified fibers using the succinimide reaction. Optical resolution of racemic phenylalanine was also investigated in ultrafiltration experiments using the immobilized BSA hollow fiber membranes.

Experimental

Materials. Membranes used for the chemical modification were commercially available polyacrylonitrile ultrafiltration hollow fibers (HC-1, Asahi Chemical Co., Ltd., molecular weight cut-off = 13,000 g/mole). The substitution ratio of the amine residue was estimated to be 1/100 from ion exchange capacity experiments of the membranes. The fibers have internal and external skin layers and two layers of finger-like macrovoids in their cross-section. The inside and outside diameters of the fibers were approximately 0.8 mm and 1.4 mm, respectively. Water flux of the fibers in mixtures with polyethylene glycol (PEG) [Mw = 20,000 g/mole] was 2.8 $m^3/m^2 \cdot$day at a pressure difference of 100 kPa. The rejection for PEG was 40% using a feed solution concentration of 0.1 wt%.

BSA (F-V, Lot. 0125R) purchased from the Armour Pharmaceutical Co. was used in this study. Phenylalanine, hydrazine hydrate, Raney nickel, ethanol, 1,6-dimethylsuberimidate dihydrochloride (DMS), polyethylene glycol, glucose, and maltose purchased from Tokyo Chemical Industry Co., Ltd., were reagent grade and were used without further purification. Ultrapure water was used throughout the experiments.

Chemical Modifications. The polyacrylonitrile hollow fibers were dipped in a solution of 75 ml ethanol, 0.25 g Raney nickel, 25 ml hydrazine hydrate at different temperatures (50 - 80°C) and reaction times (0.5 - 23 hours). The anticipated product by the reaction is shown in Figure 1. After the reaction, the modified fibers were washed in water at 50°C for 1 hour and, thereafter, washed three more times in fresh water of 50°C each for 15 minutes under ultrasonic treatment generated by an ultrasonic cleaner (200 W, 40 KHz, Kaijo Denki Co.).

BSA Immobilization. The modified fibers were immersed in DMS solution (0.015 g/50 ml of H_2O) for 5 minutes at 25°C, which was adjusted to pH 10.0 using 0.1 N NaOH solution. After the excess DMS solution on the surface of the hollow fibers was removed, the membranes were dipped in the BSA solution (0.813 g/ 50 ml of H_2O) for 30 minutes at 25°C and subsequently for 18 hours at 4°C. The solution was then adjusted to pH 8.5 by adding 0.1 N NaOH solution. The amount of BSA bound on the membranes was estimated from the decrease of BSA concentration in the solution where the membranes were immersed. The concentration of BSA was measured by HPLC using a UV detector. After BSA immobilization (Figure 2), the immobilized BSA fibers were washed in water for 1 hour and stored in water at 4°C.

Transport Measurements. The ultrafiltration apparatus used in this study was described in previous papers (*2-5*). The pressure difference across the fibers, Δp, was 0.4 kg/cm^2 or 0.8 kg/cm^2. The ultrafiltration measurements were performed at 25±0.5°C. The rejection was calculated from the concentration ratio of solute in the feed solution, C_f, and the concentration of solute in the permeate, C_p,:

$$R = (1 - C_p/C_f) \times 100 \ (\%) \tag{1}$$

The concentration of glucose (Mw = 180 g/mole), maltose (Mw = 342 g/mole), poly(ethylene glycol) (PEG) 1000 (Mw ~ 950-1,050 g/mole), PEG 2000 (Mw ~ 1,800-2,200 g/mole), PEG 4000 (Mw ~ 3,000 g/mole), PEG 6000 (Mw ~ 7,800-9,000 g/mole), and PEG 20,000 (Mw ~ 15,000 g/mole) was measured by an HPLC system using the refractive index as a detector (ERC-7510, Erma Optical Works, Ltd.) .

Optical Resolution of Amino Acid. Optical resolution of phenylalanine was also performed in this study. After the racemic phenylalanine permeated through the immobilized BSA membranes, the concentration of the D- and L-phenylalanines in the filtrate were determined using HPLC (880-PU, UV-970, Jasco Co.) with a Crownpak column [CR(+), Daicel Chemical Industries, Ltd.]. The detection limit of phenylalanine was less than 0.0001 mM in this system. The separation factor, α, is defined as, $[J_D/J_L]/[C_{feed}(D)/C_{feed}(L)]$, where $C_{feed}(L)$ and $C_{feed}(D)$ are the concentrations of the L-isomer or D-isomer in the feed, respectively, and J_L and J_D are the fluxes of the L-isomer or D-isomer, respectively. Because the feed solution used in this study was the racemic phenylalanine solution (i.e., $C_{feed}(D) = C_{feed}(L)$) and the flux of the solute was directly related to the concentration in the permeate (i.e., $J_D/J_L = C_p(D)/C_p(L)$ where $C_p(L)$ and $C_p(D)$ were the concentrations of the L-isomer or D-isomer in the permeate, respectively), α could be reduced to the concentration ratio of the D-isomer to L-isomer in the permeate. The standard deviations for the flux and rejection measurements were 10% and 20% for four independent experiments.

Results and Discussion

Chemical Modification. Polyacrylonitrile (PAN) hollow fibers were chemically modified on their surface according to the experimental procedures described above. The anticipated product of the reaction with the hollow fibers and hydrazine hydrate is shown in Figure 1. The mechanism of the reaction was postulated from the reaction of nitriles with hydrazine hydrate and Raney nickel and from the color

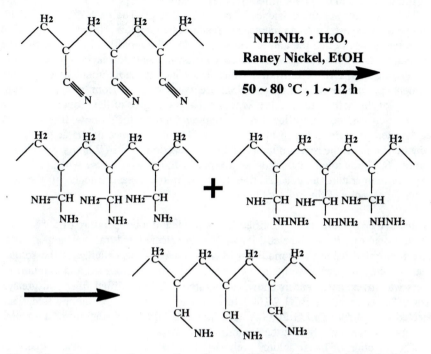

Figure 1. Reaction scheme of surface-modified polyacrylonitrile hollow fibers.

$$\boxed{\text{Modified PAN}}\!-\!NH_2 \;+\; CH_3OC\overset{\overset{\displaystyle{}^+NH_2}{\|}}{}(CH_2)_6\overset{\overset{\displaystyle{}^+NH_2}{\|}}{C}OCH_3$$

$$\Rightarrow \boxed{\text{Modified PAN}}\!-\!NH\overset{\overset{\displaystyle{}^+NH_2}{\|}}{C}(CH_2)_6\overset{\overset{\displaystyle{}^+NH_2}{\|}}{C}OCH_3 \;+\; CH_3OH$$

$$\Rightarrow \boxed{\text{Modified PAN}}\!-\!NH\overset{\overset{\displaystyle{}^+NH_2}{\|}}{C}(CH_2)_6\overset{\overset{\displaystyle{}^+NH_2}{\|}}{C}NH\!-\!\!\widehat{BSA} \;+\; CH_3OH$$

with $\widehat{BSA}\!-\!NH_2$

Figure 2. Immobilization of BSA onto the surface of chemically modified polyacrylonitrile hollow fibers.

formation of polyacrylonitrile treated with alkaline reagents as reported in the literature (23-24).

Infrared spectra of the modified fibers showed a strong peak around 3,200-3,400 cm^{-1}, which indicates the existence of an amine group. The peak height at 3,200-3,400 cm^{-1} increased with an increase in reaction temperature. It is suggested from the infrared spectra that the amine group existed on the modified fibers and that the amount of the amine segment increased with an increase in reaction temperature. The degree of the reduction ratio, which is defined as the amount of the amine segments per unit residue of polyacrylonitrile, was estimated from the ion-exchange capacity of the hollow fibers. It was found that the degree of the reduction ratio in the reaction can be controlled by the dipping time of the hollow fibers and/or reaction temperature in the reactive solution. The dependence of the degree of the reduction ratio on the reaction temperature at a reaction time of 7 hours is shown in Figure 3. The amount of the amine group was found to increase with increasing reaction temperature and/or reaction time from the measurements of the ion-exchange capacity of the hollow fibers.

Transport Experiments.　Modified fibers reacted at a temperature of 50°C and a reaction time of 7 hours, labeled MF-50-7, were selected and transport experiments were performed using these modified fibers. Rejection dependence of the solutes based on their molecular weights at $C_f = 0.1$ wt% for the modified and unmodified fibers was investigated and is shown in Figure 4. The modified fiber had greater than 95% rejection for PEG 20,000 feed solution, whereas the unmodified fiber rejected only 40% of PEG 20,000. The modified fiber had, therefore, a smaller molecular weight cut-off than the unmodified fiber.

The rejection of linear solutes such as poly(ethylene glycol) was less than that of nonlinear solutes such as glucose (Mw = 180 g/mole) and maltose (Mw = 342 g/mole) for the modified fibers. These results can be explained based on deformation and stretching of poly(ethylene glycol); thus, PEG is able to pass more easily through the pores of the fibers than glucose or maltose.

Optical Resolution.　Optical resolution of racemic phenylalanine using the immobilized BSA membranes was performed. Two types of immobilized BSA membranes were prepared in this study. One membrane, BSA-1, was prepared from unmodified fibers where BSA was immobilized on the amine residue (D.S. = 0.01). The other membrane, BSA-2, was prepared from modified fibers (MF-50-7), which had an amine group degree of 0.26. The amount of BSA immobilized on the membranes per area for BSA-1 and BSA-2 membranes was estimated to be 223 nmole/cm^2 and 154 nmole/cm^2, respectively (see Table I). A 500 ml phenylalanine solution was circulated through the immobilized BSA membranes and permeate and concentrate solutions were pumped back into the feed solution for 10-13 hours.

The time dependence of the separation factor of phenylalanine using the immobilized BSA membrane, BSA-1, was investigated for the above conditions and is shown in Figure 5. As a control measurement, experiments of optical resolution were also performed using the unmodified hollow fibers. The

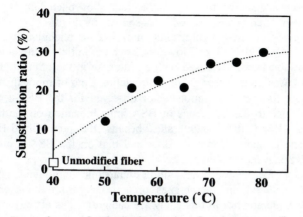

Figure 3. Dependence of substitution ratio of amine group (i.e., degree of reduction ratio) on reaction temperature for reaction time of 7 hours.

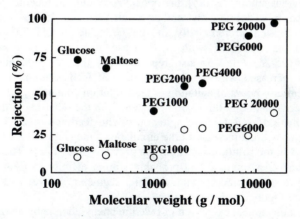

Figure 4. Dependence of rejection of solutes on their molecular weights at 20°C, $\Delta p = 0.4$ kg/cm^2 and $C_f = 0.1$ wt% for unmodified (O) and modified (●) fibers.

unmodified hollow fibers, which did not have the recognition sites of amino acid isomer (i.e., BSA) on the fibers, did not show any optical resolution of the racemic phenylalanine within the expected experimental error ($\alpha = 1.0 \pm 0.1$). On the contrary, the immobilized BSA membrane, BSA-1, provided optical resolution of the racemic phenylalanine, and the separation factor was 1.5±0.3 when the circulation time was greater than 2 hours.

The time dependence of the separation factor of phenylalanine using the immobilized BSA membrane, BSA-2, is shown in Figure 6. The BSA-2 membranes did not give any optical resolution of racemic phenylalanine during the initial 4 hours of circulation time. However, after the initial circulation time of 4 hours, the separation factor increased with an increase in the circulation time; the separation factor was greater than 3 for a circulation time of more than 11 hours, which was higher than the separation factor observed for the BSA-1 membrane. This is probably due to the difference of BSA immobilization on the membranes between the BSA-1 and BSA-2 membranes, because the immobilized BSA on the BSA-1 membrane is mainly adsorbed BSA and that on the BSA-2 membrane is covalently bound BSA on the membrane.

The mechanism for the optical resolution of racemic amino acids is due to the fact that BSA binds on the membranes preferentially to the L-isomer. The immobilized BSA membranes in this study are categorized as affinity membranes. Figures 5 and 6 suggest that BSA immobilized on the surface of the membranes with covalent binding provides an extremely slow binding characteristic with L-phenylalanine that is on the order of hours.

In our previous studies (25-26), the optical resolution of racemic phenylalanine was performed in a solution system containing BSA. It was found that D-phenylalanine existed preferentially in the permeate at pH 7.0 and that the separation factors increased with a decrease in the feed concentration of phenylalanine (25-26). It was also reported (25) that the separation factor of phenylalanine increased with the agitation time of feed solution. The separation factor was 12 after 6 hours of agitation of feed solution containing 0.60 mM BSA + 0.024 mM phenylalanine in the solution system. On the other hand, the separation factor was only 2 after 1 hour of agitation of the feed solution. The increased separation factor with an increase in the circulation time (Figure 6) shows a similar tendency found in the solution system in the previous study (25). The immobilized BSA in the BSA-2 membrane had probably similar affinity to BSA in the BSA solution due to long binding and flexible segments between BSA and the membranes (see Figure 2).

To compare the optical resolution of racemic phenylalanine by the BSA solution system previously studied (25-26) and the immobilized BSA membranes of this study, separation factors, flux, and binding effectiveness of BSA (i.e., concentration ratio of L-phenylalanine to BSA on a molar basis) are summarized in Table I. The concentration ratio of L-phenylalanine to BSA was calculated from the total amount of L-phenylalanine bound to BSA divided by the total amount of BSA immobilized on the membranes or in the BSA solution on a molar basis. The amount of L-phenylalanine bound to BSA was estimated from the concentration decrease of L-phenylalanine in the feed solution measured by HPLC.

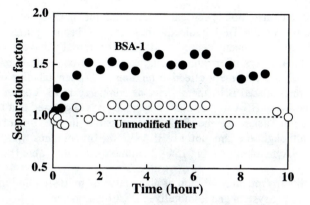

Figure 5. Time dependence of separation factor of phenylalanine for unmodified (○) and BSA-1 (●) membranes at 20°C, Δp = 0.4 kg/cm^2 and C$_f$ = 0.30 mM.

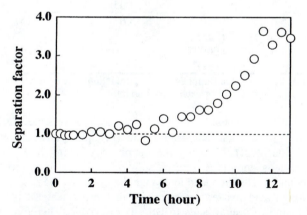

Figure 6. Time dependence of separation factor of phenylalanine for BSA-2 (○) membrane at 20°C, Δp = 0.8 kg/cm^2 and C$_f$ = 0.30 mM.

The concentration ratio of L-phenylalanine to BSA in the immobilized BSA membranes was 5.6 for the BSA-1 membrane and 60.5 for the BSA-2 membrane, while the concentration ratio in the BSA solution system was as low as 0.0007 - 0.005, as shown in Table I. The number of phenylalanine molecules bound to a BSA molecule was higher than unity for the BSA-1 and BSA-2 membranes, as shown in Table I. This phenomenon indicates phenylalanine binds to the BSA molecule not only specifically (i.e., to the binding sites of BSA that are well characterized) but also non-specifically, although the number of phenylalanine molecules bound to the BSA molecule has not yet been reported. Thus, the concentration ratio of L-phenylalanine to BSA in the immobilized BSA membranes was about 3 to 4 orders of magnitude higher than the concentration ratio in the solution system. An extremely effective binding of L-phenylalanine on BSA was found in the immobilized BSA membranes as compared to the concentration ratio of L-phenylalanine to BSA in the solution system. It is suggested that the effective binding of amino acids on the BSA was possible in the immobilized BSA membranes, although the amount of BSA in the membranes was low in the immobilized BSA membranes (e.g., 154 - 223 $nmole/cm^2$, see Table I).

Table I. Optical resolution of phenylalanine by BSA solution system and immobilized BSA membranes.

Method	C_{Phe} (mg/l)	BSA (mmole)	Phe/BSA (mole/mole)	J $(m^3/m^2 \cdot d \cdot kgcm^2)$	α
BSA	1.0	0.105	0.0007	0.70	4.8
solution	7.0	0.105	0.005	0.71	1.1
BSA-1	5.0	0.00045 (223)[a]	5.6	2.15	1.5[b]
BSA-2	5.0	0.00027 (154)[a]	60.5	0.30	3.2[c]

a) is the amount of BSA per area ($nmole/cm^2$); b) circulation time of feed is greater than 2 hours; c) circulation time of feed is greater than 2 hours.

The binding of amino acids to BSA is known to be reversible (25-28). After ultrafiltration of BSA and racemic amino acid (i.e., phenylalanine, leucine, and tryptophan) solution at pH 7.0 using the membranes having MWCO of 50,000 g/mole in the previous study (25), a buffer solution of pH 3.0 was injected into the BSA solution bound to L-amino acid to change the pH of the BSA solution. It was found that L-amino acid existed preferentially in the permeate at pH 3.0, which indicates the binding of amino acids to BSA is reversible. Therefore, the present membranes can be used for affinity separation between L-isomer and D-isomers.

Conclusions

The surface reaction of polyacrylonitrile hollow fibers with hydrazine and Raney nickel by a one-step reaction introduced NH_2 units on the surface. The modified fibers had a smaller molecular weight cut-off than the unmodified fibers. The separation factor that was determined using the immobilized BSA membranes prepared from chemically modified fibers (BSA-2) was two times higher than that of immobilized BSA membranes prepared from unmodified hollow fibers (BSA-1) for a circulation time greater than 11 hours. It can be concluded that the immobilized BSA membranes prepared from chemically modified hollow fibers efficiently showed optical resolution of racemic phenylalanine.

Literature Cited

1. Nabe, A.; Staude, E.; Belfort, G. *J. Membrane Sci.* **1997**, *133*, 57.
2. Higuchi, A.; Iwata, N.; Tsubaki, M.; Nakagawa, T. *J. Appl. Polym. Sci.* **1988**, *36*, 1753.
3. Higuchi, A.; Iwata, N.; Nakagawa, T. *J. Appl. Polym. Sci.* **1990**, *40*, 709.
4. Higuchi, A.; Nakagawa, T. *J. Appl. Polym. Sci.* **1990**, *41*, 1973.
5. Higuchi, A.; Koga, H.; Nakagawa, T. *J. Appl. Polym. Sci.* **1992**, *46*, 449.
6. Higuchi, A.; Ishida, Y.; Nakagawa, T. *Desalination* **1993**, *90*, 127.
7. Higuchi, A.; Hara, M. *J. Phys. Chem.* **1996**, *100* (6), 2183.
8. Crassous, G.; Harjanto, F.; Mendjel, H.; Sledz, J.; Schue, F.; Meyer, G.; Jozefowicz, M. *J. Membrane Sci.* **1985**, *22*, 269.
9. Stengaard, F. F. *Desalination* **1988**, *70*, 207.
10. Noshay, A.; Robeson, L. M. *J. Appl. Polym. Sci.* **1976**, *20*, 1885.
11. Hartwing, A.; Mulder, M.; Smolders, C. A. *Adv. Colloid Interface Sci.* **1994**, *52*, 65.
12. Yamagishi, H.; Crivello, J. V.; Belfort, G. *J. Membrane Sci.* **1995**, *105*, 237.
13. Breitbach, L.; Hinke, E.; Staude, E. *Angew. Makromol. Chem.* **1991**, *184*, 183.
14. Guiver, M. D.; Apsimon, J. W.; Kutowy, O. *J. Polym. Sci., Part C, Polym. Lett.* **1988**, *26*, 123.
15. Guiver, M. D.; Croteau, S.; Hazlett, J. D.; Kutowy, O. *Br. Polym. J.* **1990**, *23*, 183.
16. Kiyohara, S.; Sasaki, M.; Saito, K.; Sugita, K.; Sugo, T. *J. Membrane Sci.* **1996**, *109*, 87.
17. Yang, M.-C.; Tong, J.-H. *J. Membrane Sci.* **1997**, *132*, 63.
18. Hosch, J.; Staude, E. *J. Membrane Sci.* **1996**, *121*, 71.
19. Klein, E.; Yeager, D.; Seshadri, R.; Baurmeister, U. *J. Membrane Sci.* **1997**, *129*, 31.
20. Kubota, N.; Kounosu, M.; Saito, K.; Sugita, K.; Watanabe, K.; Sugo, T. *J. Membrane Sci.* **1997**, *134*, 67.
21. Roper, D. K.; Lightfoot, E. N. *J. Chromatogr.* **1995**, *702*, 3.
22. Thommes, J.; Kula, M.-R. *Biotech. Progress* **1995**, *11*, 357.
23. Zajac, W. W. Jr.; Siuda, J. F.; Nolan, M. J.; Santosusso, T. M. *J. Org. Chem.* **1971**, *36* (23) 3539.
24. Takata, T.; Taniyama, M. *Kobunshi Kagaku (Polymer Chemistry)* **1962**, *19*, 641.
25. Higuchi, A.; Hara, M.; Horiuchi, T.; Nakagawa, T. *J. Membrane Sci.* **1994**, *93*, 157.
26. Higuchi, A.; Hashimoto, T.; Yonehara, M.; Kubota, N.; Watanabe, K.; Uemiya, S.; Kojima, T.; Hara, M. *J. Membrane Sci.* **1997**, *130*, 31.
27. Kragh-Hansen, U. *Pharmacol. Rev.* **1981**, *33*, 17.
28. Yang, J.; Hage, D. S. *Anal. Chem.* **1994**, *66*, 2719.

Chapter 16

Facilitated Transport Membranes Incorporating Metal Affinity for Recovery of Amino Acids

C. Oxford, D. Crookston, R. R. Beitle, and M. R. Coleman[1,2]

Department of Chemical Engineering, University of Arkansas, Fayetteville, AR 72701

Facilitated transport metal affinity membranes that incorporate metal affinity ligands are being investigated as a recovery scheme for specific amino acids (i.e. histidine) and proteins. The inclusion of metal affinity ligands as fixed site carriers within poly(vinyl alcohol) gel membranes resulted in increases in the mass transfer coefficient of histidine and increases in the selectivity of histidine to a noninteracting amino acid control, phenylalanine. Membranes containing chelated Cu^{+2} or Co^{+2} exhibited the largest increase in histidine mass transfer rate and those containing Ni^{+2} or Zn^{+2} displayed no increase in mass transfer coefficient relative to the control membrane. The effect of chelated metal type on histidine transport was similar to trends exhibited by other metal affinity separation methods.

Advances in the field of biochemical engineering have allowed for the manufacturing of gene products for the development of new pharmaceuticals, industrial processes, and consumer products. In short, the cultivation of an organism via either aerobic or anaerobic fermentation leads to the biological production of the desired material. These advancements in the field of bio-process engineering require efficient methods of product recovery, as the type, number, yield, and degree of purity achieved in each isolation step can significantly impact the overall yield and purity of the final product. Indeed, the cost of amino acid and protein recovery and purification can comprise a significant portion of the total production cost. Therefore, the development of efficient recovery techniques is central to biochemical engineering.

[1]Current address: Chemical and Environmental Engineering Department, University of Toledo, 2801 West Bancroft, Toledo, OH 43606.
[2]Corresponding author.

A number of recovery methods which exploit metal affinity have been developed in the past several years including: (i) immobilized metal affinity chromatography (IMAC), (ii) metal enhanced two-phase aqueous extraction, (iii) metal affinity precipitation, (iv) magnetic immobilized metal affinity separations (MIMAS), and (v) immobilized metal affinity membrane separations (IMAMS). Common to each method is differential coordinate bond formation between a specific biomolecule and a metal ion which is bound via a chelating group to either a stationary phase (IMAC, MIMAS, or IMAMS), or a mass separating agent (two phase aqueous extraction or precipitation). Histidine is of central focus to metal affinity based recovery schemes, because the imidazole nitrogens of the (histidine) side chain easily bind to metal ions. Hence, metal affinity based separations methods are potentially useful for recovery of histidine, or for isolation of proteins which display a high histidine content. Under controlled conditions of pH and salt content, the divalent metal ion-histidine interaction may be tuned toward the recovery of a target species.

All of the aforementioned metal affinity techniques are cyclic in nature, i.e. following an adsorption event the biomolecule of interest is eluted in some fashion. Dense film membrane based recovery schemes, on the other hand, typically offer the potential to operate in a continuous recovery mode. This potential for continuous recovery along with other advantages of membrane based unit operations (low energy requirement and small modular size) have prompted our group to develop a facilitated transport membrane system incorporating metal affinity. Facilitated transport metal affinity membranes (FTMAMs) are expected to be a viable method for bioseparations, since fixed site carrier facilitated transport membranes have been shown to be useful for diverse separations.

This paper explores various aspects of FTMAM which utilizes metal affinity based interactions for the recovery of amino acids, peptides, and proteins. Specifically, the effect of incorporation of chelated metal ions on amino acid transport within a dense poly(vinyl alcohol) (PVA) membrane has been investigated. After a discussion of the pertinent biochemical principals, data will be presented which summarizes our initial efforts to characterize pure component amino acid transport. Membranes employed in this study differ by the choice of metal ion which serves as affinity ligands. Finally, potential applications will be commented upon within the context of macromolecule recovery and controlled release.

Background and Theory

Metal Affinity Chemistry.   Pseudobiospecific separation methods have been developed which rely on highly specific interactions to achieve separation, but are generic enough to be applicable to categories of biomolecules. Metal affinity, triazine-based dyes, thiophillic, lectin, immunoglobin, and many other affinity methods can be used to recover a biological molecule via a specific interaction (e.g. IgG affinity chromatography) (*1*). Several of these methods can be quite expensive because biological ligands may be required. Metal affinity is an attractive method of recovery, since the affinity ligand used to achieve the requisite separation can be quite inexpensive and stable (*2*). Clearly defined characteristics, diverse operating conditions, and many other factors contribute to the growing interest in developing metal affinity based methods.

For immobilized metal affinity based separations, a chelate group is used to fix the metal ion ligand to either a stationary phase or mass separating agent as shown in Figure 1. Chelate groups are strong Lewis acids, which in a similar fashion to ethylenediamine tetraacetic acid (EDTA), form several coordinate bonds with the metal ion through the sharing of three or more pairs of electrons. Historically, iminodiacetic acid (IDA) has been commonly employed as the chelate group, since this tridentate chelator as well as the chemistry used to prepare the metal affinity media are simple and reliable. Other chelating groups such as nitrilotriacetic acid are utilized, although the chemistry used to prepare the intermediate compound is slightly more complex (*3-5*). A variety of methods have been used to couple the chelate group to the stationary phase or mass separating agent which include radiation-induced graft polymerization, epoxide activation, bis-oxiraine coupling, and tresyl chloride reaction.

Transition metal ions are employed as affinity ligands, and the ability to choose between divalent copper, cobalt, zinc, and nickel provides a high degree of flexibility in designing recovery schemes. Histidine can display appreciable Cu^{+2} affinity, as the association constant for Cu^{+2}-histidine is relatively high (*6-10*). Other factors can attenuate the Me^{+2}-histidine interaction. For example, modifications to the mobile phase composition (e.g. inclusion of salt, organics, and/or detergents) provide an additional measure of freedom, since the inclusion of one or more attenuating material(s) can attenuate adsorption differences between competing species. The inclusion of a chaotropic salt (e.g. NaCl) performs several important functions including: suppression of ionic effects, and promotion of adsorption (*11*). Selectivity and strength of binding will depend on the type of salt employed (sodium phosphate, potassium sulfate, sodium acetate, and others), as well as its concentration (0-4 molar) (*9,12*). The addition of either organic solvents or detergents/denaturants can further mediate the adsorption event, as hydration of the chelate moiety may be affected due to increases in mobile phase hydrophobicity (*4, 9,12*).

Desorption is accomplished in several ways, including: (i) pH reduction, (ii) competing agent addition, (iii) salt concentration changes, and (iv) chelate annihilation. Elution via pH reduction occurs primarily due to protonation of the imidazole ring of histidine. The addition of competing agents such as histamine, glycine, and imidazole release proteins by vying for coordinate bonds with the chelated metal ion. Finally, if drastic measures are required, a soluble metal chelate, e.g. EDTA, is added. Metal ions will strip from the chelate group in the presence of the soluble metal chelate, and subsequently annihilate the affinity ligand's efficacy.

Separation of a mixture may be achieved by judicious choices regarding: metal ion, buffer and elution conditions, inclusion of organic solvents in mobile phases, and the exploitation of genetic techniques to enhance [metal] affinity (7-9, 11-17) Many examples have been cited which demonstrate the ability, diversity, and potential of techniques which exploit metal affinity [for a review see Arnold (6), Beitle and Ataai (18), Sulkowski (17,19)]. Although IMAC when performed in a simulated moving bed mimics a continuous process, continuous methods of recovery that employ metal affinity have not been developed.

Facilitated Transport Using Metal Affinity Fixed Site Carriers. Facilitated transport membranes use a rapid, reversible chemical reaction to augment the diffusion of a solute through a semi-permeable membrane (20,21). The carrier molecule can react selectively and reversibly with the solute, thereby, providing a means of enhancing the solute flux relative to non-reactive solutes. Pertinent to the development of FTMAMs is the use of divalent metal ions chelated within poly(vinyl alcohol) gel membranes. Chelated metal ions should act as fixed site carriers for specific amino acids (i.e. histidine) and proteins. For facilitated transport membranes, as illustrated for the FTMAM system in Figure 2, the solute first reacts at the feed side with a fixed site carrier molecule covalently bound via a spacer arm to the polymer matrix of the membrane. Through a combination of thermal motion of the chemical carrier, solute desorption, solute diffusion, and solute adsorption by a new carrier group, the reactive solute moves through the dense film and emerges on the permeate side. Therefore, this process may be viewed as an exchange reaction, with the solute flux through the facilitated transport membranes controlled by free diffusion and the facilitation process. For FTMAM systems, a metal chelating moiety (IDA) will be covalently bound to the PVA backbone via a spacer arm (1,4-butanediol diglycidyl ether) to form a membrane which will bind metals and is designated PVA-IDA. The PVA-IDA-Me^{+2} membranes rely upon the highly selective metal affinity interactions that are illustrated in Figure 1 to provide an enhanced flux for histidine or proteins. Because divalent metal ions can exhibit very different strengths of interaction with amino acids, the choice of metal ion would be expected to affect the rate of mass transfer of histidine in a PVA-IDA-Me^{+2} membrane.

A pH gradient between the feed and permeate chambers can also be established to alter the facilitated transport of histidine. Because the pKa of histidine is 6.5, a pH of larger value would augment the binding of the imidazole ring to Me^{+2}. As the pH decreases, passes through the pKa, and finally approaches 5.0, the dissociation constant for histidine and the metal ion increases. Thus, by providing a pH difference in conjunction with a concentration gradient, transport of this amino acid across the membrane may be further augmented. In this study, the effect of divalent

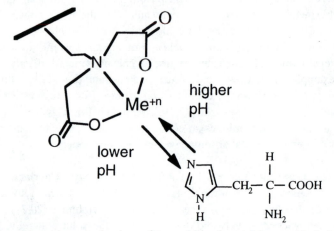

Figure 1. Interaction mechanism of immobilized metal ions with imidazole ring of histidine.

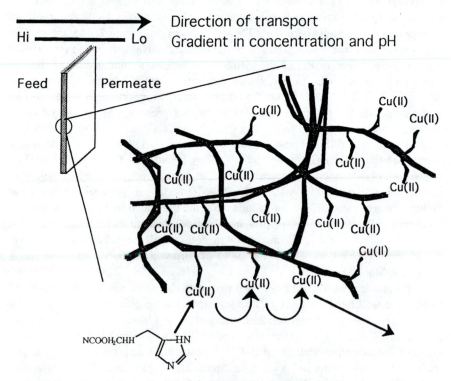

Direction of transport
Gradient in concentration and pH

Figure 2. Mechanism of transport of histidine in fixed site carrier facilitated transport metal affinity membranes.

metal ion upon the rate of mass transfer of histidine in a series of PVA-IDA-Me^{+2} membranes was investigated for a system operating at a pH of 6.5 in both the feed and permeate solutions.

Experimental

Materials. Reagents for membrane formation and functionalization were purchased from Aldrich Chemical Co. The poly(vinyl alcohol) (PVA) used in this study had an average molecular weight of 70,000 g/mol. Solutions were prepared in either deionized water or MilliQ ultrapure water where appropriate. Amino acids were purchased from Sigma Chemical at a purity of 97%.

Membrane Formation. Poly (vinyl alcohol) gel membranes (9 x 8.5 cm) were prepared from a water solution in an electrophoresis casting block using the following procedure. For each membrane, eight milliliters of a 0.125 g/ml PVA-water solution was mixed with the materials listed in Table I which included methanol, acetic acid, sulfuric acid, and glutaraldehyde. Glutaraldehyde was used as the crosslinking agent for PVA gel formation. The solution was quickly poured into a gel casting block set at a well defined thickness of 0.75 mm and allowed to crosslink at room temperature for several hours.

Table I.   Solutions for preparation of FTMAM membranes.

PVA Gel Membrane Formation

Material	Stock Concentration	Amount
Poly(vinyl alcohol)	0.125 g/ml	8.0 ml
Methanol	50 % (v/v)	300 μl
Acetic acid	10%	300 μl
Sulfuric acid	10%	900 μl
Glutaraldehyde	2.5%	900 μl

Membrane Functionalization (two separate solutions)

Material	Stock Concentration	Amount
Step 1: Functionalization with Spacer Arm (Figure 3)		
1,4 Butanediol diglycidyl ether	neat	20 ml
Sodium hydroxide	300 mM	80 ml
Sodium borohydrate	---	60 mg
Step 2: Functionalization with Chelating Ligand (Figure 3)		
Iminodiacetic acid	---	10 g
Sodium carbonate	2.0 M	100 ml

Membrane Functionalization. Using bis-oxirane coupling chemistry, the spacing element 1,4-butanediol diglycidyl ether (1,4-BDE) was used to couple iminodiacetic acid to PVA gel membranes as shown in Figure 3. First, the PVA membranes were incubated for 12 hours in a solution consisting of 1,4-BDE, sodium hydroxide, and sodium borohydride in the amounts listed in Table I. Prior to immersion of the membrane in this solution, sodium borohydride was allowed to dissolve for >10 minutes. Because at low concentrations the diepoxide could in principal react with two -OH groups to form additional crosslinks, this reaction was performed in an excess of 1,4-BDE to assure selectivity towards a single epoxide reacting with a single hydroxyl of PVA. To complete functionalization, the membranes were immersed for 12 hours in a second solution of sodium carbonate containing 0.1 g IDA per ml at 60°C (Figure 3 and Table I). Following each synthesis step, the modified membranes were washed in an excess phosphate buffer to remove residual reactant.

Membranes that were functionalized with 1,4-BDE and IDA and were not charged with metal were used as controls in mass transfer experiments (denoted PVA-IDA in Figure 3) as these membranes should not exhibit affinity for amino acids or proteins. Finally, PVA-IDA membranes were incubated in a 5 mg/ml solution of metal ion for 12 hours to immobilize metal ions (i.e. Cu^{+2}, CO^{+2}, Ni^{+2}, or Zn^{+2}). The membranes were washed in deionized water followed by a wash in a buffer solution to remove any unbound metal. A priori, these membranes should exhibit affinity towards histidine and will be designated PVA-IDA-Me^{+2}.

Metal and Water Content. The metal content of each membrane was determined using elemental analysis at Galbraith Laboratories in Knoxville, Tennessee. Gravimetric analysis of wet and dry samples was used to determine the mass percent water in a given membrane. Samples of each type of membrane were equilibrated in deionized water, patted dry to remove surface water and weighed immediately (m_{wet}). The samples were then dried in a conventional oven at 80°C for several days and finally in a vacuum oven at 100°C for at least 24 hours to remove residual water. The dried films were weighed (m_{dry}) and the weight percent water (w_{water}) was determined using equation (1)

$$W_{water} = \frac{m_{wet} - m_{dry}}{m_{wet}} \times 100 \ (\%) \qquad (1)$$

Determination of Mass Transfer Coefficients. The mass transfer coefficients of histidine were measured at room temperature using a stirred diffusion cell consisting of two 160 ml chambers (feed and permeate,) as illustrated in Figure 4 (*22*). The membrane was sandwiched between glass holders and clamped between the chambers of the diffusion cell. Magnetic stirrers were used in each chamber to keep the solutions well mixed and reduce boundary layer mass transfer resistances. The feed and permeate chambers were buffered to a pH of 6.5 using 0.05 M sodium phosphate in a solution of 0.50 M NaCl.

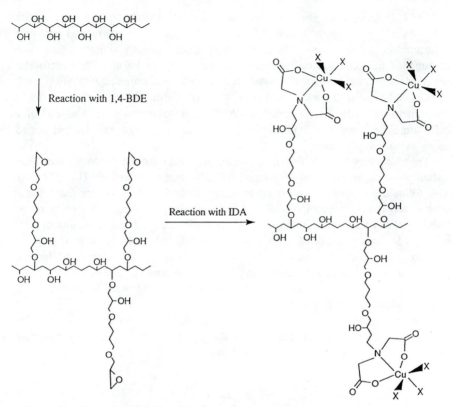

Figure 3.   Functionalization chemistry for attaching a chelating ligand (IDA) to a poly(vinyl alcohol) backbone via a spacer arm (1,4 BDE).

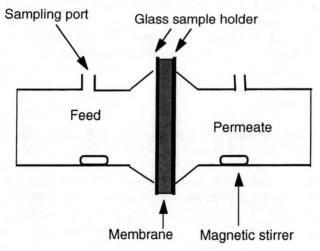

Figure 4.   Diffusion apparatus for mass transfer experiments with histidine and phenylalanine.

For a diffusion experiment, the feed and permeate chambers were charged with solutions of a given amino acid at a feed concentration of 1.5 mg/ml and permeate concentration of 0.3 mg/ml. Solutions were simultaneously pumped into each chamber of the diffusion cell with a peristaltic pump to balance the hydrostatic pressure on the membrane, and the stirrers and clock were started. Samples (800 ml) were withdrawn at predetermined time intervals during the course of diffusion experiment (~ 5 hours) and stored at –20°C prior to analysis. At the end of a diffusion experiment, the fluid was removed from the system and the pH tested using a standard pH meter.

The concentrations of amino acid in each sample were determined spectrophotometrically using a Beckman DU-640 spectrophotometer (I = 232 nm). The following procedure was used to insure that the value of the absorbance represented that of the amino acid, and was not rendered inaccurate due to (i) p-bonding of any trace metal ion in solution or (ii) pH effects. Briefly, an aliquot (200 ml) of a sample was treated with a slurry of 500 ml chelating resin (Chelating Sepharose Fast Flow, Pharmacia) in a buffered solution of 0.5 M acetate buffer at a pH 5.0 (ratio of 1:4 settled volume to total volume) to remove trace metal ions. This mixture was allowed to equilibrate for 30 minutes and was separated using centrifugation and/or filtration.

The overall mass transfer coefficients for the amino acids were obtained based on a pseudo-steady-state mass balance on the feed and permeate. The overall mass transfer coefficient (K_o) can be determined from

$$\ln\left[\frac{C_f - C_p}{C_{f0} - C_{p0}}\right] = -AK_o t\left(\frac{1}{V_f} - \frac{1}{V_p}\right) \tag{2}$$

where C_f and C_p are the bulk feed and permeate concentrations at time t, C_{f0} and C_{p0} are the initial bulk feed and permeate concentrations, A is the effective diffusion area, and V_f and V_p are the feed and permeate volumes. The facilitation factor is defined as the ratio of the overall mass transfer coefficient of a solute through a PVA-IDA-Me^{+2} membrane to the mass transfer coefficient in the PVA-IDA membrane.

Results and Discussion

Poly(vinyl alcohol) gel membranes were successfully functionalized using metal affinity chemistry and were shown to bind metals via several methods. Metal ion retention was qualitatively established by washing metal charged media with excess 1 M sodium chloride, followed by several washes of 0.05 M phosphate + 1 M sodium chloride. After decanting the final, colorless wash, 1 ml of 0.05 M EDTA + 1 M sodium chloride was added. The EDTA solutions turned color appropriate for a given FTMAM membrane (i.e. blue -Cu^{+2}, green -Ni^{+2}, pink -CO^{+2}), indicating that Me^{+2} was readily chelated, and bound Me^{+2} can be stripped from the media. Zn^{+2} chelation was determined in the same fashion, except precipitation of $Zn(OH)_2$ was used to indicate the presence of this metal ion in either a wash, or in the final EDTA wash. Elemental analysis of samples indicated that approximately 17 mM per gram dry film of each metal ion was chelated in a consistent fashion.

In an earlier study, examination of the rate of transport of phenylalanine demonstrated that a non-interactive amino acid control was quite insensitive to the presence of divalent copper in FTMAM membranes (*23*). Briefly, mass transfer coefficients of phenylalanine in PVA-IDA and PVA-IDA-Cu^{+2} were 3.42 x 10^{-5} cm²/s and 3.40 x 10^{-5} cm²/s, respectively, as determined with feed and permeate concentrations of 2.5 mg/ml and 0.0 mg/ml. Although a pH gradient was used (8.0 and 5.0), the aromatic R-group of phenylalanine is not ionizable and therefore indifferent to this range of H_3O^+ concentration.

The results obtained when the transport of histidine across a control membrane of PVA-IDA was followed are shown in Figure 5. The data is represented in the form of normalized permeate concentrations and times to account for subtle differences in the initial concentrations and between runs. The normalized permeate concentration is defined as the difference between the permeate concentration at time t and the initial permeate concentration. The raw diffusion data is shown in the inset of Figure 5 for initial permeate concentrations in the range 0.25 - 0.35 mg/ml. Parallel lines of permeate concentration versus time for these three runs indicate that consistent rates of transport were obtained for several membranes of the same thickness. An average mass transfer coefficient of 3.7 x 10^{-5} cm²/s for the histidine in the PVA-IDA membranes was calculated using Equation 2 and the data in Figure 5.

Raw and normalized histidine permeate concentration versus time are shown in Figure 6 for both the control and copper charged membranes. This plot represents the average concentration dependent data for multiple runs for each type of membrane. Data from three experiments for PVA-IDA were used to prepare the curve for histidine transport in the control system, and two runs were used for the composite of PVA-IDA-Cu^{+2}. Representatives of the raw data for one of the PVA-IDA and PVA-IDA-Cu^{+2} experiments are shown in the inset in Figure 6. Based on the disparity in the relative slopes this data demonstrates that the rate of transport of histidine through the PVA-IDA-Cu^{+2} membranes was greater than in the PVA-IDA membrane. Mass transport coefficients in the presence and absence of divalent copper were determined to be 6.3 x 10^{-5} cm²/s versus 3.7 x 10^{-5} cm²/s. Hence, the mass transfer coefficient of histidine transport is increased two-fold.

Table II summarizes the results obtained for histidine transport in PVA-IDA membranes charged with either Cu^{+2}, Ni^{+2}, CO^{+2} or Zn^{+2}. Divalent copper and cobalt increased the mass transfer coefficient approximately two-fold relative to PVA-IDA, whereas nickel and zinc showed no major effect on histidine transport. Water content (mass %) and total metal loading (mM per g dry weight) were approximately the same for each membrane type. Therefore, differences in mass

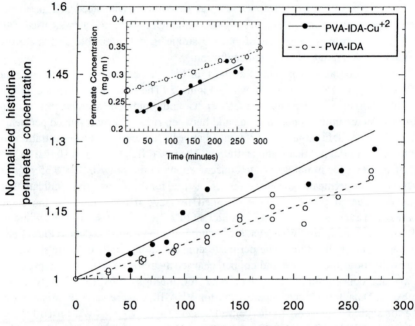

Figure 5. Normalized histidine permeate concentration for three diffusion experiments in three PVA-IDA membranes. The inset is the time dependent histidine permeate concentration for the diffusion experiments. The experiments were performed with a pH of 6.5 in the feed and permeate sides of the membrane and with an initial feed concentration of 1.5 mg/ml.

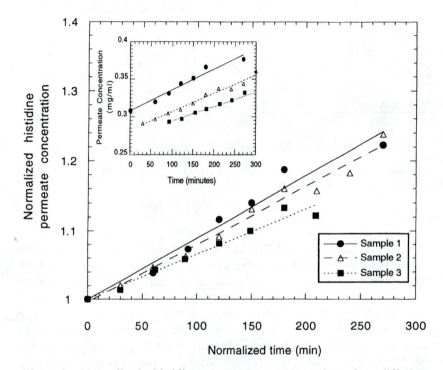

Figure 6. Normalized histidine permeate concentration for diffusion experiments in the PVA-IDA and PVA-IDA-Cu^{+2} membranes. The inset is the time dependent histidine permeate concentration for the diffusion experiments. The experiments were performed with a pH of 6.5 in the feed and permeate sides of the membrane and with an initial feed concentration of 1.5 mg/ml.

transfer coefficients were attributed to differences in interactions between the chelated metal and histidine, and not due to differences in ligand density or in diffusivity.

Table II. Characteristics of metal affinity membranes.

Membrane	Water content (wt%)[a]	Mass transfer coefficient[b] (10^{-5} cm^2/s)	Facilitation factor[c]
PVA-IDA	NA	3.7	NA
PVA-IDA-Cu^{+2}	0.87	6.3	1.7
PVA-IDA-Ni^{+2}	0.87	3.5	0.95
PVA-IDA-Co^{+2}	0.88	7.4	2.0
PVA-IDA-Zn^{+2}	0.88	3.6	0.97

[a] Average values wt% water for several films with a standard deviation of less than ± 0.02.
[b] Mass transfer coefficients reported in cm^2/s with a standard deviation of less than ±0.90 cm^2/s.
[c] Facilitation Factor defined as ratio of MTC for PVA-IDA-Me^{+2} to MTC for PVA-IDA.

Conclusions

A preliminary investigation of FTMAM indicated that chelated divalent metals within gel membranes can increase the rate of mass transfer of histidine. The choice of divalent metal had an impact on the increase in mass transfer coefficient for histidine in a similar fashion to that encountered with other immobilized metal affinity systems. Therefore, FTMAM warrants further investigation as an aqueous based recovery scheme for histidine, or other amino acids that have weak metal affinity (e.g. tryptophan). Interestingly, FTMAM could aid in the minimization of solvent use for amino acid recovery by providing a concentrated feed for the extraction step. Methods of peptide or protein recovery could also be developed based on FTMAM; however, it is important to note that the mass transport of a species larger than a single amino acid will be considerably slower. The anticipated decrease in this rate will therefore require an efficient method of thin film formation to combat this drawback.

Future research includes an investigation of the effect of pH gradient upon the rate of mass transfer in these membranes with respect to metal ion variations. Peptide and protein recovery via FTMAM will also be examined using materials of varying histidine content. Finally, this system may provide a convenient means by which controlled release may be further studied, since the method by which material is retained within the matrix may easily be tuned to achieve release rates that differ significantly.

Acknowledgements

The support of the National Science Foundation through the Presidential Faculty Fellows Program (CTS-9553267) in funding this project is gratefully acknowledged. Also, the authors are grateful to the National Science Foundation for support of Felecia Nave through the Research Experience for Undergraduates Site in Separations and the University of Arkansas for providing support for Chris Oxford.

Literature Cited

1.   Briefs, K.; Kula, M. *Chem. Eng. Sci.* **1992**, *47*, 141.
2.   Porath, J.; Carlsson, J.; Olsson I.; Belfrage, G. *Nature* **1975**, *258*, 598.
3.   Hochuli, E.; Dobeli H.; Schacher, A. *J. Chroma.* **1987**, *411*, 177.
4.   Hochuli, E.; Bannwarth, W.; Dobeli, H.; Gentz, R.; Stuber, D. *Bio./Techn.* **1988**, *6*, 1321.
5.   Hochuli, E. *J. Chroma.* **1988**, *444*, 293.
6.   Arnold, F. *Bio/Technology* **1991**, *9*, 151.
7.   Hemdan, E.; Porath, J. *J. Chroma.* **1985**, 323, 255.
8.   Hemdan, E., Zhao, Y.; Sulkowski, E.; Porath, J. *Proc. Natl. Acad Sci. USA* **1989**, *86*, 1811.
9.   Yip, T.; Nakagawa, Y.; Porath, J. *Anal. Biochem.* **1989**, *183*, 159.
10. Zhao, Y.; Sulkowski, E.; Porath, J. *Euro. J. Biochem.* **1991**, *202*, 1115.
11. Porath, J. *Biotech. Prog.* **1987**, *3*, 14.
12. El Rassi, Z.; Horvath, C. *J. Chromat.* **1986**, *359*, 241.
13. Beitle, R.; Ataai, M. *Biotech.Prog.* **1993**, *9*, 64.
14. Hemdan, E.; Porath, J. *J. Chromat.* **1985**, *323*, 265.
15. Hutchens, T; Li, C. *J. Molecular Recognition*, **1988**, *1*, 80.
16. Hutchens, T.; Yip, T. *J. Chromat.* **1990**, *500*, 531.
17. Sulkowski, E. *Trends in Biotech.* **1985** 3, 1.
18. Beitle, R.; Ataai, M. Immobilized Metal Affinity Chromatography and Related Techniques, in *New Developments in Bioseparation,* M. Ataai and S. Sikdar, Eds. **1992**, American Institute of Chemical Engineers: New York.
19. Sulkowski, E. *BioEssays* **1989**, 10, 170.
20. Noble, R.; Koval, C.; Pellegrino, J. *Chem. Eng. Prog.* **1989**, *3*, 58.
21. Noble, R. *J. of Faraday Transactions* **1991**, *87*, 2089.
22. Li, R.; Barbari, T. *J. Membr. Sci.* **1994**, *88*, 115.
23. Chai, S., Beitle, R.; Coleman, M. *Inter. J. Biochroma.* **1996**, *2*, 125.

Chapter 17

Membranes from Modified Poly(1-trimethylsilyl-1-propyne)

C. Merano[1], M. Andreotti[2], A. Turturro[1], F. Vigo[1], and G. Costa[2]

[1]Dipartimento di Chimica e Chimica Industriale, University of Genova, via Dodecaneso 31–16146 Genova, Italy
[2]Istituto di Studi Chimico-Fisici di Macromolecole Sintetiche e Naturali, IMAG-CNR, via De Marini 6–16149 Genova, Italy

The application of poly(1-trimethylsilyl-1-propyne) [PTMSP] in membrane-based gas separation processes is limited by its low selectivity and very fast physical aging which negatively affect the performance of this extraordinary polymer. To overcome these drawbacks several methods have been proposed and among them copolymerization seems to be one of the most promising. Novel copolymers with different compositions, obtained from monomers of 1-trimethylsilyl-1-propyne [TMSP] and 1-phenyl-2-trimethylsilylacetylene [$PhC\equiv CSiMe_3$] or 1-trimethylsilyl-1-pentyne [$n\text{-}PrC\equiv CSiMe_3$] were synthesized, characterized, and isotropic films were made from solution by evaporation of the solvent under controlled conditions. Permeability and selectivity measurements, carried out with air, demonstrated that the permeation properties were stable even after a week of testing, indicating that the introduction of comonomer units to the PTMSP mainchain prevents physical aging to occur. When $PhC\equiv CSiMe_3$ was used as a comonomer, the gas permeability was always lower than that of the PTMSP homopolymer and the selectivity depended on the copolymer composition. Membranes made from TMSP/n-$PrC\equiv CSiMe_3$ copolymers exhibited higher permeabilities than PTMSP and their selectivity was essentially independent of the composition.

Poly(1-trimethylsilyl-1-propyne) [PTMSP] has drawn great attention in the last decade due to its extremely high gas permeability, which, despite its glassy state, is an order of magnitude higher than that of polydimethylsiloxane, the most permeable rubber known (1). Moreover, the variation of its permeability as a function of temperature shows an opposite trend as compared to that of conventional glassy polymers, such as polysulfone, that is, by decreasing the temperature, the permeability coefficient of PTMSP increases (2). A loose packing of the chains, due to the steric hindrance of the substituents, together with a rigid backbone induced by the presence of the alternating double bonds along the main chain with both cis- and trans-configuration seem to be responsible for an extremely high free volume and, consequently, the peculiar transport properties of PTMSP (3-6).

However, when a PTMSP membrane is stored at ambient conditions or used continuously in gas separations with permanent gases or in pervaporation processes for the separation of liquids, its permeability decreases significantly over time (2,3,7-10).

The physical aging in PTMSP and other substituted polyacetylene films, such as poly(*tert*-butylacetylene) (*11*), seems to be responsible for this reduction in permeability. Many efforts have been made to overcome this drawback and different approaches such as UV irradiation (*12,13*), copolymerization (*14,16*), chemical and physical modifications (*17-21*), plasma polymerization (*22*), and grafting of a proper monomer (*23,24*) have been suggested. In these cases, an improvement in selectivity and a reduction of the gas permeability combined with a stable long-term membrane performance were achieved.

In the present work, the synthesis of some novel copolymers, obtained from TMSP and 1-trimethylsilyl-1-pentyne [n-PrC≡CSiMe$_3$] or 1-phenyl-2-trimethylsilylacetylene [PhC≡CSiMe$_3$], and the transport properties of membranes composed of these polymers are reported.

Experimental

Materials. The monomers TMSP, n-PrC≡CSiMe$_3$, and PhC≡CSiMe$_3$ (from Aldrich) were distilled over CaH$_2$ and stored under argon. All solvents were dried under reflux over LiAlH$_4$, distilled, and kept under argon atmosphere. The catalysts, TaCl$_5$ and NbCl$_5$, and cocatalysts, Ph$_3$Bi (99%) and Bu$_4$Sn (98%), were purchased from Aldrich-Fluka and were used without further purification.

Polymer Synthesis. Homopolymers and copolymers were synthesized in toluene following the procedures described in the literature (*25,26*). The polymerization was carried out at 80°C over a period of 24 hours. Conversion was determined by measuring the initial and final monomer concentrations by gas chromatography (GC) using chlorobenzene as an internal standard. The polymer yield, as a methanol-insoluble fraction, was determined by gravimetry, after the polymers were dissolved in toluene and precipitated twice in a large excess of methanol.

Polymer Characterization. Solubility tests in different solvents were made in order to ascertain the formation of copolymers. Molecular characterization of the polymers was carried out by intrinsic viscosity measurements in toluene and by size-exclusion-chromatography (SEC; Perkin-Elmer Diode Array Detector), using $CHCl_3$ as the eluent with a flow rate of 1ml/min and a set of μ-styragel columns of 10^3, 10^4, and 10^5 Å.

The thermal properties of all polymers were evaluated by thermal gravimetric analysis (Perkin Elmer TGA-7) and differential scanning calorimetry (DSC-7 Perkin Elmer). A heating rate of 10 K/min was always used. Raman spectra, both on monomers and polymers, were recorded with a Bruker RSS 100 spectrophotometer.

Membrane Preparation. Membranes were prepared from toluene solutions, typically 0.7-1.5% by weight of polymer, by evaporation at room temperature under controlled conditions over a period of 7 days and dried at 60°C under vacuum overnight. The thickness of the membranes ranged from 25 to 50 μm. Only freshly prepared membranes were used for permeability measurements.

Permeability and Selectivity Measurements. Membranes were tested by permeability measurements using gas chromatography-grade air. The apparatus used for the tests was assembled in our laboratories and described in a previous paper (24). Membranes were tested at a feed pressure of 1 MPa and a permeate pressure of 0.1 MPa. Each experiment was carried out for 140 to 180 hours in order to check the long-term stability of the membranes.

Results and Discussion

A preliminary investigation on the polymerization of n-PrC≡CSiMe$_3$ and PhC≡CSiMe$_3$ (Figure 1) was carried out to verify the possibility to obtain homopolymers. The reactions were performed at 80°C in toluene for 24 hours by using different catalysts, such as NbCl$_5$ and TaCl$_5$, with or without a cocatalyst (Ph$_3$Bi, Bu$_4$Sn). None of the reactions resulted in high molecular weight polymers, that is, methanol insoluble products, although in some cases monomer conversions up to 70% were obtained. This result suggests that in our experimental conditions the two monomers were essentially unable to form homopolymers.

Copolymers of different compositions were obtained by polymerizing TMSP with various amounts of n-PrC≡CSiMe$_3$ or PhC≡CSiMe$_3$. The results are summarized in Tables I and II, where the conversion, polymer yields, and inherent viscosities are given together with the molar ratios of the initial monomers; a molar ratio of [TMSP]/[Cat] = 50 was used in all experiments.

The most efficient catalyst system was an equimolar mixture of TaCl$_5$ and Ph$_3$Bi, which generally gave nearly 100% conversion of TMSP and lower values for the comonomers (between 30 and 80 wt%). This means that copolymer compositions differed from those of the corresponding monomer mixtures, which were used to identify the copolymers. Currently, characterizations by NMR spectroscopy are in progress to determine the real compositions of the copolymers and their chemical structures. When only $TaCl_5$ was used, conversions were fairly high, but the methanol insoluble fraction was generally rather low, indicating that high molecular weight polymers can only be obtained when a cocatalyst is present. Polymers made from $NbCl_5$ always gave very low conversions and yields.

Table I. Copolymers of TMSP and n-PrC≡CSiMe$_3$. [catalyst]/[co-catalyst] = 1; $[TMSP]_0/[TaCl_5] = 50$; solvent: toluene; T = 80°C; t = 24 h.

[TMSP]$_0$/[COM]$_0$	Catalyst	Conversion (%)		Yield (wt%)	η inh[a] (dl/g)
		TMSP	COM		
50:50	TaCl$_5$	84	28	24	-
50:50	NbCl$_5$	-	-	-	-
80:20	TaCl$_5$-Ph$_3$Bi	90	71	80	7.3
80:20	TaCl$_5$-Bu$_4$Sn	50	5	-	-
80:20	NbCl$_5$-Ph$_3$Bi	15	10	-	-
70:30	TaCl$_5$-Ph$_3$Bi	96	33	63	6.8
60:40	TaCl$_5$-Ph$_3$Bi	94	10	51	-
50:50	TaCl$_5$-Ph$_3$Bi	96	33	53	4.3

a) determined at c = 0.1 g/dl in toluene at 30°C.

Table II. Copolymers of TMSP and PhC≡CSiMe$_3$. [catalyst]/[co-catalyst] = 1; $[TMSP]_0/[TaCl_5] = 50$; solvent: toluene; T = 80°C; t = 24 h.

[TMSP]$_0$/[COM]$_0$	Catalyst	Conversion (%)		Yield (wt%)	η inh[a] (dl/g)
		TMSP	COM		
50:50	TaCl$_5$	100	36	51	-
50:50	NbCl$_5$	40	-	-	-
80:20	TaCl$_5$-Ph$_3$Bi	100	64	89	9.2
70:30	TaCl$_5$-Ph$_3$Bi	100	63	60	8.7
60:40	TaCl$_5$-Ph$_3$Bi	95	80	58	8.9
50:50	TaCl$_5$-Ph$_3$Bi	100	56	41	.
80:20	NbCl$_5$-Ph$_3$Bi	60	20	-	-

a) determined at c = 0.1 g/dl in toluene at 30°C.

Solubility tests and characterization by Raman spectroscopy, carried out on all polymers, confirmed that copolymers were obtained, although their composition was not the same as for the initial reaction mixture. The Raman spectrum of one of the copolymer samples [(TMSP/PhC≡CSiMe$_3$ (70/30)] together with that of the PTMSP homopolymer are shown in Figure 2; the presence of the aromatic substituent in the copolymer is confirmed by the absorption bands at 3,060 cm^{-1}.

256

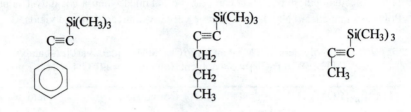

1-phenyl-2-trimethylsilylacetylene   1-trimethylsilyl-1-pentyne   1-trimethylsilyl-1-propyne

(PhC≡CSiMe₃)   (n-PrC≡CSiMe₃)   (TMSP)

Figure 1.   Structure of di-substituted acetylene-based monomers.

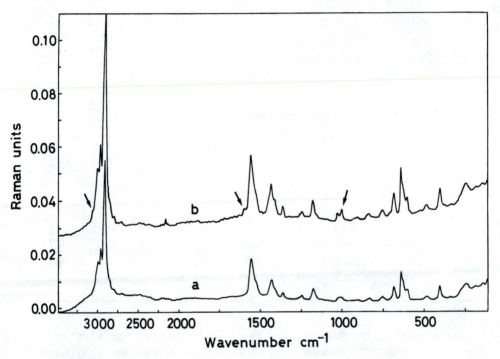

Figure 2.   Raman spectra of (a) PTMSP homopolymer [catalyst: TaCl₅], and (b) TMSP/PhC≡CSiMe₃ 70/30 copolymer [catalyst:TaCl₅/Ph₃Bi].

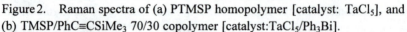

(aromatic CH), 1,600 cm^{-1} (anti-symmetric stretching of the ring), and 1,000 cm^{-1} (phenyl breathing).

Furthermore, the thermal stability of the copolymers was always higher than that of PTMSP, as shown in Figure 3. Degradation temperatures, T_d, of 330, 370, and 390°C were found for PTMSP, TMSP/PhC≡CSiMe$_3$, and TMSP/n-PrC≡CSiMe$_3$ (50/50 moles) copolymers, respectively.

Molecular characterization was performed by SEC on TMSP/n-PrC≡CSiMe$_3$ and TMSP/PhC≡CSiMe$_3$ copolymers obtained from 50/50 mixtures of the co-monomers. Wide molecular weight distributions were found. The M_{peak} values, based on a polystyrene calibration, ranged from 200,000 g/mol for TMSP/n-PrC≡CSiMe$_3$ copolymers to 500,000 g/mol for the TMSP/PhC≡CSiMe$_3$ systems. However, viscosity data and molecular weights cannot be related straightforwardly due to the limited number of samples examined so far.

The DSC analysis did not show any glass transition temperature prior to polymer degradation or any exothermic phenomenon associated with chain packing, as previously observed for PTBA and PTMSP samples (*11,27*).

The transport properties of the membranes were investigated by permeability measurements, reported in Barrer units (1 Barrer =10^{-10} cm^3(STP)·cm/cm^2·s·cm·Hg). The membrane selectivity, α, was calculated from the ratio of the O_2 and N_2 permeabilities. Permeation tests carried out on membranes prepared from PTMSP and from TMSP/PhC≡CSiMe$_3$ copolymers are reported and compared as a function of time in Figures 4 and 5. The slight increase in permeability of copolymers which occurred after a few hours of testing (Figure 4) can be ascribed to moistening of the membranes which occurred after placing them into the test cell. This effect can hardly be detected in PTMSP due to the permeability decrease which occurred rapidly with time in this polymer. The permeability trend of copolymers is in accordance with the observed decrease in selectivity (Figure 5). However, after a few hours of operation, the permeability and selectivity of the copolymers did not change anymore until the end of the experiments (140 hours). On the contrary, as is well known (*2,3,7-10*), and shown in Figure 4, the performance of PTMSP changed dramatically during testing. Moreover, it appears that the characteristics of the membranes depended on the copolymer composition; the permeability increased and selectivity decreased by increasing the TMSP content in the copolymer and, after 140 hours, membranes made from TMSP/PhC≡CSiMe$_3$ (50/50) copolymer had practically the same permeability as PTMSP.

The introduction of a co-monomer unit bearing a phenyl group provides membranes with a lower but more stable permeability, which suggests that molecular motions might be restricted and/or the excess free volume of the copolymers did not decrease during aging. When the methyl group in PTMSP, which has a lower critical surface tension compared to a phenyl ring (*28*), was replaced by an aromatic moiety, an increase in chain packing occurred. In other words, the strong chain-chain interactions combined with the planar structure of the aromatic ring allowed a tight chain packing, and, thus, a reduction in the excess free volume (*5*).

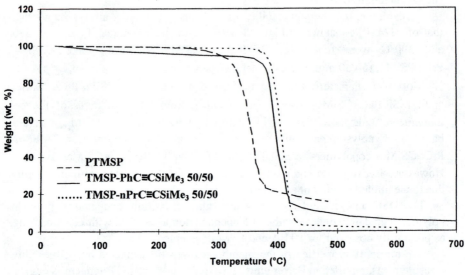

Figure 3. TGA traces of PTMSP [catalyst:TaCl$_5$] (---), TMSP/PhC≡CSiMe$_3$ 50/50 [catalyst: TaCl$_5$/Ph$_3$Bi] (—), and TMSP/n-PrC≡CSiMe$_3$ 50/50 [catalyst: TaCl$_5$/Ph$_3$Bi] (.....) copolymers.

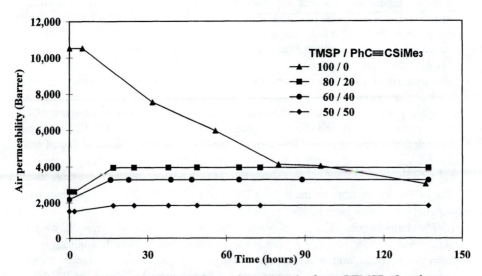

Figure 4. Air permeability of membranes made from PTMSP [catalyst: TaCl$_5$] and TMSP/PhC≡CSiMe$_3$ copolymers [catalyst: TaCl$_5$/Ph$_3$Bi] as a function of testing time.

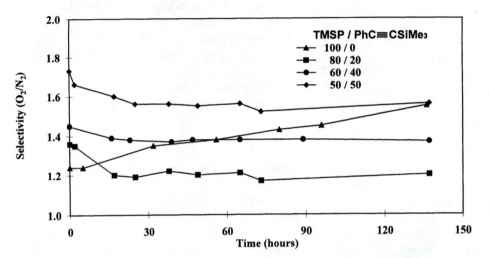

Figure 5.   Oxygen/nitrogen selectivity of membranes made from PTMSP [catalyst TaCl$_5$] and TMSP/PhC≡CSiMe$_3$ copolymers [catalyst TaCl$_5$/Ph$_3$Bi] as a function of testing time.

Similar results were recently published for membranes prepared from copolymers of TMSP and 1-phenyl-1-propyne [PP] and blends of the corresponding homopolymers (29). In this case, membranes with a high stability of gas permeability were obtained by using either copolymers of TMSP with small amounts of PP monomer or blends of PTMSP and PPP, although the permeability values were always lower than those of PTMSP.

Very interesting results are shown in Figures 6 and 7 for $TMSP/n\text{-}PrC\equiv CSiMe_3$ copolymer membranes. The air permeability of these copolymers was practically independent of the composition and was always much higher than that of PTMSP (13,000 and 3,000 Barrers, respectively, after 140 hours). On the other hand, the selectivity values of the copolymers were rather low, even lower than that of the PTMSP homopolymer. However, the most important feature is the constant value of the performance for all the copolymers membranes, at least for time period during which the measurements were carried out (0-140 hours).

The difference in the chemical structure of TMSP and $n\text{-}PrC\equiv CSiMe_3$ was only one substituent; the methyl group was replaced by a larger n-propyl side-group. It is known that longer n-alkyl side chains have higher flexibility and can increase the free volume (5); consequently, the high permeability of the copolymers of $TMSP/n\text{-}PrC\equiv CSiMe_3$ can be ascribed to a higher free volume content than that of PTMSP.

Conclusions

The introduction of a phenyl- or a propyl-group can either inhibit the physical aging of PTMSP-based copolymer membranes or drastically reduce the kinetics of the packing phenomenon (5), due to the simultaneous presence of the co-monomers of the very bulky trimethylsilyl-group and a phenyl- or n-propyl-substituent. Spectroscopic investigations are currently in progress to obtain additional insights on the copolymer composition, in order to better correlate the performance of the samples to their structure. Further experimental work is under way to synthesize novel copolymers from TMSP and other silicon-containing comonomers bearing bulky substituents to improve our knowledge of structure-property relationships in this very interesting class of polyacetylenes.

Acknowledgements.

The authors greatfully acknowledge Mr. V. Trefiletti for carrying out the thermal characterizations. This work was partially supported by MURST (60%) and by Progetto Strategico CNR, Italy.

Literature Cited

1. Masuda, T.; Isobe, E.; Higashimura, T. *J. Am.Chem. Soc.* **1983**, *105*, 7473.
2. Takada, K.; Matsuya, H.; Masuda, T.; Higashimura, T. *J. Appl. Polym. Sci.* **1985**, *30*, 1605.
3. Masuda, T.; Higashimura, T. *Adv. Polym. Sci.* **1987**, *81*, 121.

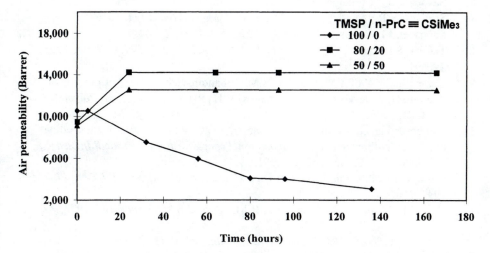

Figure 6. Air permeability of membranes made from PTMSP [catalyst: TaCl₅] and TMSP/n-PrC≡CSiMe₃ copolymers [catalyst: TaCl₅/Ph₃Bi] as a function of testing time.

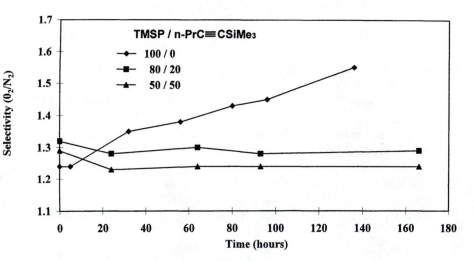

Figure 7. Oxygen/nitrogen selectivity of membranes made from PTMSP [catalyst: TaCl₅] and TMSP/n-PrC≡CSiMe₃ copolymers [catalyst: TaCl₅/Ph₃Bi] as a function of testing time.

4. Clough, S.B.; Sun, X.F.; Tripathy, S.K.; Baker, G.L. *Macromolecules* **1991**, *24*, 4264.

5. Savoca, A.C.; Surnamer, A.D.; Tien, C.-F. *Macromolecules* **1993**, *26*, 6211.

6. Srinivasan, R.; Auvil, S.R.; Burban, P.M. *J. Membr. Sci.* **1994**, *86*, 67.

7. Langsam, M.; Robeson, L.M. *Polym. Eng. Sci.* **1989**, *29*, 44.

8. Tasaka, S.; Inagaki, N.; Igawa, M. *J. Polym. Sci., Part B, Polym. Phys.* **1991**, *29*, 691.

9. Yampol'skii, Yu.P.; Shishatskii, S.M.; Shantorovich, V.P.; Antipov, E.M.; Zuzmin, N.N.; Rykov, S.V.; Khodjaeva, V.L.; Plate, N.A. *J. Appl. Polym. Sci.* **1993**, *48*, 1935.

10. Camera-Roda, G.; Bottino, A.; Capannelli, G.; Costa, G.; Sarti, G. Proceedings of 5th Int.Conference on *"Pervaporation Processes in the Chemical Industry"*, R.Bakish (Ed.), Englewood NJ, 1991.

11. Costa, G.; Chikhaoui, S.; Turturro, A. Carpaneto, L. *Macromol. Chem. Phys.* **1997**, *198*, 239.

12. Hsu, K.K.; Nataraj, S.; Thorogood, R.M.; Puri, P.S. *J. Membr. Sci.* **1993**, *79*, 1.

13. Chen, G.; Griesser, H.J.; Mau, A.H. *J. Membr. Sci.* **1993**, *82*, 99.

14. Hamano, T.; Masuda, T.; Higashimura, T. *J. Polym. Sci., Part A, Polym. Chem.* **1988**, *26*, 2603.

15. Goudong, Z.; Nakagawa, T.; Higuchi, A.; Nagai, K. *Chin. J. Polym. Sci.* **1992**, *10*, 203.

16. Nagai, K.; Higuchi, A.; Nakagawa, T. *J. Polym. Sci., Part B, Polym. Phys.* **1995**, *33*, 289.

17. Nagase, Y.; Ishihara, K.; Matsui, K. *J. Polym. Sci., Part B, Polym. Phys.* **1990**, *28*, 377.

18. Nagase, Y.; Ueda, T.; Matsui, K.; Uchikura, M. *J. Polym. Sci., Part B, Polym. Phys.* **1991**, *29*, 171.

19. Nagase, Y.; Sugimoto, K.; Takamura, Y.; Matsui, K. *J. Appl. Polym. Sci.* **1991**, *43*, 1227.

20. Nagase, Y.; Takamura Y.; Matsui, K. *J. Appl. Polym. Sci.* **1991**, *42*, 185.

21. Nakagawa, T.; Fujisaki, S.; Nakano, H.; Higuchi, A. *J. Membr. Sci.* **1994** , *94*, 183.

22. Lin, X.; Chen, J.; Xu, J. *J. Membr. Sci.* **1994**, *90*, 81.

23. Yang, J.-S.; Hsiue, G.-H. *J. Membr. Sci.* **1996**, *111*, 27.

24. Vigo, F.; Traverso, M.; Uliana, C.; Costa, G. *Sep. Sci. and Techn.* **1996**, *31*, 349.

25. Masuda, T.; Isobe, E.; Higashimura, T. *Macromolecules* **1985**, *18*, 841.

26. Masuda, T.; Isobe, E.; Hamano, T.; Higashimura, T. *Macromolecules* **1986**, *19*, 2448.

27. Costa, G. unpublished results.

28. Burton, A.F.M. *"Handbook of Solubility Parameters and other Cohesion Parameters"*, CRC Press Inc. Boca Raton (Florida) 1983.

29. Nagai, K.; Mori, M.; Watanabe, T.; Nakagawa, T. *J. Polym. Sci., Part B, Polym. Phys.* **1997**, *35*, 119.

Chapter 18

Pervaporation Properties of Surface-Modified Poly[(1-trimethylsilyl-1-propyne] Membranes

T. Uragami, T. Doi, and Tadashi Miyata

Chemical Branch, Faculty of Engineering and High-Technology Research Center, Kansai University, Suita, Osaka 564–8680, Japan

Poly[1-(trimethylsilyl)-1-propyne] (PTMSP) membranes were surface-modified using polymer additives and their pervaporation properties for an aqueous ethanol solution were investigated. The polymer additives, PFA-*g*-PDMS and PMMA-*g*-PDMS, were synthesized by graft-co-polymerization of fluoroacrylate (PFA) and oligo dimethylsiloxane (DMS), and methylmethacrylate (MMA), respectively. The contact angle of water on the surface of the air-side of the surface-modified PTMSP membranes increased by the addition of PFA-*g*-PDMS and PMMA-*g*-PDMS. These results suggest that the surface-modified PTMSP membranes were more hydrophobic. Also, X-ray photoelectron spectroscopy (XPS) measurements supported that the polymer additives were concentrated at the surface of the air-side of the modified PTMSP membranes. The ethanol/water selectivity of the PTMSP membrane modified with PFA-*g*-PDMS increased significantly without a large decrease in flux compared with that of the PTMSP membrane. On the other hand, after modification of PTMSP with PMMA-*g*-PDMS both the flux and ethanol/water selectivity had a maximum at a certain amount of PMMA-*g*-PDMS. The ethanol/water selectivity of the modified membrane was higher than that of the PTMSP membrane. High ethanol/water selectivity of the surface-modified PTMSP membranes depended on the solubility selectivity of the permeating species.

It is well known that a poly[1-(trimetylsilyl)-1-propyne] (PTMSP) membrane has a high ethanol/water selectivity and a high flux for aqueous ethanol solutions in pervaporation (PV) *(1,2)*. The high ethanol/water selectivity results from the high solubility of ethanol in the PTMSP membrane. Recent work has focused on improvement of the ethanol/water selectivity for aqueous ethanol solutions of PTMSP membranes in pervaporation applications. The ethanol/water selectivity of PTMSP membranes was improved by grafting dimethylsiloxane (*3,4*) and acrylic acid (*5*), introducing alkylsilyl (*6*) and fluoroalkyl groups on the PTMSP backbone (*7*), and blending dimethylsiloxane oligomer in PTMSP; however, the fluxes of the modified PTMSP membranes were low.

A higher ethanol/water selectivity for a modified PTMSP membrane cannot be achieved on the basis of differences in diffusivity, because ethanol is larger than water. Therefore, modifications of the membrane surface to enhance the solubility of ethanol in the membrane relative to that of water are required. As the solubility of ethanol in the membrane is improved, however, the membrane surface swells, and consequently, the solubility of water in the membrane surface also increases. Accordingly, it is very important to lower the affinity of water molecules for the membrane. There are many techniques for surface modification of polymer membranes. One of these techniques is the addition of a polymer additive to the casting solution, which readily modifies the membrane surface.

In this study, polymer additives consisting of a hydrophobic part to repel water at the membrane surface and an interacting part for the membrane matrix were synthesized for the purpose of surface modification of the PTMSP membrane. The relationship between the surface properties of the PTMSP membranes modified with the polymer additives and their permeation properties in pervaporation of aqueous ethanol solutions is discussed in detail.

Experimental

Materials. 1-(Trimethylsilyl)-1-propyne, purchased from Hüls America, Inc., was distilled twice over calcium hydride under nitrogen atmosphere. The polymerization catalyst, tantalum pentachloride ($TaCl_5$) (Wako Pure Chemical Industries, Ltd.) was used without further purification. The polymerization solvent, toluene, was washed with 5 wt.% H_2SO_4, 10 wt.% NaOH solution, and water, dried over calcium chloride for 2 days, and then distilled twice over calcium hydride under nitrogen atmosphere. Heptadecafluoro decylacrylate (FA) and oligodimethylsiloxane (DMS) macro-monomer, which has 26 units of DMS, were supplied by Dainippon Ink Chemicals, and Toray Dow Corning Silicone Co. Ltd., respectively. Methyl methacrylate (MMA) was purified by distillation under reduced pressure under nitrogen atmosphere. 2,2'-Azobis(isobutyronitril) (AIBN) was recrystallized from benzene solution and used as an initiator. The other solvents and reagents were of analytical grade obtained from standard commercial sources and used without further purification.

Synthesis of Poly[1-(trimethylsilyl)-1-propyne]. Poly[1-(trimethylsilyl)-1-propyne] (PTMSP) was prepared by the method developed by Masuda *et al.* (*8*). 1-trimethylsilyl-1-propyne (TMSP) was polymerized in toluene with $TaCl_5$ as a catalyst at 80°C for 3 h. The polymer was reprecipitated several times from a toluene solution into an excess amount of methanol, and was dried *in vacuo* at 80°C for 24 h before use. Number-average and weight-average molecular weights of PTMSP, determined by gel permeation chromatography (GPC) (Waters Associate Inc., R-400) equipped with a TSK-GEL column (Tosoh Co. Ltd; G2000HXL, G3000HXL, G5000HXL), and ultraviolet spectrophotometry (Shimadzu Co. Ltd; Spd-2A), were 6×10^5 g/mol and 4×10^5 g/mol, respectively.

Synthesis of Polymer Additives. FA and DMS macromonomer, and MMA and DMS were dissolved in toluene with AIBN (1.0 wt.% relative to the monomers) and were copolymerized at 60°C for 6 h under nitrogen atmosphere in a glass tube, respectively, as shown in Figure 1. The resulting PFA-*g*-PDMS and PMMA-*g*-PDMS polymer additives were isolated by precipitation with ethanol and a 1:2 (w/w) mixture of *n*-hexane and ethanol, respectively. Thereafter, the additives were purified by reprecipitation from toluene solution into ethanol and a 1:2 (w/w) mixture of *n*-hexane and ethanol, respectively, and dried at 40°C *in vacuo*.

Number-average and weight-average molecular weights of PFA-*g*-PDMS and PMMA-*g*-PDMS were determined by gel permeation chromatography (GPC) and ultraviolet spectrophotometry. Tetrahydrofuran was used as an eluent and the calibration was made with a polystyrene standard. The compositions of PFA-*g*-PDMS and PMMA-*g*-PDMS were determined from 400 MHz ^{1}H nuclear magnetic resonance (NMR) (JEOL; GSX-400) spectra by measuring the integrals of the peaks assigned to methylene protons (4.2 ppm) of the FA and DMS protons (0 ppm) of the DMS macromonomer, and methyl protons (3.5 ppm) of the MMA and DMS protons (0 ppm) of the DMS macromonomer, respectively, after the purified copolymer was dissolved in chloroform-*d* containing 1 vol.% tetramethylsilane (TMS). Characterizations of the PFA-*g*-PDMS and PMMA-*g*-PDMS copolymers are summarized in Table I.

Table I. Properties of polymer additives used for surface modification of PTMSP membranes.

Polymer additive	m:n	Mn (g/mol)	Mw (g/mol)	Mw/Mn
PFA-*g*-PDMS	0.64:0.36	2.0×10^5	5.9×10^4	3.3
PMMA-*g*-PDMS	0.67:0.33	2.0×10^5	7.5×10^4	2.6

Preparation of Membranes. Desired amounts of polymer additive (PFA-*g*-PDMS and PMMA-*g*-PDMS) were added to a 2 wt.% PTMSP solution in toluene at 25°C. PTMSP membranes modified with PFA-*g*-PDMS or PMMA-*g*-PDMS were made by pouring the solutions onto a rimmed glass plate and allowing the casting solvent (toluene) to evaporate at 25°C. The modified PTMSP membranes were immersed in methanol for 3 days to remove the residual solvent from the membrane matrix and then kept in pure methanol. The modified PTMSP membranes were dried completely under reduced pressure and then tested in pervaporation experiments. The resulting membranes were transparent and their thickness was about 20 μm.

Contact Angle Measurements. The contact angles of water on the surface of the air- and glass-plate-sides of the PTMSP membrane and PTMSP membranes modified with PFA-g-PDMS and PMMA-g-PDMS were measured using a contact angle meter (Erma, Model G-I) at 25°C. The contact angle, θ, was determined from equation 1,

$$\theta = \cos^{-1}\{(\cos\theta a + \cos\theta r) / 2\} \tag{1}$$

where θa and θr are the advancing contact angle and receding contact angle, respectively.

X-ray Photoelectoron Spectroscopy (XPS). Elemental analyses of the surfaces on the air- and glass-plate-sides of the PTMSP and surface-modified PTMSP membranes were measured by X-ray photoelectoron spectroscopy (XPS, Shimadzu ESCA-750).

Pervaporation Experiments. Pervaporation experiments were carried out using an apparatus reported in previous papers (9-12) at a temperature of 40°C and a permeate pressure of 1×10^{-2} Torr. The effective membrane area was 13.8 cm^2. An aqueous solution of 10 wt.% ethanol was used as the feed solution. The compositions of the feed and permeate were determined by gas chromatography (Shimadzu GC-9A). The pervaporation fluxes were determined from the weight of the permeate collected in a cold trap, permeation time, and effective membrane area.

Characterization of Membranes. The contact angles of water on the surfaces of the air- and glass-plate-sides of surface-modified PTMSP membranes are shown in Figures 2 and 3 as a function of the amount of the polymer additive. Using PFA-g-PDMS as a polymer additive, the contact angle of water on the surface of the air-side increased with an increase of the PFA-g-PDMS content in the PTMSP membrane. However, the contact angle on the glass-plate-side did not change significantly, as shown in Figure 2. The increase in the contact angle to water suggests that PFA-g-PDMS was localized on the air-side of the membrane. Hence, the surface of the modified PTMSP membrane became more hydrophobic and,

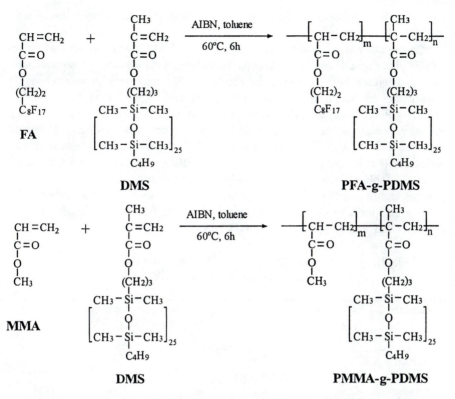

Figure 1. Synthesis of PFA-*g*-PDMS and PMMA-*g*-PDMS used as additives for modification of PTMSP membranes.

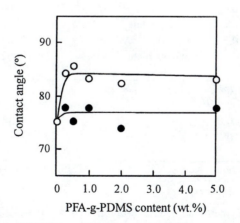

Figure 2. Effect of the PFA-*g*-PDMS content on the contact angle to water on the surface of air-side (○) and glass-plate-side (●) of surface-modified PTMSP membranes.

consequently, the surface of the air-side of the modified PTMSP membranes was water-repellent.

On the other hand, when a small amount of PMMA-*g*-PDMS was added to PTMSP, both contact angles of water on the surface of the air- and glass-plate-sides increased remarkably, as shown in Figure 3. The increases in contact angle demonstrate that both membrane surfaces became water-repellent. As the amount of PMMA-*g*-PDMS increased, the contact angles to water of both surfaces decreased and then became equal. To clarify the differences between PTMSP membranes modified with PFA-*g*-PDMS and PMMA-*g*-PDMS, elemental analyses of the surfaces of the air- and glass-plate-side of the modified membranes were performed by XPS. Table II summarizes the compositions on both surface sides of the PTMSP and PFA-*g*-PDMS- and PMMA-*g*-PDMS-modified PTMSP membranes. Table II also includes theoretical compositions of the PTMSP homopolymer and PFA-*g*-PDMS and PMMA-*g*-PDMS copolymers.

Table II. Surface compositions of PTMSP and surface-modified PTMSP membranes determined by XPS.

Sample	Additive content (wt.%)	Surface	Atomic ratio		
			F/C	O/C	Si/C
PTMSP	0	Air	-	-	0.164
	0	Glass plate	-	-	0.177
PTMSP modified with PFA-*g*-PDMS	0.1	Air	0.107	0.087	0.205
		Glass plate	0.021	0.031	0.170
	0.5	Air	0.137	0.115	0.184
		Glass plate	0.025	0.071	0.199
PTMSP modified with PMMA-*g*-PDMS	0.1	Air	-	0.085	0.200
		Glass plate	-	0.015	0.179
	0.5	Air	-	0.094	0.184
		Glass plate	-	0.029	0.161
PTMSP[a]	0	-	-	-	0.167
PFA-*g*-PDMS[a]	0	-	0.351	0.332	0.302
PMMA-*g*-PDMS[a]	0	-	-	0.397	0.355

[a]Theoretical compositions were calculated from PTMSP homopolymer and PFA-*g*-PDMS and PMMA-*g*-PDMS copolymers (Table I).

For the PTMSP membranes surface-modified with PFA-*g*-PDMS, the ratio of F/C in the membrane surface on the air-side was considerably higher than that of the membrane surface on the glass-plate-side. In addition, the F/C ratio on the membrane surface of the air-side increased with increasing PFA-*g*-PDMS content; however, the F/C ratio on the glass-plate-side did not change. A comparison of the F/C ratio of the surface-modified PTMSP with 0.1 wt.% PFA-*g*-PDMS and the theoretical ratio of F/C of PFA-*g*-PDMS suggests that PFA-*g*-PDMS occupies about 30 wt.% of the surface of PTMSP membranes.

On the other hand, the ratio of O/C on the surface of the air-side of the PTMSP membrane modified with PMMA-*g*-PDMS was higher than that on the glass-plate-side. Contact angle measurements to water on the surface of the glass-plate-side of the PTMSP membrane with additional PMMA-*g*-PDMS, however, showed that the surface of the glass-plate-side became water-repellent, as shown in Figure 3. Thus, the results of the XPS measurements are different from those of the contact angle measurements. This difference may be attributed to the fact that the contact angle measurement reflects only the surface properties of the membranes, whereas the XPS measurement provides information of the composition of the membrane for a depth of 0-100 Å. This difference has to be investigated in more detail in the future.

From the results of the contact angle and XPS measurements, the structures of surface-modified PTMSP membranes with additional PFA-*g*-PDMS and PMMA-*g*-PDMS are illustrated schematically in Figure 4. As shown in Figure 4(a), the PFA-*g*-PDMS molecules, which have a lower surface free energy than PTMSP, are localized on the surface of the PTMSP membrane on the air-side to thermodynamically stabilize the surface of the membrane. Consequently, the surface of the PTMSP membrane modified with PFA-*g*-PDMS became remarkably water-repellent. The structure of the PTMSP membrane modified with PMMA-*g*-PDMS is shown in Figure 4(b). In this case, the PMMA-*g*-PDMS molecules were also localized on the surface of the PTMSP membrane on the air-side. Furthermore, because the PMMA component in the PMMA-*g*-PDMS molecule had a lower hydrophobicity than the PTMSP molecule, the PMMA-*g*-PDMS was also localized on the surface of the glass-plate-side. However, the degree of the localization was lower than that on the air-side. Consequently, the contact angles in both membrane surfaces on the air- and glass-plate-side increased remarkably with the addition of the PMMA-g-PDMS additive.

Pervaporation Properties of Surface-Modified PTMSP Membranes. Figure 5 shows the flux and ethanol concentration in the permeate for an aqueous solution of 10 wt.% ethanol in pervaporation experiments of PTMSP membranes modified with PFA-*g*-PDMS. In all pervaporation experiments, the membrane surface of the air-side of the surface-modified PTMSP membranes was facing the feed solution. Figure 5 also includes the contact angles of the surface-modified PTMSP membranes. As can be seen from Figure 5, the flux decreased slightly but the ethanol

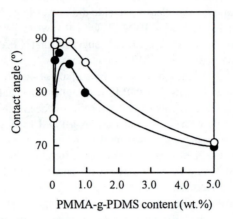

Figure 3. Effect of the PMMA-*g*-PDMS content on the contact angle to water on the surface of air-side (○) and glass-plate-side (●) of surface-modified PTMSP membranes.

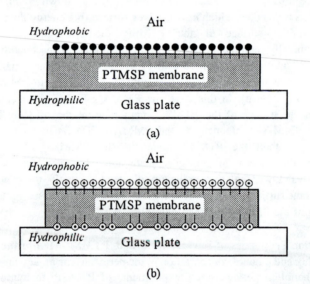

Figure 4. Structure of the surface-modified PTMSP membranes. (a) PTMSP membrane modified with PFA-*g*-PDMS: ● is PFA-*g*-PDMS, ● and | are PFA and PDMS, respectively, (b) PTMSP membrane modified with PMMA-*g*-PDMS: ☉ is PMMA-*g*-PDMS, ☉ and | are PFA and PDMS, respectively.

concentration in the permeate increased considerably with an increase of the PFA-*g*-PDMS content.

The pervaporation properties of PTMSP membranes modified with PMMA-*g*-PDMS for an aqueous solution of 10 wt.% ethanol are shown in Figure 6. In this case, both the flux and ethanol concentration in the permeate increased with an increasing amount of PMMA-*g*-PDMS and had a maximum at 0.05 wt.% of PMMA in PTMSP. The increase in the ethanol concentration in the permeate with an increase of the polymer additives, shown in Figures 5 and 6, indicates an increase of ethanol/water selectivity. The ethanol/water selectivities of the surface-modified PTMSP membrane correlate well with the contact angles to water of the membrane surface on the air-side of their membranes. That is, the ethanol/water selectivity of the PTMSP membranes modified with PFA-*g*-PDMS and PMMA-*g*-PDMS increased with an increase of the contact angle to water. The relationship between the ethanol/water selectivity and the contact angle to water of the surface-modified PTMSP membranes suggests that the ethanol/water selectivity is governed by the solubility of the components into the membrane and not by the diffusivity of the components through the membrane. This hypothesis can be supported by the illustration shown in Figure 4. The membrane surfaces of the air-side of the PTMSP membranes modified with PFA-*g*-PDMS and PMMA-*g*-PDMS become more hydrophobic and are more water-repellent than a PTMSP membrane. Therefore, the solubility of water in the surface-modified PTMSP membrane was lowered significantly relative to that of ethanol and, consequently, the ethanol/water selectivity of the surface-modified PTMSP membranes was enhanced.

Theoretical Discussion for the Pervaporation Characteristics of Surface-Modified PTMSP Membranes. The permeation mechanism of dense polymer membranes is generally explained by the solution-diffusion model (*13,14*). In this study, the surface-modified PTMSP membranes with the addition of polymer additives showed higher ethanol/water selectivity than that of the PTMSP membrane. The higher ethanol/water selectivity of the modified membranes can be attributed to a higher solubility of ethanol in the membrane. The relationship between the surface properties of a membrane and its ethanol/water selectivity can be analyzed by using the Flory-Huggins equation. Previously, Mulder *et al.* (*15-17*) compared experimental compositions sorbed into membranes for aqueous ethanol solutions in cellulose acetate, polyacrylonitrile, and polysulfone membranes with the theoretical values calculated by the Flory-Huggins equation and found good agreement with experimental data. In this study, the theoretical compositions sorbed into the surface-modified PTMSP membranes for an aqueous ethanol solution were also estimated.

The surface free energy was estimated by the contact angles (θ) to water and formamide on the surface of the surface-modified PTMSP membranes and by Owens's method (*18*) using equations (2) and (3).

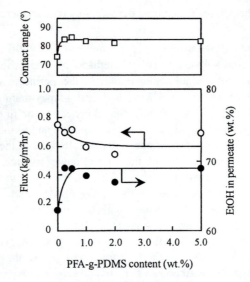

Figure 5.   Flux (○) and ethanol concentration in the permeate (●) for an aqueous solution of 10 wt.% ethanol through surface-modified PTMSP membranes in prevaporation as a function of the content of PFA-*g*-PDMS additive. The contact angle as a function of PFA-*g*-PDMS content is indicated by (□). Reproduced with permission from ref. 21. Copyright 1997. Wiley.

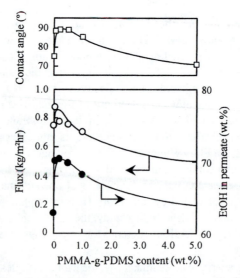

Figure 6.   Flux (○) and ethanol concentration in the permeate (●) for an aqueous solution of 10 wt.% ethanol through surface-modified PTMSP membranes as a function of the content of PMMA-*g*-PDMS additive. The contact angle as a function of PMMA-*g*-PDMS content is indicated by (□).

$$(1+\cos\theta)\cdot\gamma_l/2 = (\gamma_s^d\cdot\gamma_l^d)^{1/2}\cdot(\gamma_s^P\cdot\gamma_l^P)^{1/2} \qquad (2)$$

$$\gamma_s = \gamma_s^d + \gamma_s^P \qquad (3)$$

where γ_l and γ_s are the surface free energies of the liquid and the surface-modified PTMSP membranes, and γ_l^d and γ_l^P, and γ_s^d and γ_s^P are the dispersion force components and polar force components of the surface free energy of the liquid and the surface-modified PTMSP membranes, respectively.

The solubility parameter of the surface-modified PTMSP membranes, δ_P, was calculated from equation (4).

$$\delta_P = e_{coh}^{1/2} \qquad (4)$$

where e_{coh} is the cohesion energy density, which was determined by equation (5).

$$\gamma = 0.75\ e_{coh}^{2/3} \qquad (5)$$

where γ is the surface free energy, which was determined by measuring the contact angles and using equation (2). Using equations (2) and (5), the value of δ_P in equation (4) could then be determined.

The interaction parameters, χ_{iP}, for water and ethanol and the surface-modified PTMSP membranes were calculated from equation (6).

$$\chi_{iP} = V_P\ (\delta_P - \delta_i)/RT + 0.34 \qquad (6)$$

where R is the gas constant, T is the absolute temperature, and δ_i is the solubility parameter of component i.

Furthermore, the interaction parameter between water and ethanol, χ_{12}, was calculated from equation (7) (*17*).

$$\chi_{12} = \{x_1\ln (x_1/v_1)+x_2\ln (x_2/v_2)+\Delta G^E/RT\}/x_1v_2 \qquad (7)$$

where x_1 and x_2 are the weight fractions of water and ethanol in the feed solution, v_1 and v_2 are the volume fractions of water and ethanol, respectively, and ΔG^E is the excess free energy of mixing calculated from equation (8) (*19*).

$$\Delta G^E/RT = x_1\ln\gamma_1 + x_2\ln\gamma_2 \qquad (8)$$

where γ_1 and γ_2 are the activity coefficients of water and ethanol, which were calculated using the Wilson equation. The Wilson parameters were obtained from vapor-liquid equilibrium data (*20*).

From the above results, the theoretical ethanol concentration in the surface-modified PTMSP membranes was determined by the Flory-Huggins equation (9) (*17*).

$$\ln(\phi_1/\phi_2)-\ln(v_1/v_2) = (m-1)\ln(\phi_2/v_2)-\chi_{12}(\phi_2-\phi_1)-\chi_{12}(v_1-v_2)-\phi_P(\chi_{1P}-m\chi_{2P}) \quad (9)$$

$$m=V_1/V_2 \quad (10)$$

where ϕ_1, ϕ_2, and ϕ_P are the volume fractions of water, ethanol, and the surface-modified PTMSP membrane, respectively, and V_1 and V_2 are the molar volumes of water and ethanol, respectively. A schematic representation for the Flory-Huggins equation is illustrated in Figure 7.

In Figure 8, the effect of (i) PFA-*g*-PDMS and (ii) PMMA-*g*-PDMS on the interaction parameters χ_{1p} and χ_{2p}, calculated using equation (6) is shown. As can be seen from Figure 8, the interaction parameter, χ_{1p} (water-polymer), of the PTMSP membranes modified with PFA-*g*-PDMS and PMMA-*g*-PDMS was higher than the interaction parameter, χ_{2p} (ethanol-polymer), in those membranes. This results supports the hypothesis that ethanol can interact more easily with the surface-modified PTMSP membranes.

In Figure 9, the theoretical ethanol concentrations sorbed in the surface of the air-side of the modified PTMSP membranes with (i) PFA-*g*-PDMS and (ii) PMMA-*g*-PDMS, calculated from equation (9), are compared with those in the permeate of the membranes. The theoretical ethanol concentrations for the PFA-*g*-PDMS and PMMA-*g*-PDMS surface-modified PTMSP membranes were greater than that for the unmodified PTMSP membrane. This result is in agreement with the permeation experiments using an aqueous solution of 10 wt.% ethanol. However, there was a big difference between the absolute values of the theoretical ethanol concentrations based on penetrant solubility and those in the permeate of the permeation experiment. This difference is due to the fact that permeation depends on the solubility of the permeants in the membrane and also the diffusivity of the permeants in the membrane. However, the amount of PFA-*g*-PDMS and PMMA-*g*-PDMS added to the PTMSP membrane was extremely small and the additives were essentially localized on the membrane surface. Therefore, it is reasonable to assume that the internal structure of the surface-modified PTMSP membrane was almost the same as that of the unmodified PTMSP membrane. Consequently, it can be concluded that the improvement of the ethanol/water selectivity of the PTMSP membrane was due to the surface modification and was influenced by the solubility of the permeants relative to the diffusivity of the permeants in the membrane.

The above results show that the solubility of ethanol in the surface-modified PTMSP membrane was improved by modifying the membrane surface with an extremely small amount of polymer additive. The ethanol/water selectivity was

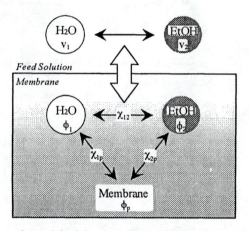

Figure 7.   Schematic model of the interactions between water, ethanol, and the polymer membrane.

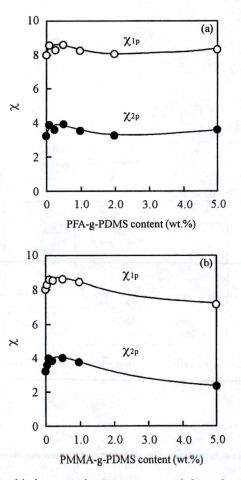

Figure 8. Relationship between the χ parameter and the polymer additives: (a) PTMSP membrane modified with PFA-*g*-PDMS, (b) PTMSP membrane modified with PMMA-*g*-PDMS.

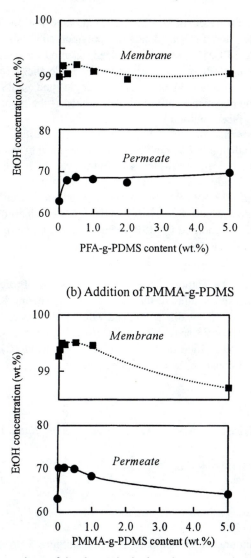

Figure 9. Comparison of the theoretical ethanol concentration sorbed in the surface at the air-side of the surface-modified PTMS membranes calculated from equation (9) and those in the permeate of the membranes.

enhanced without decreasing the permeability of the PTMSP membrane after modification.

Conclusions

To improve the performance of a PTMSP membrane for an aqueous ethanol solution for pervaporation (PV) applications, polymer additives such as PFA-g-PDMS and PMMA-g-PDMS were synthesized as modifiers for the membrane surface. The addition of these polymer additives made the PTMSP membrane surface remarkably water-repellent. Contact angle and XPS measurements demonstrated that the polymer additives were localized at the membrane surface on the air-side of the surface-modified PTMSP.

When the surface-modified PTMSP membranes were tested with an aqueous solution of 10 wt.% ethanol in PV, the flux was almost equal to that of the PTMSP membrane; however, the ethanol/water selectivity was higher than that of the PTMSP membrane. In addition, the relationship between the surface properties of the surface-modified PTMSP membrane and the ethanol/water selectivity was investigated using the surface free energy and Flory-Huggins equation. It was shown that the surface-modified PTMSP membrane had higher ethanol solubility than the PTMSP membrane.

Acknowledgements. The authors express their thanks to Professor Katsuhiko Nakamae and Dr. Takashi Nishino of the Faculty of Engineering, Kobe University, for measurements of the X-ray diffraction and valuable discussion. This work was supported by the following Kansai University Research Grants: Grant-in-Aid for Joint Research, 1996 and a Grant-in-Aid for Scientific Research (C) (No.09651005) from the Ministry of Education, Science, Sports and Culture.

Literature Cited

1. Ishihara, K.; Nagase, Y.; Matsui. K. *Makromol.Chem. Rapid Commun.* **1986**, *7*, 43.
2. Masuda, T.; Tang, B.-Z.; Higashimura, T. *Polym. J.* **1986**, *18*, 565.
3. Nagase, Y.; Ishihara, K.; Matsui, K. *J. Polym. Sci. Part B:* Polym. *Phys.* **1990**, *28*, 377.
4. Nagase, Y.; Ueda, T.; Matsui, K.; Uchikura, M. *J. Polym. Sci. Part B: Polym. Phys.* **1991**, *29* 171.
5. Masuda, T.; Kotoura, M.; Tsuchihara, K.; Higashimura, T. *J. Appl. Polym.Sci.* **1991**, *43*, 423.
6. Nagase, Y.; Takamura, Y.; Matsui, K. *J. Appl. Polym.Sci.* **1991**, *42*, 185.
7. Nagase, Y.; Sugimoto, K; Takamura, K. *J. Appl. Polym.Sci.* **1991**, *43*, 1223.
8. Masuda, T.; Isobe, E.; Higashimura, T. *Macromolecures* **1985**, *18*, 841.

9. Uragami, T.; Morikawa, T. *Makromol.Chem.* **1989**, *190*, 399.

10. Uragami, T.; Morikawa, T.; Okuno, H., *Polym. J.* **1989**, *30*, 1117.

11. Inui, K.; Miyata, T.; Uragami, T. *Angew. Makromol.Chem.* **1996**, *240*, 241.

12. Miyata, T.; Takagi, T.; Uragami, T. *Macromolecules* **1996**, *29*, 7787.

13. Binning, R. C.; Lee, R. J.; Jennings, J. F.; Martin, E. C. *Ind. Eng. Chem.* **1961**, *53*, 47.

14. Aptel, P.; Cuny, J.; Jozefonvicz, J.; Movel, G.; Neel, J. *J. Appl. Polym.Sci.* **1974**, *18*, 365.

15. Mulder, M. H.; Smolders, C. A. *J. Memb. Sci.* **1984**, *17*, 289.

16. Mulder, M. H.; Franken, A. C. M.; Smolders, C. A. *J. Memb. Sci.* **1985**, *23*, 41.

17. Mulder, M. H. Thermodynamic Principles of Pervaporation, in *Pervaporation Membrane Separation Processes*; Huang, R. Y. M., Ed.; Membrane Science and Technology Series, 1; Elsevier; Amsterdam-Oxford-New York-Tokyo, 1991, pp. 225-251.

18. Owens, D. K.; Wendent, R. C. *J. Appl. Polym.Sci.* **1969**, *13*, 1741.

19. Pesek, S. C.; Koros, W. J. *J. Memb. Sci.* **1993**, *81*, 71.

20. Gmehling, J.; Onken, U.; Arlt, W. *Vapor-Liquid Equilibrium Data Collection; Chemistry Data Series,* Vol. I, Part 6a, Dechema, Frankfurt, 1980.

21. Uragami, T. *Macromolecul. Symp.* **1997**, *118*, 419.

Chapter 19

Simple Surface Modifications of Poly(dimethylsiloxane) Membranes by Polymer Additives and Their Permselectivity for Aqueous Ethanol Solutions

Takashi Miyata, Yuichi Nakanishi, and Tadashi Uragami

Faculty of Engineering and High-Technology Research Center, Kansai University, Suita, Osaka 564–8680, Japan

This paper describes simple surface modifications of polydimethylsiloxane (PDMS) membranes by adding hydrophilic and hydrophobic block copolymers containing a PDMS component before the formation of PDMS networks on stainless-steel plates. The relationship between the surface characteristics of the surface-modified PDMS membranes and their permselectivity for aqueous ethanol solutions by pervaporation are discussed. Contact angle measurements demonstrated that the addition of a hydrophilic block copolymer made the stainless-steel-side surface of the PDMS membrane very hydrophilic. On the other hand, addition of a hydrophobic block copolymer produced a hydrophobic surface at its air contact side. Making the membrane surface hydrophilic gave the PDMS membrane a higher water-ethanol selectivity. On the other hand, introducing a very hydrophobic surface improved its ethanol-water selectivity without lowering its flux. Our results suggest a novel concept for designing the structure of high-performance, tailor-made pervaporation membranes, that is, a method for controlling the permselectivity of a pervaporation membrane without lowering its permeability.

Pervaporation is a promising membrane process for the separation of azeotropic and close-boiling point mixtures, which cannot be adequately separated by distillation. Therefore, recent research has focused on the characteristics of permeation and separation for liquid mixtures through a variety of membranes by pervaporation (1). We have focused on ethanol-selective polydimethylsiloxane (PDMS) membranes for the separation of aqueous ethanol solutions by pervaporation. Furthermore, we investigated the effect of the structures of multicomponent polymer membranes containing PDMS on their selectivity for aqueous ethanol solutions (*2-5*). Our previous studies showed that the phase structures of multicomponent polymer membranes such as block copolymer, graft copolymer, and polymer blend membranes strongly influence their permeability and selectivity. These results suggested that designing the phase structures of multicomponent polymer membranes is of great importance in improving their performance. Because the permeation mechanism of organic liquid mixtures by pervaporation is based on the solution-diffusion theory, designing a membrane material in which ethanol is more preferentially sorbed than water is an approach for enhancing the ethanol-selectivity of membranes.

Generally, the properties of multicomponent polymer surfaces are quite different from those of the bulk polymer because of the surface localization of a component (*6-21*). For example, a fluorine-containing component in a multicomponent polymer is preferentially concentrated to its surface to minimize the surface free energy of the system. Therefore, polymer additives which can be localized to the polymer surface might enable simple surface modification of pervaporation membranes.

This paper describes a simple surface modification of the PDMS membrane by addition of hydrophilic and hydrophobic block copolymers and the selectivity for aqueous ethanol solutions through the resulting membranes by pervaporation. The block copolymers containing a PDMS component as the anchor part were synthesized by the polymerization of hydrophilic and hydrophobic monomers via a PDMS macro-azo-initiator (*22-24*). Furthermore, the relationship between the surface characteristics of the resulting surface-modified PDMS membranes and their selectivity for aqueous ethanol solutions is discussed in detail.

Experimental Section

Materials. A PDMS macro-azo-initiator (PASA) (*22-24*) (Figure 1a), which has 59 units of a PDMS block as shown below, was supplied by Wako Pure Chemical Industries, Ltd. Diethylacrylamide (DEAA) (Figure 1b) and nonafluorohexyl methacrylate (NFHM) (Figure 1c) supplied by Kohjin Co., Ltd. and Daikin Industries, Ltd., respectively, were used without further purification. The solvents and other reagents were of analytical grade obtained from standard commercial sources, and were used without further purification.

Synthesis and Characterization of Block Copolymers. The appropriate amount of DEAA or NFHM was dissolved together with PASA in benzene to make a 40 wt% solution in a glass tube. The polymerization was carried out at 60°C for 6 h under nitrogen atmosphere. The resulting PDMS-*b*-PDEAA and PDMS-*b*-PNFHM copolymers were isolated by slow precipitation with a mixture of *n*-hexane and ethanol. The polymers were purified by reprecipitation from benzene solution in a mixture of *n*-hexane and ethanol, and dried at 40°C in vacuum.

Average molecular weights of PDMS-*b*-PDEAA and PDMS-*b*-PNFHM were determined by gel permeation chromatography (GPC) (Waters Associate Inc.; R-400), equipped with a TSK-GEL column (Tosoh Co. Ltd.; G2000HXL,

G3000HXL, G5000HXL) and ultraviolet spectrophotometry (Shimadzu Co. Ltd.; SPD-2A). Tetrahydrofuran was used as an eluent and the calibration was made with polystyrene standards. The composition of the resulting block copolymers was determined from 270 MHz [1]H-Nuclear Magnetic Resonance (NMR) (JEOL; GSX-270) spectra by measuring the integrals of the assigned peaks, after the purified copolymer was dissolved in chloroform-d. The compositions and molecular weights of the resulting block copolymers are summarized in Table I.

Table I. Results of polymerization of diethylacrylamide (DEAA) and nonafluorohexyl methacrylate (NFHM) by PDMS macro-azo-initiator (PASA).

Monomer	Monomer DMS content (mole%)	Polymer DMS[a] content (mole%)	Molecular weight[b]		
			Mn	Mw (g/mole)	Mw/Mn
DEAA	30	47	7,000	13,000	1.88
	50	68	9,900	13,000	1.35
	70	82	4,200	10,000	2.35
NFHM	30	52	22,000	52,000	2.26
	50	65	42,000	60,000	1.42
	70	79	28,000	51,500	1.37

[a] determined by ^{1}H-NMR
[b] determined by GPC

Membrane Preparation. PDMS-b-PDEAA or PDMS-b-PNFHM was dissolved in benzene at 25°C at a concentration of 4 wt.%. PDMS-b-PDEAA and PDMS-b-PNFHM membranes were prepared by pouring the casting solutions onto rimmed glass plates and allowing the solvent to evaporate completely at 25°C. The resulting membranes were transparent and their thickness was about 40 μm.

Furthermore, surface-modified membranes were prepared by polycondensation of the PDMS base polymer in the presence of block copolymers as follows: the block copolymers were dissolved in tetrahydrofuran (THF) solutions of the PDMS base polymer (4 wt.%, with tetraethyl orthosilicate as a cross-linker (excess to the PDMS base polymer) and stannous dibutyl diacetate as a catalyst (1.0 wt.% relative to the PDMS base polymer) in the polycondensation reaction. After the resulting mixture was cast on a stainless-steel plate, the PDMS network was formed by evaporating the THF at 25°C for 8 h. The resulting PDMS-b-PDEAA/PDMS and PDMS-b-PNFHM/PDMS membranes were transparent.

Contact Angle Measurements. Contact angles of water on the air-side and stainless-steel-side surfaces of the block copolymer membranes and surface-modified PDMS membranes were measured by using a contact angle meter (Erma Model G-I) at 25°C.

Transmission Electron Micrographs (TEM). The block copolymer membranes were vapor-stained with an aqueous solution of 5 wt.% RuO_4 in glass-covered dishes (25). The stained membranes were embedded in epoxy resin and cross-sectioned into thin films with a thickness of approximately 60 nm using a microtome (Leica; Reichert Ultracut E). The morphological features were observed with a transmission electron microscope (TEM) (JEOL JEM-1210) at an accelerating voltage of 80 kV.

Pervaporation Experiments. Pervaporation experiments were carried out using the apparatus described in previous papers (2-5). The feed temperature was 40°C and the permeate pressure was 0.01 Torr. The effective membrane area was 13.8 cm^2. An aqueous solution of 10 wt.% ethanol was used as the feed solution. The compositions of the feed and permeate were determined using gas chromatography (Shimadzu GC-9A) equipped with a flame ionization detector (FID) and a capillary column (Shimadzu Co. Ltd.; Shimalite F) heated to 200°C. The fluxes of the membranes for an aqueous ethanol solution were determined from the weight of the permeate collected in a cold trap, permeation time, and effective membrane area.

Results and Discussion

Hydrophilic and Hydrophobic Block Copolymer Membranes. Figures 2 and 3 show the effect of copolymer compositions on the contact angle of water on the PDMS-*b*-PDEAA and PDMS-*b*-PNFHM membranes, and the ethanol concentration in the permeate and normalized flux for an aqueous solution of 10 wt.% ethanol through the membranes. The normalized flux is the product of the flux and the membrane thickness. The contact angle of water on the PDMS-*b*-PDEAA membrane decreased with increasing PDEAA content. Accordingly, the ethanol concentration in the permeate decreased strongly because an introduction of the PDEAA component made the membrane surface hydrophilic. On the other hand, the contact angle measurements for the PDMS-*b*-PNFHM membranes revealed that an increase in the PNFHM content leads to a more hydrophobic surface in contrast to that of PDEAA. However, the ethanol-selectivity of the PDMS-*b*-PNFHM membrane decreased in spite of its hydrophobic surface. These results suggest that both membranes changed from being ethanol-selective to water-selective by the introduction of the PDEAA and PNFHM component although the effects of the introduced PDEAA and PNFHM component on their surface characteristics are quite different.

The phase structures of the PDMS-b-PNFHM membranes were investigated to reveal the reason why the membranes became water-selective in spite of their very hydrophobic surfaces. Figure 4 shows the transmission electron micrographs of the cross-sections of the PDMS-*b*-PNFHM membranes with PNFHM contents of 35 and 52 mole%. In the micrographs, the dark region stained by RuO_4 represents the PDMS component. These micrographs demonstrate that both membranes showed distinct microphase separation consisting of a PDMS phase and a PNFHM phase.

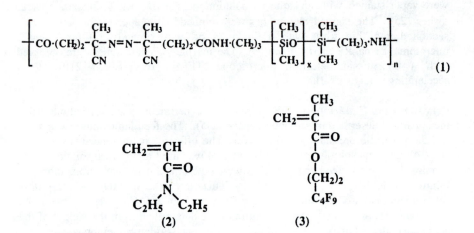

$$\left[CO\cdot(CH_2)_2-\underset{\underset{CN}{|}}{\overset{\overset{CH_3}{|}}{C}}-N=N-\underset{\underset{CN}{|}}{\overset{\overset{CH_3}{|}}{C}}-(CH_2)_2\cdot CONH\cdot(CH_2)_3-\left[\underset{\underset{CH_3}{|}}{\overset{\overset{CH_3}{|}}{SiO}}\right]_x\underset{\underset{CH_3}{|}}{\overset{\overset{CH_3}{|}}{Si}}-(CH_2)_3\cdot NH\right]_n \quad (1)$$

Figure 1. Structures of monomers used for membrane preparation. (**a**) PDMS macro-azo-initiator, (**b**) diethylacrylamide, and (**c**) nonafluorohexyl methacrylate (NFHM).

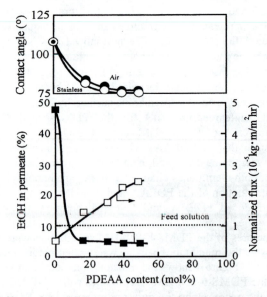

Figure 2. Effect of the PDEAA content on the ethanol concentration in the permeate (■) and normalized flux (□) for an aqueous solution of 10 wt% ethanol through PDMS-*b*-PDEAA membranes by pervaporation at 40°C and contact angles of water on their surfaces: O stainless-steel-side; ● air-side.

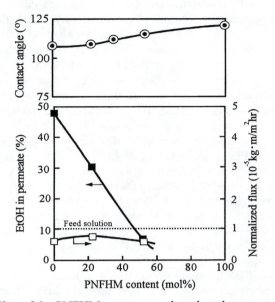

Figure 3.   Effect of the PNFHM content on the ethanol concentration in the permeate (■) and normalized flux (□) for an aqueous solution of 10 wt% ethanol through PDMS-*b*-PNFHM membranes by pervaporation at 40°C and contact angles of water on their surfaces: O stainless-steel-side; ● air-side.

PNFHM content; 35 mol%   **PNFHM content; 52 mol%**

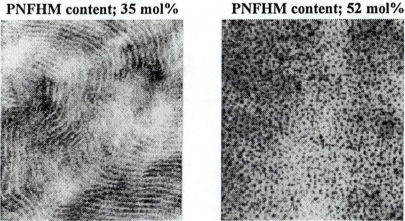

Figure 4.   Transmission electron micrographs of the cross-sections of PDMS-*b*-PNFHM membranes. The dark region stained by RuO$_4$ represents the PDMS component.

In previous papers (2-4), we reported that membranes having a continuous PDMS phase were ethanol-selective and that those having a discontinuous PDMS phase were water-selective. Similarly, the reason why the PDMS-b-PNFHM membranes became water-selective in spite of their very hydrophobic surface can be explained by the morphological changes with the copolymer composition. The PDMS component in the membranes with low PNFHM content forms a continuous phase in the direction of the membrane thickness. At higher PNFHM content, the membrane becomes ethanol-selective because the penetrant diffuses mainly through the continuous PDMS phase. On the other hand, the membranes with high PNFHM content had a discontinuous PDMS phase and a continuous PNFHM phase, and then the penetrant diffuses preferentially through the continuous, rigid PNFHM phase having higher Tg. The membranes with high PNFHM content prefer ethanol to water in the sorption step but the diffusivity of ethanol is much lower than that of water in the rigid PNFHM phase. Consequently, in spite of the very hydrophobic surface, the membranes with high PNFHM content were water-selective due to the very low diffusivity of ethanol in the continuous PNFHM phase. Hence, the membrane surface must be modified without changing the bulk structure of the PDMS membrane.

Surface-Modified PDMS Membranes. In general, the characteristics of multicomponent polymer surfaces are quite different from those of the bulk polymer, because a component is preferentially localized at the surface due to the minimization of the surface free energy. Therefore, membrane surfaces can easily be modified by using such surface localization. In this study, surface-modified PDMS membranes were prepared by adding a small amount of hydrophilic and hydrophobic block copolymers before the formation of PDMS networks on stainless-steel plates. PDMS-b-PDEAA with a PDMS content of 53 mole% and PDMS-b-PNFHM with a PDMS content of 48 mole% were used as the hydrophilic and hydrophobic polymer additives, respectively. As shown in the previous section, both block copolymer membranes were water-selective.

Figure 5 shows the relationship between the block copolymer content and the contact angle of water on the PDMS-b-PDEAA/PDMS and PDMS-b-PNFHM/PDMS membranes. The contact angle of water on the stainless-steel-side of the PDMS-b-PDEAA/PDMS membrane decreased after the addition of the hydrophilic block copolymer; however, the air-side contact angle did not change. On the other hand, the contact angle on the air-side surface of the PDMS-b-PNFHM/PDMS membrane increased; in this case, the stainless-steel-side contact angle remained constant. These results indicate that the hydrophilic PDMS-b-PDEAA and hydrophobic PDMS-b-PNFHM were localized at the stainless-steel and air-side surface, respectively, in order to minimize the surface free energy of the modified membranes. Consequently, we can conclude that the PDMS membranes having hydrophilic and hydrophobic surfaces without changing their bulk structures can easily be prepared by adding a small amount of the block copolymers as polymer additives.

The effect of the block copolymer content on the ethanol concentration in the permeate and normalized flux for an aqueous solution of 10 wt.% ethanol through the surface-modified PDMS membranes are shown in Figure 6. In these pervaporation experiments, the surface-modified PDMS membranes were placed in a pervaporation cell with the modified surfaces facing the feed side. The ethanol concentration in the permeate passing through the hydrophilic PDMS-*b*-PDEAA/PDMS membranes decreased gradually with increasing hydrophilic PDMS-*b*-PDEAA content. On the other hand, the ethanol concentration in the permeate through the PDMS-*b*-PNFHM/PDMS membranes showed a maximum at a PDMS-*b*-PNFHM content of 5 wt.%. It is interesting to note that the ethanol/water selectivity of the PDMS membrane can be enhanced by the addition of the water-selective PDMS-*b*-PNFHM. Furthermore, when the block copolymer content was increased up to 5 wt.%, the behavior of the ethanol concentration in the permeate corresponded to that of the contact angles, regardless of the kind of the block copolymer. This fact suggests that the selectivity of the surface-modified PDMS membranes was associated with their surface characteristics. Estimating the ethanol concentration in the membrane surfaces by the Flory-Huggins theory can provide important information about the relationship between membrane surfaces and their selectivity for aqueous ethanol solutions.

Relationship Between Membrane Surface and Selectivity. The solubility of a penetrant into a membrane can be described using the Flory-Huggins equation (*26*). Therefore, the theoretical ethanol concentration in a membrane can be determined if the polymer-solvent interaction parameter, which can be estimated from the solubility parameter is known. At first, we tried to determine the solubility parameter of membrane surfaces from the surface free energy, which can be derived from the contact angles by equation 1 (*27,28*)

$$\frac{(1+\cos\theta)\cdot\gamma_l}{2} = \left(\gamma_s^d \cdot \gamma_l^d\right)^{1/2} + \left(\gamma_s^p \cdot \gamma_l^p\right)^{1/2} \tag{1}$$

$$\gamma_s = \gamma_s^d + \gamma_s^p$$

where θ is the contact angle, and γ_s and γ_l are the surface free energy of the solid and the liquid, and γ_s^d γ_s^p, γ_l^d, and γ_l^p are the dispersion force components and polar force components of the surface free energy of the solid and the liquid, respectively.

Furthermore, the cohesive energy density, e_{coh}, of a polymer surface is correlated with the surface free energy by equation 2:

$$\gamma \approx 0.75 e_{coh}^{2/3} \tag{2}$$

Therefore, the solubility parameter, δ, can be calculated from the cohesive energy density by equation 3:

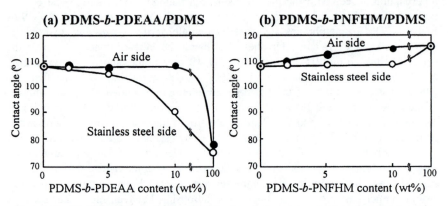

Figure 5. Relationship between the block copolymer content and contact angles of water on the PDMS-*b*-PDEAA/PDMS (**a**) and PDMS-*b*-PNFHM/PDMS (**b**) membranes; ○ stainless-steel-side; ● air-side.

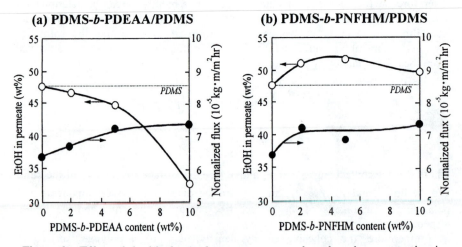

Figure 6. Effect of the block copolymer content on the ethanol concentration in the permeate (○) and normalized fluxes (●) for an aqueous solution of 10 wt% ethanol through (**a**) PDMS-*b*-PDEAA/PDMS and (**b**) PDMS-*b*-PNFHM/PDMS membranes by pervaporation at 40°C.

$$\delta = e_{coh}^{1/2} \tag{3}$$

The solubility parameter determined by this method is an indicator for the hydrophobicity of a polymer surface. The surface properties of the block copolymer/PDMS membranes are summarized in Table II.

Table II. Surface properties of block copolymer/PDMS membranes.

Block copolymer	Block copolymer content (wt.%)	Surface	Contact angle (°)		γ (erg/cm^2)	δ (J$^{1/2}$/cm$^{3/2}$)
			θ_{water}	θ_{CH2I2}		
PDMS-*b*-DEAA	0	stainless	86.7	62.0	28.6	15.4
	2	stainless	85.8	60.9	29.4	15.7
	5	stainless	84.5	60.1	30.1	16.0
	100	stainless	60.0	56.4	42.9	20.8
PDMS-*b*-NFHM	0	air	86.7	62.0	28.6	15.4
	2	air	88.6	63.2	27.7	15.0
	5	air	89.2	64.4	27.0	14.7
	100	air	90.4	65.3	26.4	14.5

The state of a solvent in a membrane can be represented by the Flory-Huggins equation (*26,29,30*). Therefore, the ethanol concentration in the membrane surface can be determined from the polymer-solvent interaction parameter, χ, by equation 4, after the parameter is calculated from the solubility parameters using equation 5:

$$\ln\left(\frac{\phi_1}{\phi_2}\right) - \ln\left(\frac{v_1}{v_2}\right) = (l-1)\ln\left(\frac{\phi_2}{v_2}\right) - \chi_{12}(\phi_2 - \phi_1) - \chi_{12}(v_1 - v_2) - \phi_3(\chi_{13} - l\chi_{23}) \tag{4}$$

$$\chi_{i3} = \frac{v_3}{RT}(\delta_3 - \delta_i)^2 + 0.34 \tag{5}$$

The resulting theoretical ethanol concentrations in the PDMS-*b*-PDEAA/PDMS and PDMS-*b*-PNFHM/PDMS membrane surfaces are shown in Figure 7. With increasing block copolymer content, the ethanol concentration in the PDMS-*b*-PDEAA/PDMS membrane surface decreases but that in the PDMS-*b*-PNFHM/PDMS membrane increases gradually. These results indicate that the surface modifications by the hydrophilic PDMS-*b*-PDEAA and hydrophobic PDMS-*b*-PNFHM enhance the relative solubility of water and ethanol in the membrane surfaces, respectively. Furthermore, the solubility of the penetrant in surface-modified membranes is closely related to the surface characteristics of the membrane. Therefore, the reason why the selectivity of the PDMS membrane changed by the addition of the hydrophilic and hydrophobic block copolymers can

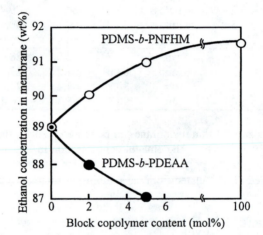

Figure 7. Effect of the block copolymer content on the theoretical ethanol concentration in the PDMS-*b*-PDEAA/PDMS (●) and PDMS-*b*-PNFHM/PDMS (O) membranes calculated on the basis of the Flory-Huggins equation.

be explained on the basis of the solubility of the penetrant in their membrane surfaces (Figure 8).

When the hydrophilic PDMS-*b*-PDEAA is added to the PDMS membrane, the membrane surface becomes hydrophilic due to the surface localization of PDMS-*b*-PDEAA. Because such a hydrophilic surface prefers water to ethanol in the sorption process, the PDMS-*b*-PDEAA/PDMS membranes showed low ethanol selectivity. On the other hand, the addition of the hydrophobic PDMS-*b*-PNFHM to the PDMS membrane makes its air-side surface very hydrophobic by surface localization of the hydrophobic component. Ethanol is more preferentially incorporated into such a hydrophobic surface than water. Then, the diffusion of the penetrant in the PDMS-*b*-PNFHM/PDMS membranes is not affected by the addition of a very small amount of the PDMS-*b*-PNFHM, because the bulk structure of the PDMS membranes does not change. Thus, the preferential sorption of ethanol in the very hydrophobic surface combined with a constant diffusivity of the penetrant in the membrane enhances the ethanol-selectivity of the PDMS-*b*-PNFHM/PDMS membranes.

Consequently, the selectivity of the PDMS membrane can be controlled by simple surface modification using various polymer additives. Especially, the addition of hydrophobic fluorine-containing block copolymers enhanced the ethanol-selectivity of the PDMS membranes without lowering their fluxes. These results provide a novel concept for designing the structure of high-performance pervaporation membranes and are first step in preparing "tailor-made" membranes for ethanol/water separation.

Conclusions

PDMS membranes were simply surface-modified by addition of hydrophilic and hydrophobic block copolymers. This work focuses on the relationship between the surface properties of modified PDMS membranes and their permselectivity for aqueous ethanol solutions by pervaporation. Although the PDMS-*b*-PDEAA and PDMS-*b*-PNFHM membranes had hydrophilic and hydrophobic surfaces, both membranes were water-selective in separating aqueous ethanol solutions. This is due to the fact that the morphology of a PDMS phase in microphase separation of the membranes strongly influences the permselectivity for aqueous ethanol solutions. Therefore, the membrane surface was modified without changing the bulk structure of the PDMS membrane by the addition of hydrophilic PDMS-*b*-PDEAA and hydrophobic PDMS-*b*-PNFHM segments into the membrane. The addition of PDMS-*b*-PDEAA made the stainless-steel-side surface of the PDMS membrane very hydrophilic, and that of PDMS-*b*-PNFHM produced a hydrophobic surface at its air-side surface. Making the surface hydrophilic and hydrophobic made the PDMS membrane water-selective and ethanol-selective, respectively, without changing its permeate flux. A theoretical discussion using the Flory-Huggins equation revealed that the preferential sorption of ethanol in the hydrophobic surfaces combined with a constant diffusivity of the penetrant in the membrane enhanced the ethanol-selectivity in the PDMS-*b*-PNFHM membrane. In

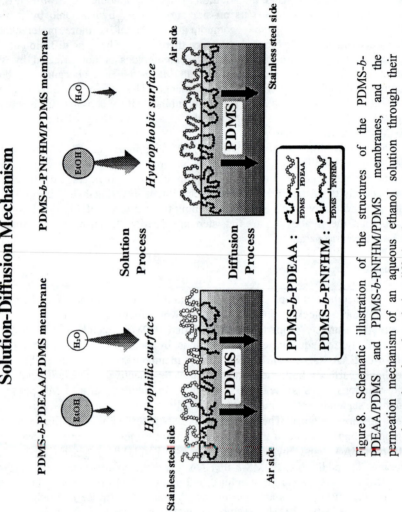

Figure 8. Schematic illustration of the structures of the PDMS-*b*-PDEAA/PDMS and PDMS-*b*-PNFHM/PDMS membranes, and the permeation mechanism of an aqueous ethanol solution through their membranes based on the solution-diffusion theory.

summary, our results suggest a novel concept for membrane design in controlling the permselectivity for liquid organic/water mixtures without lowering permeability.

Acknowledgments

This research was financially supported by the Kansai University Research Grants: Grant-in-Aid for Joint Research, 1996.

Literature Cited

1. Huang, R. Y. M. *Pervaporation Membrane Separation Processes*; Elsevier: Amsterdam, 1991.
2. Miyata, T.; Takagi, T.; Kadota, T.; Uragami, T. *Macromol. Chem. Phys.* **1995**, *196*, 1211.
3. Miyata, T.; Higuchi, J.; Okuno, H.; Uragami, T. *J. Appl. Polym. Sci.* **1996**, *61*, 1315.
4. Miyata, T.; Takagi, T.; Uragami, T. *Macromolecules* **1996**, *29*, 7787.
5. Miyata, T.; Nakanishi, Y.; Uragami, T. *Macromolecules* **1997**, *30*, 5563.
6. Andrade, J. D. *Surface and Interfacial Aspects of Biomedical Polymers; Surface Chemistry and Physics, Vol. 1*; Plenum Press: New York, 1985.
7. Owens, M. J.; Kendrick, J. C. *Macromolecules* **1970**, *3*, 458.
8. Gaines, G. L.; Bender, G. W. *Macromolecules* **1972**, *5*, 82.
9. Ratner, B. D.; Weathersby, P. K.; Hoffman, A. S.; Kelly, M. A.; Scharpen, L. H. *J. Appl. Polym. Sci.* **1978**, *22*, 643.
10. Thomas, H. R.; O'Malley, J. J. *Macromolecules* **1979**, *12*, 323.
11. Pennings, J. F. M.; Bosman, B. *Colloid Polym. Sci.* **1980**, *258*, 1099.
12. Chujo, R.; Nishi, T.; Sumi, Y.; Adachi, Y.; Naito, H.; Frentzel, H. *J. Polym. Sci., Polym. Lett. Ed.* **1983**, *21*, 487.
13. Hasegawa, H.; Hashimoto, T. *Macromolecules* **1985**, *18*, 589.
14. Yoon, S. C.; Ratner, B. D. *Macromolecules* **1986**, *19*, 1068.
15. Ratner, B. D. *Makromol. Chem., Makromol. Symp.* **1988**, *19*, 163.
16. Bhatia, Q. S.; Pan, D. H.-K.; Koberstein, J. T. *Macromolecules* **1988**, *21*, 2166.
17. Nakamae, K.; Miyata, T.; Matsumoto, T. *J. Membrane Sci.* **1992**, *69*, 121.
18. Nakamae, K.; Miyata, T.; Matsumoto, T. *J. Membrane Sci.* **1992**, *75*, 163.
19. Jones, R. A. L.; Kramer, E. J. *Polymer* **1993**, *34*, 115.
20. Gowie, J. M. G.; Devlin, B. G. ; Mcwen, I. J. *Polymer* **1993**, *34*, 501.
21. Nakamae, K.; Miyata, T.; Ootsuki, N. *Macromol. Chem. Phys.* **1994** ,*195*, 2663.
22. Inoue, H.; Ueda, A.; Nagai, S. *J. Polym. Sci., Part A, Polym. Chem.* **1988**, *26*, 1077.
23. Inoue, H.; Ueda, A.; Nagai, S. *J. Appl. Polym. Sci.* **1988**, *35*, 2039.
24. Inoue, H.; Matsumoto, A.; Matsukawa, K.; Ueda, A. *J. Appl. Polym. Sci.* **1990**, *40*, 1917.

25. Trent, J. S.; Scheinbeim, J. I.; Couchman, P. R. *Macromolecules* **1983**, *16*, 589.

26. Flory, P. J. *Principles of Polymer Chemistry*; Cornell University Press, NY, 1953.

27. Owens, D. K.; Wendt, R. C. *J. Appl. Polym. Sci.* **1969**, *13*, 1741.

28. Miyata, T; Ootsuki, N; Nakamae, K; Okumura, M; Kinomura, K. *Macromol. Chem. Phys.* **1994**, *195*, 3597.

29. Mulder, M. H. V.; Franken, T.; Smolders, C. A. *J. Membrane Sci.* **1985**, *22*, 55.

30. Okuno, H; Nishida, T; Uragami, T. *J. Polym. Sci. Polym. Phys.* **1995**, *33*, 299.

Chapter 20

Carbon Molecular Sieve Membranes: Preparation, Characterization, and Gas Permeation Properties

H. Suda and K. Haraya

Department of Chemical Systems, National Institute of Materials and Chemical Research, 1–1 Higashi, Tsukuba 305, Japan

Gas permeation properties of a series of carbon molecular sieve (CMS) membranes prepared by pyrolysis of a Kapton polyimide precursor were investigated. Pyrolysis at high temperatures and low heating rates produced the most selective membranes. The experimental observation that gas permeabilities were not always in the order of the kinetic gas diameters originated from differences in sorption and/or diffusivity. Kapton-based CMS membranes whose pore openings were slightly enlarged by moderate activation or those prepared from other polyimide precursor types provided a series of CMS membranes that had different permeation properties. Some of the membranes were effective even for alkene/alkane separation. Imidization and pyrolysis of capillary polyamic acid membranes resulted in CMS membranes with an asymmetric structure, which also exhibited excellent performance.

A carbon molecular sieve (CMS) membrane is one of the most promising inorganic membranes for gas separation applications. A distinctive feature of CMS membranes is that controlled pyrolysis of a polymeric precursor can yield a series of membranes that possess micropores of desired molecular dimensions. Gas permeation properties of CMS membranes are affected by the polymeric precursor, membrane formation method, pyrolysis method, and post-treatment conditions. This chapter focuses on factors that determine the permeation properties of flat, dense, and capillary CMS membranes.

Background

Molecular sieves have pores of molecular dimensions through which molecules with different sizes or shapes can be separated. Typical molecular sieve materials are aluminosilicates (zeolites) and CMS. Although these materials have been used in adsorption processes, recent studies have focused on their use as molecular sieve membranes for gas separation.

Separation of gases in a CMS membrane results mainly from differences in molecular gas dimensions. A summary of previous fabrication methods and gas separation performance of CMS membranes is given in Table I. Polymeric precursors used so far include polyfurfuryl alcohol (4,5), thermosetting polymers (6-8), and phenol formaldehyde (9,10). More recently, polyimide precursors (11-24) have been considered for CMS membranes. The only exception to carbon molecular sieve membranes is the work by Rao and Sircar, who prepared so-called "nanoporous" carbon membranes from polyvinylidene chloride. In these carbon membranes, large hydrocarbons permeate much faster than smaller gases, such as H_2, by competitive adsorption and surface diffusion of the more strongly adsorbed components (25-28).

A CMS membrane can be prepared either by coating of a polymeric precursor on a porous support or by the phase inversion technique, both followed by pyrolysis steps, or by one-step pyrolysis of a precursor dense film. In some of the past studies (9,10,13-15,23,24,29,30) asymmetric membranes were used as precursors, because the thin selective skin layer formed on a porous support is expected to exhibit high permeance.

The pyrolysis temperature and the heating rate have significant effects on the gas permeation properties of a CMS membrane. The choice of the pyrolysis atmosphere is also an important parameter, because the atmosphere can change the pore size and pore geometry or even the nature of the surface by sintering or activation effects.

CMS membranes can be modified to improve their permeation properties or to solve several problems inherent to their structure. The first modification methods were proposed by Koresh and Soffer (31), who changed pore openings by oxidation and sintering. The oxidized membrane type exhibited increased permeability, whereas the sintered membrane type had lower permeability. A moderately modified CMS membrane was reported by Suda et al. which exhibited 1 to 2 orders of magnitude higher permeability than that of the original membrane (22). The modified membrane was found to separate even relatively large alkene/alkane gas pairs. A similar modification was reported by Hayashi et al. (32) who controlled the pore openings by chemical vapor deposition of propylene (33). Changes in separation properties during permeation of gas mixtures were reported by Jones and Koros (34,35), because CMS membranes are vulnerable to adverse effects from exposure to organic contaminants or water vapor. Jones and Koros improved the properties of the contaminated CMS membrane by regeneration with propylene and by coating of the CMS membrane with a water-resistant polymer.

Table I. Precursor, membrane formation, and pyrolysis conditions for fabrication
of CMS membranes and their permeation performance.

Precursor	Membrane formation	Pyrolysis conditions	Representative permeation performance*	Ref.
Polyfurfuryl alcohol	unsupported or supported on silica frits	973 K, < 7 K/min	$D(H_2) = 1.2 \times 10^{-3}$ $D(N_2) = 0.2 \times 10^{-3}$ at 350 K	(4)
Polyfurfuryl alcohol	coated on graphite disk	773 K, 1.5 K/min	$D(CH_4) = 2.4 \times 10^{-8}$ $D(C_2N_6) = 0.3 \times 10^{-8}$ at 297 K	(5)
Thermosetting polymers	hollow fiber	1073-1223 K, activation	$P(O_2)=1,710$, $\alpha O_2/N_2=7.1$	(6-8)
Phenol formaldehyde	asymmetric flat-sheet	1073-1223 K N_2 and activation	$P(O_2)=2,300$ $\alpha O_2/N_2=10.7$	(9,10)
6F-polyimide (polypyrrolone)	flat dense film	1073 K, 5 K/min, N_2	$P(C_3H_6)=170$ $\alpha C_3H_6/ C_3H_8=78$ at 308 K	(11)
PMDA-ODA polyimide (Kapton)	flat dense film	1073 K, 3 K/min	$P(O_2)=1.2$ $\alpha O_2/N_2=4.6$ at 293 K	(12)
6F-containing polyimide copolymer	asymmetric hollow fiber	773-1073 K, 13.3 K/min, vacuum or He, Ar, CO_2	$(P/l)O_2=1.6-306$ $\alpha O_2/N_2=8.5-14$ for gas mixtures	(13-15)
BPDA-ODA polyimide	dip-coated on alumina tube	773-1173 K, 5 K/min, N_2	$P(CO_2)=7-330$ $\alpha CO_2/N_2=40$ at 298 K	(16,17)
PMDA-ODA polyimide (Kapton) and other polyimides	flat dense film or asymmetric capillary	773-1273 K, 1.33-13.3 K/min, Ar or vacuum	$P(H_2)=19-1,600$ $\alpha H_2/N_2=20-4,700$ $P(O_2)=0.15-383$ $\alpha O_2/N_2=4.7-36$ at 308 K	(18-24)
Poly(vinylidene chloride)	coated on graphite disk "nanoporous" membrane	1273 K, 1 K/min, N_2	$P(C_4H_{10})=112$ $\alpha C_4H_{10}/H_2=94$ at 295 K for gas mixtures	(25-28)

* P: permeability in Barrers (1 Barrer = 10^{-10} cm^3(STP)•cm/cm^2•s•cmHg); P/L:
permeance or pressure-normalized flux in GPU (1 GPU = 10^{-6} cm^3(STP/cm^2•s•cmHg);
D: diffusivity (cm^2/s); α = selectivity (-)

Experimental

Dense CMS membranes were fabricated by pyrolysis of polyimide precursor films between graphite blocks at 873-1273 K for 2 hours at a heating rate of 10 K/min under a vacuum of 10^{-5} Torr, as shown in Figure 1. The heating rate (1.33-13.3 K/min) and pyrolysis atmosphere (under vacuum or Ar flow) were also varied to study the effects of pyrolysis conditions on the CMS membrane properties. A capillary-type polyamic acid (PA) membrane with an asymmetric structure was prepared by a phase inversion technique. The microstructure of the CMS membranes was investigated by FT-IR, XPS, elemental analysis, XRD, SEM, high-resolution TEM, and gas sorption measurements. Pore size distributions were obtained by a molecular probe method from the analysis of sorption isotherms at 298 K of probe gas molecules with different kinetic diameter, σ, by application of the Dubinin-Astakhov equations (36,37). Gas permeabilities were measured at 308-373 K with a high vacuum time-lag method (38,39) under a pressure difference of 1 atm. Diffusivity values were calculated by the time-lag method. In the present study, selectivity is defined as the ratio of the permeability of a chosen gas over that of nitrogen. Details for the experimental methods are described in an earlier paper (20).

Flat Type CMS Dense Membranes Derived from Kapton Polyimide (18-20).

Pyrolysis Process of a Kapton Polyimide Film.   Upon pyrolysis, the color of a Kapton film changed from transparent yellow to dark brown, and then black due to carbonization (40). The weight and the film diameter decreased abruptly at around 800-900 K and then gradually at higher temperatures. The weight loss can be attributed to release of O_2, CO, CO_2, H_2, and N_2 during thermal degradation of Kapton polyimide. It is characteristic of the Kapton polyimide that pyrolysis of the film yields a dense CMS membrane without pinholes in spite of the rather large weight loss of about 40% and shrinkage of about 25% at 1273 K. The micropores of CMS membranes are the result of gases released from the membrane matrix during pyrolysis. Thus, the pyrolysis conditions are expected to strongly affect the pore structure of the membrane. Both the CMS membrane obtained at 1273 K under a vacuum of 10^{-5} Torr and the membrane prepared at 1223 K under Ar flow consisted essentially only of carbon (> 95%), as determined by elemental analysis, infrared spectra, and XPS spectra. The carbon prepared by pyrolysis of organic materials is known to have a turbostratic structure (41), in which layer-planes of graphite-like microcrystallines are dispersed in a non-crystalline carbon, as shown in Figure 2. High-resolution transmission electron micrographs demonstrated that such a graphite-like structure starts to appear and further develops at higher pyrolysis temperature. Therefore, evolution of gases in the early stages of pyrolysis through either imide ring cleavage or C-N bond cleavage processes (42) and nitrogen

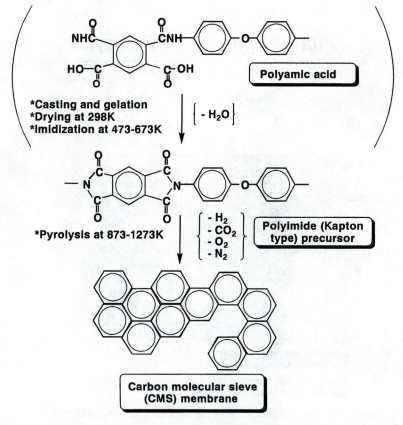

Figure 1.   Gelation, imidization, and pyrolization steps for preparation of CMS membranes.

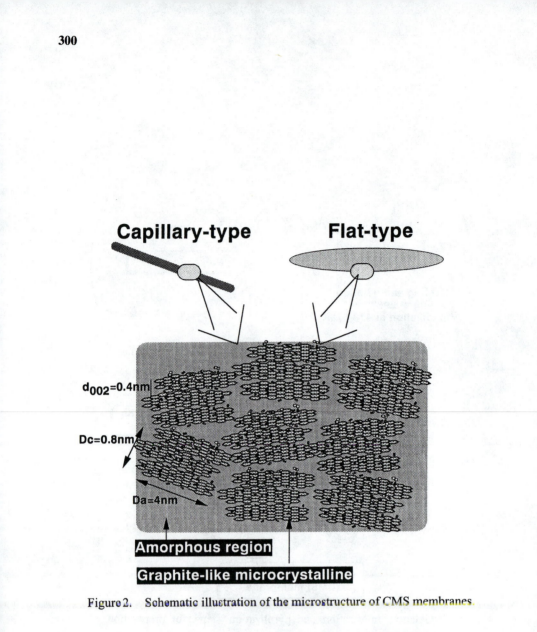

Figure 2. Schematic illustration of the microstructure of CMS membranes.

evolution in the later pyrolysis steps initiate the growth of layer-planes of graphite-like microcrystallines.

Pore Analysis of the Vacuum Pyrolized CMS Membranes. Sorption isotherms of probe molecules revealed how the pore size changed with increase in pyrolysis temperature. A membrane prepared at lower temperature (873 K) sorbed to some extent even i-C_4H_{10}, the largest sorbate (σ=5.0 Å), whereas the membrane prepared at 1273 K sorbed only CO_2, the smallest sorbate studied (σ=3.3 Å). Plots of the limiting micropore volume, W_0, against the kinetic diameters of probe molecules showed that with increasing the pyrolysis temperature both the limiting pore volume and pore size decreased. The plots also indicated that pore size distributions became sharper when the pyrolysis temperature increased. The W_0 values for CO_2 increased from 0.15 to 0.22 ml/g with decreasing pyrolysis temperature, all of which were larger than that of a microporous silica membrane (W_0=0.11 ml/g) (*43*). The microstructure (pore size, pore volume, etc.) can therefore be tailored by controlling the pyrolysis temperature.

Gas Permeation Properties of the Vacuum Pyrolized CMS Membranes. Permeation properties were compared for CMS membranes pyrolized under various conditions (pyrolysis temperature, heating rate, and pyrolysis atmosphere). The pyrolysis temperature had the greatest effect on the microstructure and the gas permeation behavior for the vacuum-pyrolized CMS membranes, i.e., the permeabilities of the gases decreased, whereas the gas selectivities increased with increasing the pyrolysis temperature. For H_2/N_2 and CO_2/N_2 gas pairs, the selectivities were 1,450 and 101 at 308 K (PH_2=59 Barrer, PCO_2=4.2 Barrer), respectively, for the membrane pyrolized at 1273 K. Such extremely high gas selectivities imply that the CMS membrane showed a molecular sieving effect. Thus, larger molecules, such as N_2, are more restricted to diffuse through narrower pores than smaller gases, such as H_2. The membrane pyrolized at lower temperature (873 K) exhibited higher permeability (PH_2=1,600 Barrer at 308 K) and had relatively low selectivity (H_2/N_2=20). This is because the average pore size of the membrane (4.55 Å) was large enough for N_2 molecules (σ=3.64 Å) to permeate through the pores. Thus, the pyrolysis temperature can be used as a parameter for modification of the pore size with respect to the target gases to be separated.

Effects of Other Pyrolysis Conditions on Gas Permeation Properties. The permeability also decreased with decreasing the heating rate from 13.3 to 1.33 K/min. The reason for this trend may be that by decreasing the heating rate pyrolysis proceeds very slowly which may result in smaller pores. The longer total time for pyrolysis may also contribute to the pore size reduction due to a sintering effect. The membrane pyrolized at the slowest heating rate of 1.33 K/min under Ar flow at 1223 K exhibited the highest selectivity of all membranes studied

(H_2/N_2=4,700; CO_2/N_2=122; and O_2/N_2=36 at 308 K, PH_2=19 Barrer), indicating the possibility of using CMS membranes for hydrogen recovery or oxygen enrichment from air.

It is worthwhile to mention the effects of the pyrolysis atmosphere. From a recent study, Geiszler and Koros (15) concluded that even residual trace levels of oxygen (0.3 ppm) in the inert gas atmosphere affected the CMS membrane properties. They also reported that traces of oxygen made the pores larger; however, a reduction in the purge flow rate from 200 to 20 cm^3(STP)/min resulted in a reduction of pore size presumably due to deposition of carbon. Thus, in this study to investigate the threshold for oxygen level, pyrolysis was carried out under high purity Ar flow (purity: >99.9995%, flow rate: 500 cm^3(STP)/min). Comparing the difference in the pyrolysis atmosphere (vacuum or Ar) for two membranes, the trace of oxygen (< 0.2 ppm), and higher flow rates did not exhibit a severe effect on the permeation properties. This result suggests that the threshold for oxygen level that would alter the CMS membrane properties may be around 0.3 ppm as reported.

Order of Gas Permeabilities vs. Kinetic Diameters of Gas Molecules. The permeabilities of the selected gases were in the order of $H_2 > He > CO_2 > O_2 > N_2$ for all the CMS membranes except for the membrane prepared at 873 K. Similar results have been obtained for our asymmetric CMS membranes (23) and for nanoporous carbon membranes (25-28). On the other hand, microporous silica membranes with a pore size of less than 20 Å used by Shelekhin et al. (44) exhibited an ordinary trend, i.e., the permeability decreased in the order of increasing kinetic gas diameter. It is valuable to give an interpretation to the order of gas permeabilities against kinetic diameter, because this kind of correlation is related to the estimation of molecular dimensions from sorption kinetics (45) and the theoretical calculation of critical pore dimensions (46).

Therefore, a detailed study was performed based on the solution-diffusion theory (47,48) for the membrane prepared at 1223 K at a heating rate of 1.33 K/min under Ar flow to elucidate the order of permeability. In the solution-diffusion model, the permeability P is given as the product of the apparent diffusivity coefficient, D, and solubility, S, as follows:

$$P = P_0 \exp\left(\frac{-E_p}{RT}\right) = D \times S \qquad (1)$$

$$P = D_0 \exp\left(\frac{-E_d}{RT}\right) \times S_0 \exp\left(\frac{-H_s}{RT}\right) \qquad (2)$$

$$P = D_0 S_0 \exp\left\{\frac{-(E_d + H_s)}{RT}\right\}$$
(3)

The apparent activation energy for permeation, E_p, is the sum of the apparent activation energy for diffusion, E_d, and the apparent heat of sorption, H_s.

Figures 3 (a)-(c) show that the permeabilities and the diffusivities decreased from 10^1 to 10^{-3} Barrer and 10^{-7} to 10^{-12} cm^2/s at 373 K, respectively, in the order of the kinetic diameter, σ, of the penetrants studied. On the other hand, the solubility increased from 10^{-3} to 1 cm^3(STP)/cm^3(CMS)·cmHg in the order of the Lennard-Jones potential, ε/K (25). The greater change in the diffusivity compared to that of the solubility implies that the selectivity of the CMS membranes can be attributed to a molecular sieving effect that results mainly from the dependence of diffusivity on the pore size. Figures 4 (a)-(c) show the temperature dependence of permeability, diffusivity, and solubility. The E_d and H_s values were derived from the Arrhenius expression and could be correlated linearly with a positive slope, and the square of the kinetic diameter and ε/K, respectively. Thus, it can be implied that gases with larger cross-sectional area need greater activation energy for diffusion and that more condensable gases are more easily sorbed.

The anomalous behavior that the larger H_2 gas (σ=2.89Å) permeates faster through our CMS membranes than the smaller He gas (σ =2.60 Å) must result from a higher solubility of H_2. The present study revealed that the diffusivity of He at 373 K is about two times higher than that of H_2, whereas the solubility of He was about one-third of that of H_2. The phenomenon that CO_2 (σ =3.30 Å) permeates faster than the smaller He (σ =2.60 Å) through our membrane and for Hayashi's CMS membrane both pyrolized at 873 K (16) can be explained in a similar manner. However, the present study reports only single-gas selectivities and not mixed-gas selectivities. Considering that heavier hydrocarbons permeated faster than the smaller H_2 or CH_4 by selective surface flow through nanoporous carbon membranes (25), the present CMS membranes may also exhibit a different permeation behavior in gas mixtures. Such measurements are currently carried out and may give further insight into the permeation mechanism.

Flat-Type CMS Dense Membranes Prepared from Other Precursors (21)

The excellent gas separation performance of the Kapton polyimide derived CMS membranes can be attributed to the change in the pore size upon heating. It seems possible to further improve the gas separation performance (particularly permeabilities rather than selectivities) by the choice of the polymeric precursor. Here, the gas separation properties and the relationship to their microstructure are presented for CMS membranes prepared by pyrolysis of several types of polyimide films, including Kapton, Upilex, and Cardo types. An important criterion

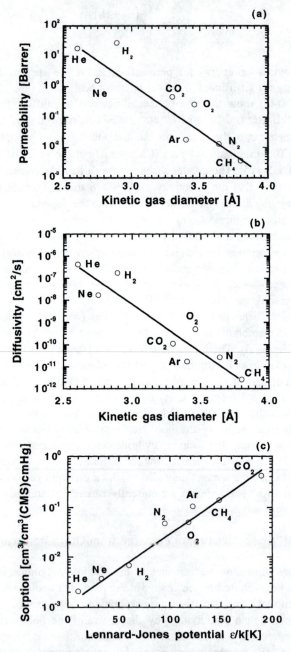

Figure 3. Relationship between (a) permeability and kinetic gas diameter, (b) diffusivity and kinetic gas diameter, and (c) solubility and L-J potential for a CMS membrane (measured at 373 K) prepared by pyrolysis of a Kapton film at 1223 K for 2 hours at a heating rate of 1.33 K/min under Ar flow.

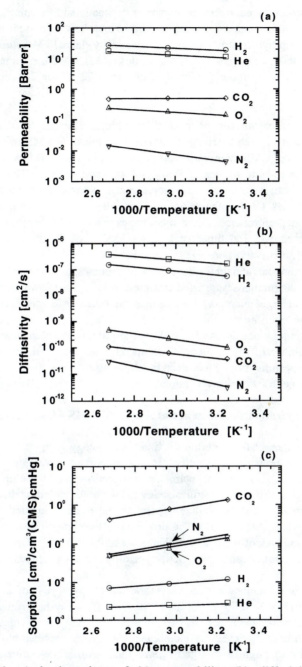

Figure 4. Arrhenius plots of (a) permeability (b) diffusivity, and (c) solubility for a CMS membrane prepared by pyrolysis of a Kapton film at 1223 K for 2 hours at a heating rate of 1.33 K/min under Ar flow.

in choosing an appropriate polymeric precursor is its fractional free volume (FFV). A Cardo-type polyimide is one of the best candidate materials, because it has very bulky groups in the polymer structure and, consequently, large FFV values (0.14-0.20) compared to those of Kapton (0.11) and Upilex (0.12). CMS membranes were fabricated by pyrolysis of these polyimide films at 1273 K for 2 hours under vacuum.

Structure-Gas Permeation Property Relationships. The gas permeabilities decreased with increasing the kinetic gas diameter for almost all CMS membranes. For the Kapton-derived CMS membranes, the permeabilities decreased, whereas the selectivities increased with increasing the pyrolysis temperature. The Upilex-derived CMS membranes exhibited similar separation performance as the Kapton-based CMS membranes. On the other hand, the Cardo-derived CMS membranes had different separation performance. These membranes exhibited higher permeability (e.g., PCO_2 increased over 100 times) with some loss in selectivity (e.g. CO_2/N_2 decreased 50 %). The effective pore sizes were around 3.7 and 4.1 Å for Kapton or Upilex-derived membranes and for Cardo-derived membranes, respectively.

The pore size distributions suggested that the Cardo-derived CMS membranes had bigger pores and larger micropore volume than the other CMS membranes. The pore sizes were in good agreement with those expected from the permeability-kinetic gas diameter relationship. As the FFV of the polyimide precursors became larger, the permeability of the resultant CMS membranes increased, as expected. Consequently, permeability can be increased by choosing an appropriate polymeric precursor with a larger fractional free volume.

Alkene/Alkane Separation Performance of Activated CMS Membranes (*22*)

Separation of light alkene/alkane mixtures is a key technology in the petrochemical industry. Propylene/propane separation is an important example, because propylene is used as an important petrochemical feedstock and useful for the production of a wide variety of chemicals. Several methods including distillation, adsorption, absorption, and membrane separation have been applied for the separation of propylene/propane (*49*). The drawbacks of polymeric membranes are a low selectivity and lack of thermal and chemical stability (*50-52*). On the other hand, inorganic membranes can exhibit better performance for propylene/propane separation (*53,54*).

The original CMS membrane used was fabricated by pyrolysis of a Kapton polyimide film at 1273 K for 2 hours under vacuum. Because the original CMS membrane had pores of about 3.7 Å, which are too small for permeation of hydrocarbons such as propylene, the membrane was further calcined under mild activation condition to slightly enlarge its pore dimensions. The activation was carried out as follows; the original membrane was heated to 673 K at a heating rate of 10 K/min with an Ar purge (100 cm³/min), and then the Ar was replaced with He (100 cm³/min) containing water vapor of about 20 mmHg. After a period of 10 min, the He was exchanged with Ar again, and then the sample was allowed to cool to room temperature. The sorption isotherms for probe molecules were determined for both the CMS membranes before and after the mild activation step. The sorption isotherms showed that the former adsorbed only the small CO_2 molecule (σ=3.3 Å), whereas the latter adsorbed even the larger C_2H_6 molecule (σ=3.9 Å). This result suggests that the mild activation step is effective for gradual opening of pore dimensions. The average pore size increased from about 3.7 A to 4.0 Å due to the mild activation step.

The single-gas permeabilities of the original CMS membrane were in the order of the kinetic gas diameters. The high selectivities of the membrane resulted from tailored pore dimensions suitable for separation of those gas pairs. On the other hand, the permeabilities of the modified CMS membrane were ten to several hundred times higher than those of the original membrane made at 373 K, which is mainly due to larger pore dimensions.

Additional permeation tests showed that there was appreciable permeation of even larger hydrocarbons. In particular, the selectivities for light alkene/alkane gas pairs with the same carbon number were rather high. The selectivity for propylene/propane, for example, was over 100 at 308 K (PC_3H_6 = 4.2 Barrer). This value is much higher than those reported for polymeric membranes. A rather high selectivity of about 6 was also obtained at 308 K for the ethylene/ethane gas pair (PC_2H_4 = 17 Barrer). The differences in diffusivity and/or sorption are the reasons why the CMS membrane exhibited excellent selectivity for alkene/alkane gas pairs. The CMS membrane activated under mild conditions appears to be one of the most promising candidates for alkene/alkane separation. The other candidate inorganic materials for alkene/alkane separation are sol-gel derived silica membranes (53).

Capillary-Type CMS Membranes with Asymmetric Structure (23,24)

The permeation properties of flat-sheet CMS membranes are controlled by micropores ranging from 3 to 4 Å. For practical applications, however, a capillary-type membrane is preferable to a flat-sheet membrane, because the former has a larger membrane area per unit module volume. The capillary CMS membrane must have a controlled asymmetric structure, consisting of a dense, selective surface layer with a molecular sieving ability and a porous supporting layer to attain both high selectivity and permeance. Although there have been some attempts to prepare capillary CMS membranes (6,29,30), it appears difficult to control the membrane structure. We developed a novel method for preparing asymmetric capillary CMS membranes. The method involves the formation of a capillary type polyamic acid (PA) membrane followed by an imidization and pyrolysis step, as shown in Figure 1.

A capillary-type PA membrane with an asymmetric structure was prepared by a phase inversion technique. A 16 wt% solution of PA in dimethylacetamide (DMAC) was used for casting. The PA solution was coated on the surface of a polytetrafluoroethylene (PTFE) tube of 1.8 mm outside diameter. After two seconds of exposure in air, the coated PTFE support was immersed continuously into a coagulant bath of water or ethanol at 276 K. The composite membranes were washed with water for 1 day. The PTFE tube was removed after drying under ambient conditions for 1 day. The dried capillary PA membrane was further dried under a vacuum of 10^{-5} Torr, and then imidized to a Kapton-type polyimide with two heating steps at 473 K for 30 min and at 673 K for 1 hour under vacuum. The polyimide membrane was pyrolized under vacuum at a heating rate of 20 K/min up to 1223 K for 1 hour.

SEM photographs of the pyrolized capillary CMS membranes revealed an asymmetric structure consisting of a dense surface layer and porous substructure, as shown in Figure 5. This asymmetric structure is characteristic for polymer membranes prepared by the phase inversion technique. The structure was formed in the gelation step of the polyamic acid and was also maintained in the imidization and pyrolization steps. Generally, the dense surface layer became thinner and the pore dimensions became larger with acceleration in the exchange rate of solvent with the coagulant. Using ethanol as the coagulant, the gelation process proceeded slowly, resulting in a thicker surface layer of about 10 μm. On the other hand, the dense surface layer became thinner using water as the coagulant, because the exchange of a polar solvent such as DMAC with water proceeded rather rapidly. The gas permeances were in the order of $H_2 > He > CO_2 > O_2 > N_2$, similar to those of flat-sheet CMS membranes. The selectivities were very high for the CMS membrane prepared in ethanol as the coagulant. For H_2/N_2, the selectivity value reached up to 1,080 (permeance of $H_2 = 7.4 \times 10^{-7}$ cm^3(STP)/cm^2·s·cmHg), which was comparable to that found in our flat-sheet CMS membranes (18-20). The capillary CMS membrane prepared using water as the coagulant increased the permeances by two to three orders of magnitude. It was difficult, however, to control the thickness of the dense surface layer because of the rather rapid exchange of the polar solvent (DMAC) with water. As a result, a reproducible membrane performance in the water coagulant system was difficult to achieve. The poor selectivities were caused by defects in the thin surface layer.

Capillary CMS membranes were modified to improve their separation performance (24). Modification was carried out by several methods including a modification in the starting polymer concentration and composition and using a dual-coagulation method. Polyamic acid solutions of up to 19 wt% in DMAC coagulated in water produced an open structure with macrovoids and a skin layer with defects. Concentrated solutions of up to 25 wt% yielded a more sponge-like structure with less macrovoids, but a less permeable skin layer (~2 μm thick). A very dense skin layer of about 2 μm with an open substructure was obtained by a

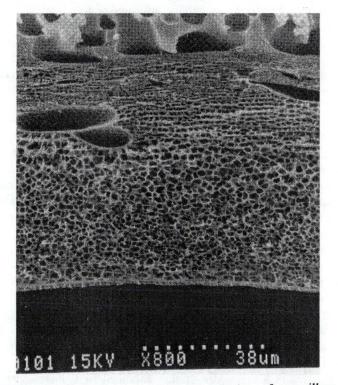

Figure 5. SEM photograph of a cross-sectional view of a capillary CMS membrane with an asymmetric structure prepared using water as the coagulant.

dual-coagulation method, where membranes were first coagulated in ethanol and then immersed in water. Optimal results were obtained by using a 15 wt% PA solution in 55 wt% DMAC and 30 wt% glycerol and by using water as the coagulant. The resultant membrane consisted of a thin skin layer (0.3-0.5 μm) and a homogeneous substrate with a regular sponge-like structure.

The permeances of smaller gases like H_2 and He for this membrane increased with an increase in temperature (form 273 to 523 K) in an Arrhenius-type plot. However, the slopes of other gases like CO_2 became larger at higher temperatures. This observation implies that competing transport mechanisms, that is, activated diffusion through micropores or surface diffusion or Knudsen diffusion through defect macropores contributed to this behavior. Thus, a further modification consisted of coating the CMS membranes with a thin layer of polydimethylsiloxane to prevent gas flow through possible defects. The modified CMS membrane displayed almost a straight line in the Arrhenius plot for all the gases studied. Furthermore, the membrane had selectivities of about 2,000 for He/N_2 and 1,000 for H_2/N_2 at 273 K, respectively (H_2 permeance=7.2 x 10^{-7} cm^3(STP)/cm^2·s·cmHg). The H_2/N_2 selectivity reached 136 even at a temperature of 523 K (H_2 permeance = 8.3 x 10^{-5} cm^3(STP)/cm^2·s·cmHg). In addition, mixed-gas measurements showed that the permeation behavior for H_2/CO and H_2/CO_2 was not different from that observed in single-gas measurements. This observation implies that each gas molecule can diffuse almost freely through the slit-like pores of the CMS. Consequently, the separation performance for mixed-gases of CMS membranes may be close to those obtained in single-gas permeation experiments.

Conclusions

The present study showed that a series of flat- and capillary-type CMS membranes with tailored microstructure (pore size, pore volume, etc.) can be obtained by controlling the pyrolysis conditions (pyrolysis temperature, heating rate, and pyrolysis atmosphere) and post-treatment conditions. Of the pyrolysis conditions, the heating temperature had the greatest effect on the microstructure and gas permeation properties of the membranes. The membrane pyrolized at the slowest heating rate of 1.33 K/min under Ar flow exhibited the highest selectivity of all membranes studied. The high selectivity of the CMS membrane can be attributed to a molecular sieving effect that resulted mainly from the dependence of the diffusivity of penetrants on the size of the micropores ranging from 3.7 to 4.6 Å. Furthermore, the higher permeabilities of H_2 and/or CO_2 compared with that of the smaller He can be ascribed to the larger solubilities of the former two gases. Other membranes whose pore openings were enlarged by moderate activation or those prepared from other polyimides with larger fractional free volume provided a series of CMS membranes that exhibited different permeation properties. In some cases, the membranes were effective even for alkene/alkane separation. It has been recognized that there exists a "tradeoff" relationship between selectivity and permeability in polymer membranes, that is, a membrane with higher permeability tends to exhibit a lower selectivity, and vice versa, as shown in Figure 6 for O_2/N_2 separation (55,56). Similar relationships can also be depicted for other gas pairs.

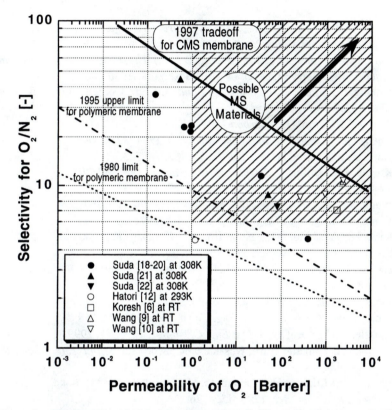

Figure 6. Trade-off relationship between O_2/N_2 selectivity and O_2 permeability for polymeric membranes (data for polymers from Koros *et al.* (*55,56*)) and for presently available CMS membranes.

Progress in the last decade resulted in better polymeric membranes for O_2/N_2 separation; however, there are essentially no polymeric membranes above the tradeoff line. On the other hand, molecular sieve membranes are expected to lie above the upper bound lines for polymers and their separation performance may lie in the cross-hatched region, shown in Figure 6. However, it appears that there exists a new tradeoff line for CMS membranes.

Literature Cited

1. Breck, D.W. in *Zeolite Molecular Sieves-Structure, Chemistry, and Use*, p.636, John Wiley & Sons, New York, 1974.
2. Reid, R.C.; Prausnitz, J.M.; Poling, B.E. in *The Properties of Gases and Liquids*, 4th ed., p.733, McGraw-Hill, New York, 1986.
3. Pauling, L.; *Nature of the Chemical Bond*, 3rd Ed., Cornell University Press, Ithaca, New York, 1960,
4. Bird, J.E.; Trimm, D.L. *Carbon* **1983**, *21*, 177.
5. Chen, Y.D.; Yang, R.T. *Ind. Eng. Chem.Res.* **1994**, *33*, 3146.
6. Koresh, J.E.; Soffer, A. *Sep. Sci. Tech.* **1983**, *18*, 723.
7. Koresh, J.E.; Soffer, A. *J. Chem. Soc. Faraday I* **1986**, *82*, 2057.
8. Koresh, J.E.; Soffer, A. *Sep. Sci. Tech.* **1987**, *22*, 973.
9. Wang, S.; Zeng, M.; Wang, Z. *J. Membr. Sci.* **1996**, *109*, 267.
10. Wang, S.; Zeng, M.; Wang, Z. *Sep. Sci. Tech.* **1996**, *31*, 2299.
11. Kita, H.; Yoshino, M.; Noboriro, K.; Tanaka, K.; Okamoto, K. Proc. Euromembrane '97, Netherlands, 1997, p. 20.
12. Hatori, H.; Yamada, Y.; Shiraishi, M.; Nakata, H.; Yoshitomi, S. *Carbon* **1992**, *30*, 305.
13. Jones, C.W.; Koros, W.J. *Carbon* **1994**, *32*, 1419.
14. Jones, C.W.; Koros, W.J. *Ind. Eng. Chem. Res.* **1995**, *34*, 158.
15. Geiszler,V.C.; Koros, W.J. *Ind. Eng. Chem.Res.* **1996**, *35*, 2999.
16. Hayashi, J.; Yamamoto, M.; Kusakabe, K.; Morooka, S. *Ind. Eng. Chem. Res.* **1995**, *34*, 4364.
17. Hayashi, J.; Mizuta, H.; Yamamoto, M.; Kusakabe,K.; Morooka, S.; Suh, S-H. *Ind. Eng. Chem. Res.* **1996**, *35*, 4176.
18. Haraya, K.; Itoh, N. Proc. ICOM'93, Heidelberg, 1993, p.2.5.
19. Suda, H.; Haraya, K. *J. Chem. Soc. Chem. Commun.* **1995**, *11*, 1179.
20. Suda, H.; Haraya, K. *J.Phys.Chem.B* **1997**, *101*, 3988.
21. Suda, H.; Kazama, S.; Haraya, K. Proc. ICOM'96, Yokohama, 1996, 460.
22. Suda, H.; Haraya, K. *J. Chem. Soc. Chem. Commun.* **1997**, 93.
23. Haraya, K.; Suda, H.; Yanagishita, H.; Matsuda, M. *J. Chem. Soc. Chem. Commun.* **1995**, *17*, 1781.
24. Petersen, J.; Matsuda, M.; Haraya, K. *J. Membr. Sci.* **1997**, *131*, 85.
25. Rao, M.B.; Sircar, S. *J. Membr. Sci.* **1993**, *85*, 253.
26. Rao, M.B.; Sircar, S. *J.Membr.Sci.* **1996**, *110*, 109.
27. Anand, M.; Langsam, M.; Rao, M.B.; Sircar, S. *J.Membr.Sci.* **1997**, *123*, 17.
28. Naheiri, T.; Ludwig, K.A.; Anand, M.; Rao, M.B.; Sircar, S. *Sep. Sci. Tech.* **1997**, *32*, 1589.

29. Braymer, T.A.; Coe, C.G.; Farris, T.S.; Gaffney, T.R.; Schork J.M.; Armor, J.N. *Carbon* **1994**, *32*, 445.
30. Linkov, V.M.; Sanderson, R.D.; Jacobs, E.P. *J. Membr .Sci.* **1994**, *95*, 93.
31. Koresh, J.E.; Soffer, A. *J. Chem. Soc. Faraday Trans. I* **1980**, *76*, 2457.
32. Hayashi, J.; Yamamoto, M.; Kusakabe, K.; Morooka, S. *Ind. Eng. Chem. Res.* **1997**, *36*, 2134.
33. Hayashi, J.; Mizuta, H.; Yamamoto, M.; Kusakabe, K.; Morooka, S. *J. Membr. Sci.* **1997**, *124*, 243.
34. Jones, C.W.; Koros, W.J. *Carbon* **1994**, *32*, 1427.
35. Jones, C.W.; Koros, W.J. *Ind. Eng. Chem. Res.* **1995**, *34*, 164.
36. Dubinin, M.M. Chem.Rev. **1960**, *60*, 235.
37. Astakhov, V.A.; Dubinin, M.M.; Romankov, P.G. *Theor.Osn.Khim.* **1969**, *3*, 292.
38. Barrer, R.M. *Trans.Faraday Soc.* **1964**, *35*, 628.
39. Barrer, R.M. Proc.Int.Symp. Surface Area Determination, p.227,Bristol,1970.
40. Inagaki, M.; Harada, S.; Sato, T.; Nakajima, T.; Horino, Y.; Morita, K. *Carbon* **1989**, *27*, 253.
41. Franklin, R.E. *Proc. Roy. Soc.* **1951**, *A209*, 196.
42. Hatori, H.; Yamada, Y.; Shiraishi, M.; Yoshihara, M.; Kimura, T. *Carbon* **1996**, *34*(2), 201.
43. Bhandarkar, M.; Shelekhin, A.B.; Dixon, A.G.; Ma, Y.H. *.J. Membr .Sci.* **1992**, *75*,221.
44. Shelekhin, A.B.; Dixon, A.G.; Ma, Y.H. *J.Membr. Sci.* **1992**, *75*, 233.
45. Koresh, J.E.; Soffer, A. *J. Chem. Soc. Faraday Trans.* I **1980**, *76*, 2472.
46. Rao,M.B.; R.G.Jenkins,R.G.; Steele,W.A. *Langmuir* **1985**, *1*, 137.
47. Koros, W.J.; Chern, R.T. in *Handbook of Separation Process Technology* (Ed., Rousseau,R.W.), Chapter 20, John Wiley & Sons, New York, 1987.
48. Crank,J. in *The Mathematics of Diffusion*, 2nd ed., p.51, Clarendon, Oxford, 1975.
49. Eldridge, R.B. *Ind. Eng. Chem.Res.* **1993**, *32*, 2208.
50. Ilinitch, O.M.; Semin, G.L.; Chertova, M.V.; Zamaraev, K.I. *J. Membr. Sci.* **1992**, *66*, 1.
51. Ito, A.; Hwang, S.T. *J. Appl. Polym. Sci.* **1989**, *38*, 483.
52. Lee, K.R.; Hwang, S.T. *J. Membr. Sci.* **1992**, *73*, 37.
53. Asaeda, M.; Yamamichi,A.; Satoh, M.; Kamakura, M. Proc. 2nd. Int.Conf. Inorg. Membr., 1993, 315.
54. Nair, B.N.; Keizer, K.; Verweij, H.; Burggraaf,A.*J. Sep. Sci. Tech.* **1996**, *31*, 1907.
55. Koros, W.J. *Chem. Eng. Prog.* **1995**, 68.
56. Singh, A.; Koros, W.J. *Ind. Eng. Chem. Res.* **1996**, *35*, 1231.

Chapter 21

Preparation of Carbon Molecular Sieve Membranes and Their Gas Separation Properties

Ken-ichi Okamoto, Makoto Yoshino, Kenji Noborio, Hiroshi Maeda, Kazuhiro Tanaka, and Hidetoshi Kita

Department of Advanced Materials Science and Engineering, Faculty of Engineering, Yamaguchi University, Ube, Yamaguchi 755, Japan

Carbon molecular sieve (CMS) membranes were prepared by pyrolyzing flat membranes of 3,3',4,4'-diphenylhexafluoro-isopropylidene tetracarboxylic dianhydride (6FDA)-based polyimide and polypyrrolone and composite membranes of phenolic resin coated on alumina tubes at 500-800°C in N_2. The gas permeability and selectivity depended significantly on the precursor polymers and pyrolysis conditions. With increasing pyrolysis temperature, the permeability decreased and the selectivity increased, because the membranes became denser and the effective pore size for molecular sieving became smaller, whereas gas adsorption increased. CMS membranes from the polyimide and polypyrrolone displayed high performance for gas separation applications such as CO_2/CH_4, O_2/N_2, and propylene/propane. CMS membranes made from a phenolic resin displayed high performance for CO_2/N_2 separation.

In adsorption applications, carbon molecular sieve membranes produced by the pyrolysis of polymeric materials have been studied extensively (*1*). There is growing interest in using molecular sieve membranes to separate gas mixtures with very similar molecular dimensions or organic vapor mixtures. However, it is very difficult to prepare carbonized membranes free of pinholes and with a molecular sieving effect (*2*). Some recent work reported on lab-scale carbon molecular sieve (CMS) membranes with excellent performance for gas separation applications (*3-17*). However, it is not clear from previous work how much the precursor polymers and pyrolysis conditions affected the gas separation performance of CMS membranes. Here, we report on the preparation of CMS membranes from three different precursor polymers, that is, polyimide, polypyrrolone, and a phenolic resin, and their characterization, gas adsorption and gas separation properties.

CMS Membranes from Polyimide and Polypyrrolone

Preparation and Characterization. Figure 1 shows the chemical structures of the precursor polymers used in this work. Polyimide was prepared from 3,3',4,4'-diphenylhexafluoroisopropylidene tetracarboxylic dianhydride (6FDA) and m-phenylenediamine (mPD) by a chemical imidization method. Polypyrrolone was prepared from 6FDA and 3,3'-diaminobenzidine (DABZ) by a two-step synthesis. Poly(amic acid) films were thermally imidized at 150°C for 12 h in vacuum, followed by thermal cyclodehydration to the polypyrrolone at 300°C for 2h. The glass transition temperatures of 6FDA-mPD polyimide and that of 6FDA-DABZ polypyrrolone were 298°C and >470°C, respectively. The membranes were heated up to the pyrolysis temperature (500-800°C) at a heating rate of 5°C/min under nitrogen atmosphere and held for 0-5 h to yield carbon membranes (40-50 μm in thickness and 3 cm in diameter). The weight loss amounted to 20 and 40% for the polyimide carbonized for 1 h at 500 and 700°C, respectively, and 15 and 31% for the polypyrrolone, respectively.

TG-MS analysis indicated that carbonization began to occur at 500°C (Figure 2) with the evolution of carbon monoxide, carbon dioxide, hydrogen fluoride, and carbon tetrafluoride. The O and F atomic contents decreased significantly and the C content increased at 500-600°C, whereas the N and H contents changed rather slightly (Figure 3). From these results, it can be assumed that the initial pyrolysis process for the polypyrrolone occurred as shown in Figure 4. The cleavage of hexafluoroisopropylidene groups leads to evolution of HF and CF_4. The cleavage of cyclic imide leads to evolution of CO and formation of cyclic imidazole. The cleavage of cyclic imidazole leads to evolution of CO and formation of nitrile groups. The presence of nitrile groups in the membranes pyrolyzed at 500-600°C was confirmed from the IR spectra. A similar cleavage of hexafluoro-isopropylidene groups and cyclic imide can be assumed for the polyimide.

The d-spacing values of the carbonized membranes were determined from the diffraction peak angles and are listed in Table I together with the density values determined from the weight and apparent volume. Both the density and d-spacing hardly changed with heating up to 500°C. On the other hand, with heating above 700°C, the density increased and the d-spacing decreased, indicating that the membranes became much denser with increasing pyrolysis temperature above 700°C.

Gas Adsorption and Permeation Properties. Adsorption and desorption isotherms of N_2 for the membranes belonging to Type 1 sorption isotherms, without hysteresis in the desorption isotherms, indicated no capillary condensation. Figure 5 shows adsorption isotherms of N_2, CH_4, and CO_2 for the polypyrrolone and its carbonized membranes. The adsorption amounts were larger for the membranes pyrolyzed at a higher temperature. The carbonized polyimide

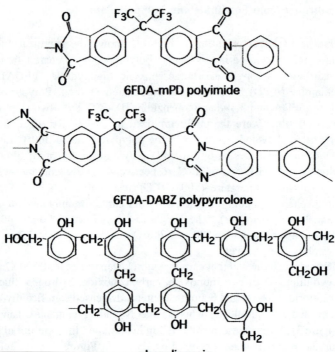

Figure 1. Chemical structures of 6FDA-mPD polyimide and 6FDA-DABZ polypyrrolone and a phenolic resin.

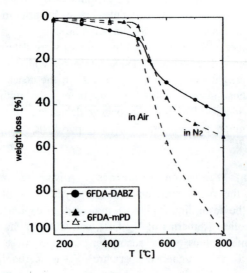

Figure 2. Thermograms of 6FDA-DABZ polypyrrolone and 6FDA-mPDpolyimide.

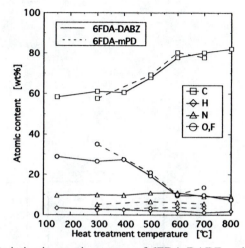

Figure 3. Variation in atomic content of 6FDA-DABZ polypyrrolone and 6FDA-mPD polyimide with heat treatment.

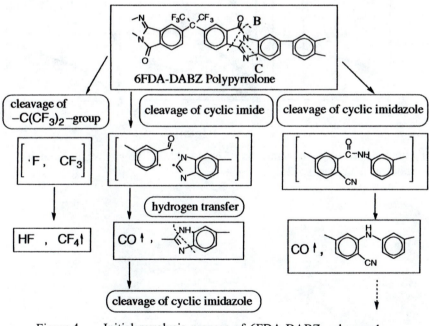

Figure 4. Initial pyrolysis process of 6FDA-DABZ polypyrrolone.

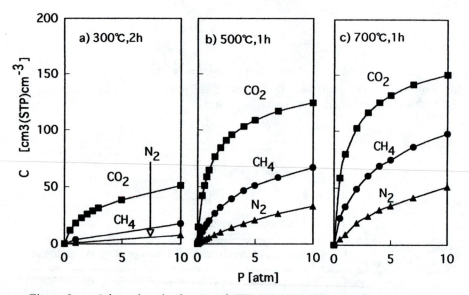

Figure 5.   Adsorption isotherms of 6FDA-DABZ polypyrrolone and its carbonized membranes at 35°C.

membranes showed a similar adsorption behavior, although their levels of adsorption were smaller compared with those of the carbonized polypyrrolone.

Table I. Physical properties of carbonized membranes.

Precursor polymer	Pyrolysis conditions	Density (g/cm^3)	d-spacing (nm)
6FDA-mPD	200°C, 20h	1.475	
	500°C, 1h	1.411	
	600°C, 1h	1.472	
	700°C, 1h	1.525	
6FDA-DABZ	300°C, 2h	1.401	0.59
	400°C, 10h	1.412	
	500°C, 1h	1.398	0.59
	550°C, 1h		0.58
	700°C, 1h	1.581	0.49
	800°C, 1h	1.643	0.44

The micropore volumes were determined from CO_2 adsorption using the Dubinin-Radushkevich equation and are listed in Table II together with the data of the carbonized membranes of other polyimides (*10,18*).   The micropore volume increased slightly with increasing pyrolysis temperature. This result may be due to the shift of the pore size distribution towards smaller pores.   The carbonized membranes had similar values of micropore volumes, indicating that the micropore volume depends rather weakly on the chemical structure of the precursor polymers.

Table II. Micropore volumes (W_0) evaluated from D-R equation of CO_2 adsorption.

Precursor polymer	Pyrolysis conditions	W_0 (cm^3/g)	Ref.
6FDA-mPD	500°C, 1h	0.21	this work
	700°C, 1h	0.22	this work
6FDA-DABZ	500°C, 1h	0.22	this work
	700°C, 1h	0.24	this work
BPDA-ODA	600°C	0.22	10
	700°C	0.25	10
	800°C	0.36	10
Kapton	600°C, 1h	0.24	18
	700°C, 1h	0.27	18
	800°C, 1h	0.25	18

Gas permeation experiments were carried out by a vacuum time-lag method. The permeability coefficient, P, (1 Barrer = 1×10^{-10} cm^3(STP)·cm/(cm^2·s·cmHg))and the permeance, P/l, (1 GPU = 1×10^{-6} cm^3(STP)/(cm^2·s·cmHg) used here are common units for gas permeation properties.

Figure 6 shows the relationship between the permeability coefficients and kinetic diameters of penetrant gases for the polypyrrolone and its carbonized membranes. Pyrolysis at 500°C for 1 h enhanced the gas permeability about 100-fold. With increasing the pyrolysis temperature to 700°C, the gas permeability decreased, but its dependence on the kinetic diameter became more pronounced. Pyrolysis at 800°C decreased the gas permeability significantly, and the membranes became permeable essentially only to H_2. A similar behavior was observed for carbonized polyimide membranes. The gas permeation properties, the density, and d-spacing data mentioned above indicate that the effective pore size of molecular sieve membranes became smaller with increasing pyrolysis temperature from 500°C to 800°C. It should be noted that the effect of the pyrolysis temperature on the gas permeability was quite different from that of gas adsorption, that is, the gas permeability decreased significantly with increasing temperature from 500°C to 700°C, whereas both gas adsorption and the micropore volume increased. Although any open pore can take part in gas adsorption, only penetrating micropores can take part in gas permeation. Moreover, bottle-neck type micropores can control the gas permeation and separation properties of the CMS membranes.

Figure 7 shows plots of the permeability ratio of CO_2/CH_4 versus CO_2 permeability for pure-gas measurements for the 6FDA-mPD polyimide and 6FDA-DABZ polypyrrolone and their carbonized membranes together with the reference data of the carbonized membranes of polyimide from 3,3'4,4'-biphenyl-tetracarboxylic acid dianhydride (BPDA) and 4,4'-oxydianiline (ODA). Figure 7 also shows the upper bound line reported by Robeson (19). This line denotes the upper limit of the CO_2/CH_4 separating ability of polymers known up to 1991. The data for 6FDA-mPD polyimide and 6FDA-DABZ polypyrrolone are below and above the upper bound line, respectively. On the other hand, the data for the carbonized membranes are much above the upper bound line, indicating that the carbonized membranes have much better performance than the precursor polymers.

The membrane performance depended significantly on the pyrolysis conditions such as temperature and time. With increasing the pyrolysis temperature from 500°C to 700°C, the CO_2 permeability decreased, but the selectivity increased, indicating that the membranes acted more like a "molecular sieve" for the CO_2/CH_4 gas pair. Increasing the heating time from 1 h to 2 or 5 h resulted in a significant decrease in the selectivity. The 6FDA-DABZ polypyrrolone membranes carbonized at 500 or 700°C for 1 h had better gas separation properties than those of the carbonized polyimide membranes.

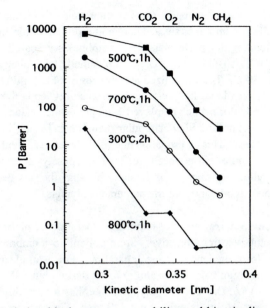

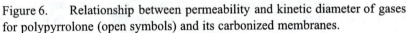

Figure 6.   Relationship between permeability and kinetic diameter of gases for polypyrrolone (open symbols) and its carbonized membranes.

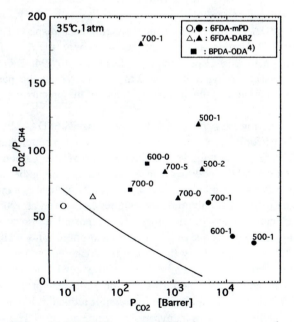

Figure 7.   Plots of PCO_2/PCH_4 versus PCO_2 for precursor polymers (open symbols) and carbonized membranes. The solid line is the upper bound line (*19*). The numerals refer to pyrolysis temperature and time.

Tables III and IV show comparisons of membrane performance toward several gas pairs among CMS membranes studied here and reported in the literature. Jones and Koros reported on the performance of membranes prepared by pyrolyzing hollow-fibers of 6FDA/BPDA-TrMPD (2,4,6-trimethyl-1,3-phenylene diamine) polyimide at 500-550°C. The carbonized membranes of 6FDA-mPD polyimide and 6FDA-DABZ polypyrrolone had very high permeability and, therefore, the flat membranes of around 50-μm-thick displayed high permeance values, comparable to those of the carbonized hollow fiber or tubular membranes having an active layer as thin as a few μm or less. This is probably due to their less efficient polymer chain packing and the more pronounced effect of pyrolysis because of the hexafluoroisopropylidene group. It is noteworthy that the gas permeability and selectivity depended significantly on the precursor polymers.

Propylene/Propane Separation. Membrane separation of olefins and paraffins is another important research objective. Solid polymeric membranes have rather poor performance for propylene/propane separation. For example, 6FDA-TrMPD polyimide has a fairly high C_3H_6 permeability of 30 Barrers but has a relatively low selectivity of 10 (*20*). 6FDA-DABZ polypyrrolone has a fairly high selectivity of 45, but its C_3H_6 permeability is only about 1 Barrer. Furthermore, the selectivity decreased significantly to 15 in gas mixture experiments. Therefore, in this study we also investigated the propylene/propane separation properties of carbonized membranes. Table V lists the performance of carbonized membranes for propylene/propane separation.

Carbonized 6FDA-DABZ polypyrrolone membranes displayed a high C_3H_6 permeability of 170 Barrers and a pure-gas selectivity of 78. In mixed-gas permeation experiments, the permeability decreased to 94 Barrers and the propylene/propane selectivity, α, decreased to 44. However, the performance of the carbonized membrane was much better than that of the precursor polypyrrolone membrane.

A similar behavior was observed for carbonized 6FDA-mPD polyimide membranes. The performance of these membranes was much better than that of the carbonized BPDA-ODA polyimide membranes and of the carbonized Kapton polyimide membrane mildly modified with water steam. This is because their less dense structure was more suitable to sieving by gases with larger molecular size differences. The 6FDA-mPD polyimide membranes carbonized at 700°C displayed a significant reduction in gas permeability after exposure to propane, probably because of micropore clogging due to irreversible adsorption. Only a slight recovery of the permeability was observed by the desorption treatment at 150°C in vacuum. After exposure to propane, the membrane reached a newly organized state and, thereafter, gave a reproducible performance for propylene/propane separation (Figure 8), which is listed in Table V. The irreversible reduction in gas permeability became much smaller with lower pyrolysis temperature. The 6FDA-mPD

Table III. Permeability (P) and selectivity of carbon molecular sieve membranes.[a]

Membrane	Ref.	Pyrolysis conditions	T (°C)	Permeability (Barrers)			Selectivity			
				H_2	CO_2	O_2	H_2/CH_4	CO_2/N_2	CO_2/CH_4	O_2/N_2
6FDA-DABZ[b]	this study	500°C, 1h	35	6,600	3,000	670	270	40	120	9.1
		700 °C, 1h		1,700	250	68	1,200	39	180	11
6FDA-mPD[b]	this study	600°C, 1h	35	19,000	16,000	3,100	39	26	32	5.2
		700°C, 1h		11,000	4,600	950	140	39	60	8.0
Kapton[b]	4	800°C, 1 h	20	-	13	1.2	-	50	-	4.6
Kapton[b]	7,8	800°C, 2h	35	670	130	35	-	42	-	12
		950°C, 1h		53	3.5	0.92	-	83	-	22
BPDA-ODA[c]	10	600°C, 0h	25	-	330	-	-	30	80	-

[a] Single-gas permeation; P is given in Barrer (1 Barrer=1x10^{-10}cm^3(STP)•cm/cm^2•s•cmHg);
[b] dense and flat-sheet; [c] tubular (effective thickness of 2 μm).

Table IV. Permeance (P/l) and selectivity of carbon molecular sieve membranes.[a]

Membrane	Ref.	Pyrolysis conditions	T (°C)	Permeance (GPU)			Selectivity			
				H_2	CO_2	O_2	H_2/CH_4	CO_2/N_2	CO_2/CH_4	O_2/N_2
6FDA-DABZ[b]	this study	500°C, 1h	35	130	59	14	270	40	120	9.1
		700 °C, 1h		34	5.0	1.4	1,200	39	180	11
unspecified[c]	3	950°C	RT	-	-	190	-	-	-	8.0
BPDA-PI[d]	13	700°C, 4 min.	50	760	130	43	200	20	33	6.7
BPDA/6FDA-TrMPD[d]	5	500°C, 2h	RT	-	(100)	(20-50)	-	(55)	-	(8.5-12)
		550°C, 2h		(170)	(73)	(15-40)	(450)	-	(190)	(11-14)
BPDA-ODA[e]	10,12	600°C, 0h	25	-	160	22[f]	-	30	80	-
		700°C, 0h		-	79		-	55	57	7.5[f]

[a] data in parenthesis are for mixed-gas permeation; feed composition: H_2/CH_4: 50 mol% H_2, CO_2/N_2: 15 mol% CO_2, CO_2/CH_4: 50 mol% CO_2, O_2/N_2: 20 mol% O_2; Permeance is given in GPU (1 GPU=1x10^{-6}cm^3(STP)/cm^2•s•cmHg);
[b] dense and flat-sheet, 50 μm thickness; [c] precursor polymer unspecified; hollow fiber (effective thickness of 6±1 μm), activated; d hollow fiber, [e] tubular; [f] at 65°C.

polyimide and 6FDA-DABZ polypyrrolone membranes carbonized at 500°C displayed reproducible performance for propylene/propane separation even at room temperature.

Table V. Performance of carbon molecular sieve membranes for
propylene/propane separation at 35°C.[a]

Membrane	Reference	Pyrolysis conditions	C_3H_6 Permeability (Barrers)		Selectivity C_3H_6/C_3H_8	
			pure-gas	mixed-gas	pure-gas	mixed-gas
6FDA-mPD[b]	this study	500°C, 1h	1,600	1,100	25	13
		600°C, 1h	1,200	660	33	16
		700°C, 1h	53	26	530	130
6FDA-DABZ[b]	this study	500°C, 1h	170	94	78	44
Modified Kapton[c]	9	1,000°C	10[e]	-	20[e]	-
BPDA-ODA[d]	11	700°C, 0h	14	14	54	46

[a]mixed-gas permeation performed with a propylene/propane mixture (50:50 mol%); [b]dense and flat-sheet (50 μm thick); [c]dense and flat-sheet (113 μm thick); [d]tubular; [e]at 100°C.

Figure 8 shows the temperature dependence of the permeability and selectivity for a propylene/propane mixture. With increasing temperature, the permeability increased, whereas the selectivity decreased. The CMS membranes should be used at higher temperature to eliminate the complicated effects of adsorption of coexisting components and contaminants.

Figure 9 shows the time dependence of adsorption of propylene and propane in the carbonized polyimide membranes. For membranes carbonized at 500°C, the equilibrium sorption amount is only a little larger for propylene than for propane, whereas the adsorption rate was much faster for propylene. For membranes carbonized at 700°C, equilibrium adsorption was attained for propylene in about 5 hours. On the other hand, equilibrium adsorption for propane required at least 100 hours. Separation of propylene/propane mixtures through the carbonized membranes is clearly due to a molecular sieving effect rather than a difference in adsorption properties. The mixed-gas, steady-state permeation properties were obtained within a much shorter time compared with the pure-component sorption and pure-gas permeation of propane. This result indicates that the presence of propylene enhanced the diffusion and permeability of propane.

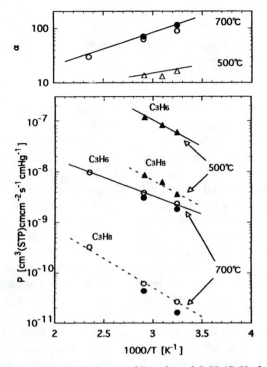

Figure 8. Temperature dependence of P and α of C_3H_6/C_3H_8 for 6FDA-mPD polyimide membranes carbonized at 500 and 700°C.

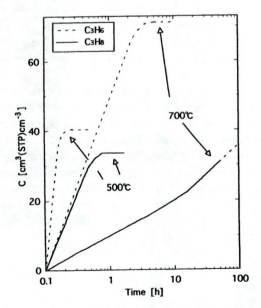

Figure 9. Time-dependence of C_3H_6 and C_3H_8 adsorption (0.5 atm, 75°C) for 6FDA-mPD polyimide membranes carbonized at 500 and 700°C.

CMS Membranes from Phenolic Resin

A phenolic resin was also selected as a precursor polymer on the basis of its high carbon yield and thermosetting properties without deforming the deposition layer on a substrate during heating and pyrolysis.

Preparation and Characterization. A phenolic resin (Bellpearl S-895) used as the starting material was supplied by Kanebo Ltd. Porous α-alumina tubes, supplied by Mitsui Grinding Wheel Co., had a mean pore diameter of 1 μm with a porosity of about 50%, and their dimensions were 1 cm in outer diameter, 0.2 cm in thickness, and 5 cm in effective length. Colloidal silica (EG-ST, supplied by Nissan Chemical Ind.) was dip-coated onto the α-alumina tubes in order to reduce the pore size of the substrate. The silica-coated alumina tubes thus prepared were used as substrate. In addition, another type of porous α-alumina tubes supplied by NOK Co. was used. These support membranes had a mean pore diameter of 0.15 μm with a porosity of 44% and their dimensions were 0.23 cm in outer diameter, 0.03 cm in thickness, and were 3.3 cm in effective length. The membranes from phenolic resin were formed by a dip process in which the porous substrate capped with silicone rubber was dipped into a 40 wt% (or 20 wt% as noted below) phenolic resin solution in methanol for 20 min at room temperature. The coated substrate was then pulled out at a rate of 1 or 20 cm/min. After air-drying at 50°C for one day and vacuum-drying at 50°C for one day, the membranes were heated up to 500-700°C at a heating rate of 5°C/min under nitrogen atmosphere and held at the temperature for 1 h. The coating-pyrolysis cycle was repeated two or three times. After the second cycle, the coating was carried out using a 20 wt% solution in methanol.

Thermogravimetry showed that a significant weight loss occurred around 200, 380, and 555°C, respectively, followed by gas evolution. The evolved gases were carbon monoxide, carbon dioxide, water, and larger molecules. The weight loss was about 45% at 800°C.

SEM analysis indicated a nonporous carbon layer on the substrate with a thickness of several micrometers. To reduce the thickness of the carbon layer, the 20 wt% phenolic resin solution was used even for the first coating. In this case, the surface carbon layer was so thin that the porous structure of the substrate could be observed from the surface view of the SEM. The cross-section view of the SEM indicated that the carbon layer was formed on the substrate in a depth of about 3 μm from the surface (Membrane ND2-3 in Table VI). The micropore volume determined from nitrogen adsorption for the phenolic resin membrane carbonized at 600°C for 2h was 0.23 cm^3/g, which was smaller than that of the 6FDA-DABZ polypyrrolone membrane (0.30 cm^3/g).

Gas Separation Properties. Gas permeation experiments were carried out by the vacuum time-lag method. Single-component or mixed-gas was fed to the outer side of the membrane in a permeation cell. The membrane was sealed in the cell module with a fluoro-rubber o-ring and the effective membrane area was 15.2 or 2.4 cm^2.

For the silica-coated substrate (Type S), the permeance decreased sharply with the second coating and started to level off after the third coating. Thus, three coatings gave highly selective membranes. For the alumina substrate having pores of 1 μm, more than 4 coatings were necessary to give a pinhole-free membrane. On the other hand, for the alumina substrate having pores of 0.14 μm (Type N), only two coatings gave highly selective membranes.

Table VI shows a comparison of membrane performance toward several gas pairs of CMS membranes from a phenolic resin. The gas permeation properties depended significantly on the type of substrate.

Table VI. Permeance (P/L) and selectivity of carbon molecular sieve membranes from phenolic resin at 35°C.[a]

Membrane	Pyrolysis conditions	Permeance (GPU)			Selectivity			
		H$_2$	CO$_2$	O$_2$	H$_2$/CH$_4$	CO$_2$/N$_2$	CO$_2$/CH$_4$	O$_2$/N$_2$
SD3-1	700°C, 1h	210	150	35	110	24	63	5.2
			(150)			(30)		
ND2-1	500°C, 1h	380	83	19	370	34	82	7.9
			(84)			(51)		
ND2-2	600°C, 1h	410	73	15	390	34	70	7.3
			(63)			(51)		
ND2-3	600°C, 1h	330	210	40	63	32	38	6.2
			(210)			(36)		
ND2-4	700°C, 1h	110	20	4.2	690	44	140	8.9
			(17)			(55)		

[a]data in parenthesis are for mixed-gas permeation; 20 mol% CO$_2$. Membrane code refers to substrate (S = silica-coated alumina tube, N = smaller size alumina tube; coating process of phenolic resin (D = dip), coating times and sample identification. ND2-3: 20 wt% phenolic resin solution was used even for the first coating.

Membrane ND2-4 had a larger penetrant-size dependence of the permeance and a higher selectivity than the membrane SD3-1. With increasing the pyrolysis temperature from 500 to 700°C, the permeance decreased, but the selectivity increased. This is similar to the behavior of the CMS membranes made from the polyimide and polypyrrolone. The permeance of N$_2$ and CH$_4$ increased with increasing temperature. The permeances of H$_2$, He, and O$_2$ were essentially independent of temperature. On the other hand, the CO$_2$ permeance decreased with increasing temperature. This behavior suggests contribution of surface diffusion for CO$_2$ permeation. As compared with the single-component system, the CO$_2$

permeance in the mixed-component system was essentially constant, but the N_2 permeance decreased. As a result, the selectivity increased by about 30-50%. The decrease in the N_2 permeance may be due to preferential adsorption of CO_2 in the micropores of the membrane and, therefore, hindrance by adsorbed CO_2 molecules to N_2 transport through the pores of the membrane. The CMS membranes from the phenolic resin displayed fairly high performance for CO_2/N_2 separation, that is, the CO_2 permeance was 84 GPU and the CO_2/N_2 selectivity was 51 at 35°C.

Table VII shows the performance for propylene/propane separation of the carbonized membranes. The carbonized phenolic resin membranes displayed fairly high performance for single-gas permeation, but in mixed-gas permeation experiments, the propylene permeance decreased significantly and the propylene/propane selectivity decreased. As a result, the performance of this membrane type was inferior to that of the carbonized BPDA-ODA polyimide membranes (11). Both membranes had a similar composite structure composed of the carbonized polymer layer on a porous α-alumina tube (NOK). However the membranes showed a very different behavior for gas mixtures, indicating that the adsorption behavior of the hydrocarbon vapor was dependent on the precursor polymer and the pyrolysis conditions. Further investigations are necessary for a more detailed evaluation for propylene/propane separation of the CMS membranes made from phenolic resin.

Table VII. Performance of carbon molecular sieve membranes for propylene/propane separation at 35°C.[a]

Membrane	Pyrolysis conditions	(P/L) propylene (GPU) pure-gas	(P/L) propylene (GPU) mixed-gas	C_3H_6/C_3H_8 selectivity pure-gas	C_3H_6/C_3H_8 selectivity mixed-gas
6FDA-mPD[b]	500°C; 1 h	28	(19)	25	(13)
6FDA-DABZ[b]	500°C; 1 h	3.1	(1.7)	78	(44)
BPDA-ODA[c]	700°C; 0 h	2.6	(2.4)	54	(46)
		9.3	(8.7[d])	29	(33[d])
Phenolic resin					
ND2-3	600°C; 1 h	19	(11[e])	54	(11[e])
ND2-5	500°C; 1 h	7.7	(0.19)	30	(27)
			(2.0[e])		(18[e])

[a] data in parenthesis are for mixed-gas permeation (50%C_3H_6/50% C_3H_8); [b] dense and flat-sheet, 50 μm thick. [c] tubular (ref. 11); [d] at 100°C; [e] at 90°C.

Conclusions

With increasing pyrolysis temperature, the permeability decreased and the selectivity of heat-treated polymeric membranes increased, because the membranes became denser and the effective pore size for molecular sieving became smaller, whereas gas adsorption increased slightly. CMS polyimide and polypyrrolone membranes displayed high performance for certain gas separation applications, such as CO_2/CH_4, O_2/N_2, and propylene/propane. CMS membranes made from a phenolic resin displayed high performance for CO_2/N_2 separation.

Acknowledgments

This work was supported partly by Grand-in-aid for Developmental Scientific Research (No. 08455367) from the Ministry of Education, Science, and Culture of Japan, and also by the Petroleum Energy Center and Japan High Polymer Center at the R & D project for membrane for petroleum component separation.

Literature Cited

1. Keller, G. E.; *Chem. Eng. Progr.* **1995**, Oct., 56.
2. Damle; A. S.. *Gas. Sep. Purification* **1994**, *8*, 137.
3. Koresh, J. E.; Sofer, A.; *Sep. Sci. Technol.* **1983**, *18*, 723.
4. Hatori,Y.; Yamada; Shiraishi, M.; Nakata, H.; Yoshiyomi, S.; *Carbon* **1992**, *30*, 305.
5. Jones, C. W.; Koros, W.J.; *Carbon* **1994**, *32*,1420.
6. Jones, C. W.; Koros W.J.; *Ind. Eng. Chem. Res.* **1995**, *34*, 158.
7. Suda, H.; Haraya, K. *J. Chem. Soc., Chem. Commn.* **1995**, 1179.
8. Suda, H.; Haraya, K., *J. Phys Chem. B* **1997**, *101*, 3988.
9. Suda; H.; Haraya, K.; *J. Chem. Soc., Chem. Commn.* **1997**, 93.
10. Hayashi, J.; Yamamoto, M.; Kusakabe, K; Morooka, S.; *Ind. Eng. Chem. Res.* **1995**, *34*, 4364.
11. Hayashi, J.; Mizuta,H.; Yamamoto,M; Kusakabe,K; Morooka, S. *Ind. Eng. Chem. Res.* **1996**, *35*, 4176.
12. Hayashi, J.; Mizuta,H.; Yamamoto, M.; Kusakabe, K.; Morooka, S. *J. Membr. Sci.* **1997**, *124*, 243.
13. Yoshinaga,T.; Shimazaki, H.; Kusuki, Y.; Japanese Patent, H4-11933 (1992).
14. Kusuki, Y.; Shimazaki, H, Tanihara, T, Nakanishi, J.; Yoshinaga, T. *J. Membr. Sci.* **1997**, *134*, 245.
15. Rao, M. B; Sircar, S. *J. Membr. Sci.* **1996**, *110*, 109.
16. Kita, H.; Yoshino, M.; Tanaka, K.; Okamoto, K. *Chem. Commn.* **1997**, 1051.
17. Kita, H.; Maeda, H.; Tanaka, K.; Okamoto, K. *Chemistry Lett.* **1997**, 179.
18. Hatori, H.; Yamada, Y.; Shiraishi, M.; Nakada, H.; Yoshitomi, S.; Yoshihira, S.; Kimura, T. *Tanso* **1995**, 94.
19. Robeson, L.M.; *J. Membr. Sci.* **1991**, *62*, 165.
20. Tanaka, K.; Taguchi, A.; Hao, J.; Kita,H.; Okamoto, K.; *J. Membr. Sci.* **1996**, *121*, 197.

Chapter 22

Preparation and Pervaporation Properties of X- and Y-Type Zeolite Membranes

Hidetoshi Kita, Hidetoshi Asamura, Kazuhiro Tanaka, and Kenichi Okamoto

Department of Advanced Materials Science and Engineering, Yamaguchi University, Tokiwadai, Ube 755, Japan

NaX and NaY zeolite membranes were prepared hydrothermally on the surface of a porous, cylindrical mullite support and were characterized by X-ray diffraction and scanning electron microscopy. The surface of a membrane prepared at 100°C for 5 hours was completely covered with randomly oriented, intergrown NaX or NaY zeolite crystals and the thickness of the membrane was about 20-30 μm. NaX and NaY zeolite membranes preferentially permeated water in pervaporation experiments of water/alcohol mixtures. The membrane was also highly alcohol-selective in pervaporation of alcohol/benzene, cyclohexane, or methyl *tert*-butyl ether (MTBE). The high selectivity of the NaY zeolite membrane can be attributed to the selective sorption of methanol into the membrane. A pervaporation process using a NaY zeolite membrane may be an alternative method for the industrial purification of MTBE.

Pervaporation has gained some acceptance in the chemical industry as an effective process for the separation of azeotropic mixtures (*1,2*). For example, pervaporation has been applied to the dehydration of organic liquids (ethanol, *i*-propanol, ethylene glycol etc.). Application of pervaporation for the separation of organic liquid mixtures, such as aliphatic-aromatic hydrocarbons or alcohols-ethers, however, is still very limited because of the low selectivity of most polymer membranes. On the other hand, membranes made from inorganic materials are generally superior to those made from polymeric materials in thermal and mechanical stability as well as chemical resistance. Among inorganic materials, zeolites are promising candidate materials for high-performance pervaporation membranes because of the unique characteristics of zeolite crystals such as molecular sieving, ion exchange, selective adsorption, and catalytic properties. Recently, we reported on the excellent pervaporation properties of NaA zeolite membranes for the separation of water/organic liquid mixtures *(3,4)*. Although a lot of research regarding water/alcohol separation by pervaporation has been carried out, membranes with both high selectivity and high flux are not commonly available. The pervaporation performance of the NaA zeolite membrane is one of the best reported so far and it may be feasible to use this membrane in industrial pervaporation systems *(5)*.

In this work, we report on the synthesis of NaX and NaY zeolite membranes and their pervaporation properties.

Experimental

Membrane Preparation. NaX and NaY zeolite membranes were grown hydrothernally on the surface of a porous, cylindrical mullite support (Nikkato Corp., 12 mm outer diameter, 1.5 mm thickness, 1.3 μm average pore size). Figure 1 shows the synthetic procedure and molar compositions of the starting gels of NaX and NaY zeolite membranes. According to the literature *(6)*, the aluminosilicate gel used in the synthesis of the zeolite membrane was prepared by mixing an aqueous sol (Aldrich, NaOH 14% and SiO_2 27%) and an alkaline aluminate solution prepared by dissolving sodium aluminate (Wako Chemical, Al/NaOH = 0.81) and sodium hydroxide (Wako Chemical) in distilled water. After formation of the gel, the reaction mixture was stored at ambient temperature for 12 hours in case of the synthesis of NaY zeolite. The porous support was coated by the water-slurry of seed crystals of NaX or NaY zeolite (Tosoh Co.) and then dried at 60°C. The gel was poured into a reaction vessel fitted with a condenser and a heater and then the support was placed in the gel. After hydrothermal treatment at 100°C for a specified reaction time, the support was taken out, washed with water, and dried at reduced pressure. The zeolite membrane was characterized by X-ray diffraction (XRD) with Cu-Kα radiation using a Shimadzu XD-3 instrument. The surface morphology of the zeolite membrane was examined by scanning electron microscopy (SEM) using a Hitachi S-2300.

Pervaporation Experiments. Pervaporation experiments were carried out using the membrane module illustrated schematically in Figure 2. Liquid mixtures were preheated to a constant temperature and were then fed by a liquid pump from a liquid reservoir to the outer side of the zeolite membrane in the stainless-steel module that was placed in a thermostated air-bath. The zeolite membrane was sealed in the cell module with fluoro-rubber o-rings. Throughout the experiments, the flow rate of the feed was held at 30-37 cm^3min^{-1}. The permeate-side of the membrane was evacuated with a vacuum pump. The gaseous permeate was collected in a liquid nitrogen trap. The effective membrane area was 47 cm^2 and the downstream pressure was maintained below 13.3 Pa. The membrane properties were evaluated by measuring the permeation flux (Q in $kgm^{-2} \cdot h^{-1}$) and the separation factor (α). The

332

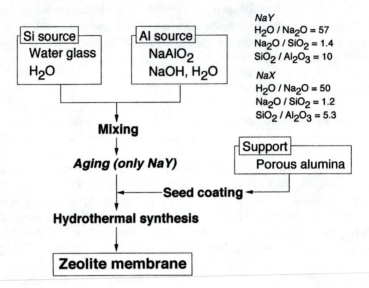

Figure 1. Schematic illustration of the synthetic methods to prepare NaX and NaY zeolite membranes.

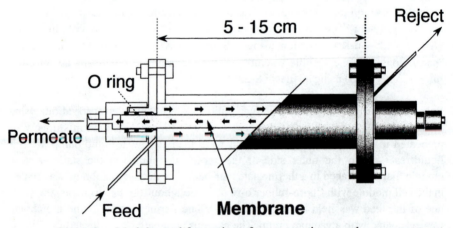

Figure 2. Module used for testing of pervaporation membranes.

permeation flux was calculated by weighing the liquid collected in the liquid nitrogen trap. The separation factor was determined as: $\alpha_{A/B} = (Y_A/Y_B)/(X_A/X_B)$, where A is the preferentially permeating component, and X_A, X_B, Y_A, and Y_B denote the weight fractions of component A and B in the feed and permeate, respectively. The compositions of feed and permeate were analyzed by gas chromatography.

Results and Discussion

Membrane Characterization. The X-ray diffraction (XRD) patterns of membranes grown hydrothermally on the porous support consist of peaks corresponding to those of the support and NaX or NaY zeolite, respectively. Figure 3 shows XRD patterns of the (i) support, (ii) NaY zeolite membranes prepared on the porous support at a crystallization time of 1 hour and 2 hours, respectively, and (iii) commercially available NaY zeolite powder. NaY crystals could be formed after 1 hour. The intensities of the diffraction peaks corresponding to NaY zeolite increased with an increase in the crystallization time. Crystals recovered from the bottom of the reaction vessel also showed the XRD patterns of FAU zeolite (X- or Y-type). The Si/Al ratio of the NaX and NaY crystals determined by atomic absorption spectrophotometry was 1.3 and 1.9, respectively.

The photomicrographs of the porous support and the surface of the NaY zeolite membrane after crystallization times of 1 hour and 2 hours, respectively, are shown in Figure 4. It was observed that small zeolite crystals were formed and accumulated on the surface of the porous support after 1 hour. These crystals grew further from 1 to 2 hours and a membrane-like phase was observed on the surface although void space between the crystals was still visible. After 5 hours, the outer surface of the porous support was completely covered with randomly oriented, intergrown NaX or NaY zeolite crystals, as shown in Figure 5. The membrane was about 20-30 μm thick based on the cross-sectional view of the photomicrograph. On the other hand, no continuous membrane phase was formed on the inner surface of the support because there were no seed crystals placed on the inner surface of the support. A prolonged crystallization time resulted in the growth of P-type zeolite, as described in the literature (7).

Pervaporation Through Zeolite Membranes. The fluxes and separation factors for liquid mixtures through NaX and NaY zeolite membranes prepared at 100°C for 5 hours together with the pervaporation performance of a NaA zeolite membrane previously reported are shown in Table I (3,4). The NaA zeolite membrane was highly selective for water permeation with a high flux because of micropore filling of water in the zeolite pores and/or the intercrystalline pores between the zeolite crystals. NaX and NaY zeolite membranes also preferentially permeated water in water/alcohol mixtures. However, both the separation factor and the flux of the

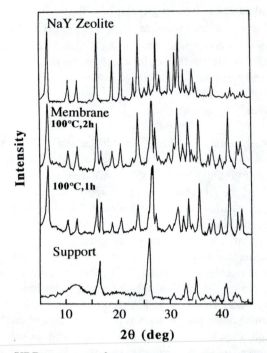

Figure 3. XRD patterns of porous support, NaY zeolite membranes, and NaY zeolite powder.

Figure 4. SEM micrographs of the surface of the porous support and a NaY zeolite membrane.

**NaY Zeolite
Membrane
(100°C, 1h)**

**NaY Zeolite
Membrane
(100°C, 2h)**

Figure 4.   *Continued.*

Surface

Cross section

Figure 5.   SEM micrographs of the surface and cross-section of a NaY zeolite membrane prepared at 100 °C for 5 hours.

zeolite membranes were lower than those of the NaA zeolite membrane, probably due to the less hydrophilic properties of the NaX and NaY zeolites.

Table I.   Pervaporation performance of NaX, NaY, and NaA zeolite membranes.

Zeolite type	Feed solution (A/B) (wt% of A)	T (°C)	Separation factor (A/B)	Flux (kg/m²·h)
NaX	Water/ethanol (10)	75	360	0.89
	Methanol/benzene (10)	50	24	1.25
	Methanol/MTBE (10)	50	320	0.26
NaY	Water/ethanol (10)	75	130	1.59
	Methanol/benzene (10)	50	3,800	0.93
	Methanol/MTBE (10)	50	3,300	1.11
		60	5,300	1.70
	Ethanol/ETBE (10)	65	1,300	0.48
	Ethanol/benzene (10)	60	930	0.22
	Ethanol/cyclohexane (10)	60	1,000	0.27
NaA	Water/methanol (10)	50	2,100	0.57
	Water/ethanol (10)	75	10,000	2.15
	Water/2-propanol (10)	75	10,000	1.76
	Water/acetone (10)	50	5,600	0.91

On the other hand, NaX and NaY zeolite membranes showed a high alcohol selectivity for several feed mixtures with methanol or ethanol. Especially high separation factors were observed for methanol/methyl *tert*-butyl ether (MTBE) mixtures (8). The methanol concentration in the permeate as a function of the methanol concentration in the feed mixture at 50°C is shown in Figure 6. Pervaporation using the NaY zeolite membrane can break the vapor-liquid azeotrope of methanol and MTBE. Furthermore, the membrane is more selective than distillation. High methanol/MTBE separation factors for the NaY zeolite membrane were recently reported over a wide range of feed compositions (8). The flux increased from 0.83 kg·m^{-2}·h^{-1} at 2.7 wt% methanol to 1.27 kg·m^{-2}·h^{-1} at 20 wt% methanol in the feed. Over the range of 20 to 50 wt% methanol, the flux was essentially constant. It has been reported that a polycrystalline silicalite membrane also permeates methanol preferentially (9). For comparison, the pervaporation performance for a silicalite membrane reported by Sano *et al.* (9) is also shown in Figure 6. The flux and the methanol/MTBE selectivity of the NaY membrane were considerably higher than those of the silicalite membrane. For organic liquid mixtures investigated in this study, pervaporation using the silicalite membrane does not appear feasible because the flux was very low due to the highly hydrophilic properties resulting from the strong adsorption of organic molecules (10).

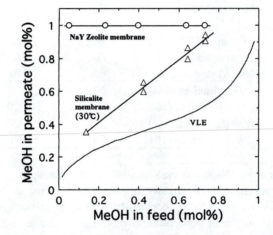

Figure 6. Methanol concentration in permeate of a (i) NaY zeolite membrane, (ii) silicalite membrane (*9*) and (iii) vapor-liquid equilibrium (*13*) for the methanol/MTBE system as a function of methanol feed concentration.

The production of MTBE as a gasoline additive has gained significant importance during the past decade (*11*). MTBE is usually produced by reacting *iso*-butene from a mixed-C$_4$ stream with excess methanol in order to achieve high conversion and to minimize side reactions. The unreacted methanol is subsequently distilled off and recycled to the process (*12*). However, methanol forms azeotropes with both MTBE and unreacted C$_4$s (*13*), which are difficult to separate by distillation. Thus, pervaporation of methanol/MTBE mixtures has attracted recent interest in industry. Although methanol/MTBE mixtures can be separated by pervaporation using polymer or inorganic membranes as shown in Table II, membranes that have both high selectivity and a high permeation flux are currently not available. On the other hand, the performance of the NaY zeolite membrane may be good enough to be economically feasible in commercial pervaporation systems for MeOH/MTBE separation.

Table II. Properties of various pervaporation membranes for the separation of methanol/MTBE mixtures.

Membrane type	MeOH in feed (wt%)	T (°C)	Separation factor (MeOH/MTBE)	Flux (kg/m^2·h)	Ref.
Cellulose acetate	0.83-6.9	23-49	14-454	0.05-1.41	14
Poly(vinyl alcohol)	5-30	45	3-4	0.12-0.16	15
BPDA polyimide	4.1	60	1,400	0.6	16
Poly (phenylene oxide)	1-20	22	8-24	0.18-0.65	17
Nafion 417	3.2-5.3	50	35	0.64	18
Poly(styrene-co-styrene-sulfonic acid) Mg ion form)	5.0-14.3	25	25,000-35,000	0.001-0.02	19
Silicalite/ stainless steel	5-50 (vol%)	30-50	4-9	0.08-0.12	9
NaY/alumina	1-50	35-60	2,000-24,000	1.0-2.3	This work

It is well known that the overall selectivity of a pervaporation process is determined by the mobility selectivity and the sorption selectivity. In the case of methanol/MTBE mixtures, the sorption process presumably determines the pervaporation performance. Methanol sorption into the membrane increases with increasing methanol concentration in the feed mixture. Increasing sorption causes an increase in the pervaporation flux. When the sorption of methanol approaches saturation, the flux becomes constant as reported previously (8). In order to clarify the hypothesis that the high selectivity of the NaY zeolite membrane can be attributed to the selective sorption of methanol into the membrane, the fluxes of the individual components are compared in Figure 7. In this figure, the temperature dependence of the pure-methanol flux and that of a mixture (10 wt% methanol/90 wt% MTBE) are compared. The pure-methanol flux and the methanol flux obtained in the mixture were almost same. On the other hand, the mixture-MTBE flux decreased remarkably compared with that of the pure-MTBE flux. Thus, permeation of MTBE was hindered by the presence of methanol. From these results, it can be concluded that methanol molecules adsorbed in zeolite pores obstructed the permeation of MTBE and the high selectivity of the NaY zeolite membrane can be attributed to the selective sorption of methanol into the membrane.

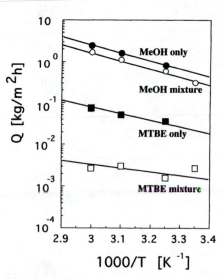

Figure 7. Arrhenius plots of the pure-component pervaporation fluxes of methanol and MTBE of a NaY zeolite membrane and mixed-pervaporation fluxes of methanol and MTBE in a methanol/MTBE mixture (feed concentration: 10 wt% methanol/90 wt% MTBE).

Acknowledgments

The present work was partially supported by a Grant-in-Aid for Scientific Research from the Ministry of Education of Japan.

Literature Cited

1. Fleming, H. L. *Chem. Eng. Progr.,* **1992**, July, 46.
2. *Pervaporation Membrane Separation Processes;* Huang, R. Y. M., Ed.; Elsevier, Amsterdam, 1991.
3. Kita, H.; Horii, K.; Ohtoshi, Y.; Tanaka, K.; Okamoto, K. *J. Mater. Sci. Lett.,* **1995***, 14,* 206*;* Kita, H.; Horii, K.; Tanaka, K.; Okamoto, K.; Miyake, N.; Kondo, M.; *Proceedings of 7th International Conference on Pervaporation Processes in the Chemical Industry;* Bakish, R., Ed.; Bakish Materials, Englewood, USA, 1995, 364.
4. Kita, H. *Maku(Membrane)* **1995***, 20,*169.
5. Kondo, M.; Komori, M.; Kita, H.; Okamoto, K. *J. Membr. Sci.* **1997***, 133,* 133.
6. *Zeolite Molecular Sieves;* Breck, D. W., Ed.; Wiley, New York, 1974, 245.
7. Vaughan, D. E. W. *Chem. Eng. Progr.* **1988**, February, 25.
8. Kita, H.; Inoue, T.; Asamura, H.; Tanaka, K.; Okamoto, K.; *J. Chem. Soc., Chem. Commun.* **1997***,45.*
9. Sano, T.; Hasegawa, M.; Kawakami, Y.; Yanagishita, H., *J. Membr. Sci.* **1995***,107,* 193.
10. Kita, H.; Tanaka, K.; Okamoto, K.; *Proceedings of ICOM96,* 1996*, 1102.*
11. Kirschner, E. M. *Chem. Eng. News,* 1996, April 8, 16.
12. Zhang, T.; Datta, R. *Ind. Eng. Chem. Res.* **1995***, 34,* 730.
13. Alm, K.; Ciprian,M. *J. Chem. Eng. Data* **1980***, 25,* 100.
14. Shah, V. M.; Bartels, C. R.; Pasternak, M.; Reale, J.; *AIChE Symp. Ser.* **1989***, 85,* 93.
15. Cen, Y.; Wesslein, M.; Lichtenthaler, R. N. *Proceedings of 4th International Conference on Pervaporation Processes in the Chemical Industry;* Bakish, R., Ed.; Bakish Materials, Englewood, USA, 1989, 522.
16. Nakagawa, K.; Matsuo, M.; *U.S. Patent* 5,292,963, 1994.
17. Doghieri, F.; Nardella, A.; Sarti, G. C.; Valentini, C. *J. Membr. Sci.* **1994***, 91,* 283.
18. Farnand, B. A.; Noh, S. H. *AIChE Symp. Ser.* **1989***, 85,* 89.
19. Chen, W-J.; Martin, C. R. *J. Membr. Sci.* **1995***,104,* 101.

Author Index

Andreotti, M., 252
Asamura, Hidetoshi, 330
Batarseh, M. T., 23
Beitle, R. R., 238
Berghmans, S., 23
Berwald, S., 189
Brack, H. P., 174
Büchi, F. N., 174
Cassel, F., 189
Cho, In-Sok, 42
Chowdhury, G., 125
Coleman, M. R., 205, 238
Costa, G., 252
Crookston, D., 238
Dioszeghy, Z., 189
Doi, T., 263
Fouda, A. E., 87
Freeman, B. D., 1
Giuver, Michael D., 137
Guo, Qunhui, 162
Hachisuka, Hisao, 65
Haraya, K., 295
Hashimoto, T., 228
Higuchi, A., 228
Huslage, J., 174
Ikeda, Kenichi, 65
Jonnason, K., 87
Kang, Y. S., 110
Kawakami, Hiroyoshi, 79
Kim, Sung Soo, 42
Kim, U. Y., 110
Kita, Hidetoshi, 314, 330
Li, K., 96
Lloyd, D. R., 23
Lunkwitz, K., 189
Maeda, Hiroshi, 314
Matsuura, T., 87, 125
Matsuyama, H., 23
Meier-Haack, J., 189
Merano, C., 252
Miyata, T., 263, 280

Moon, Y. S., 110
Müller, M., 189
Myler, U., 205
Nagaoka, Shoji, 79
Nakanishi, Yuichi, 280
Noborio, Kenji, 314
Nohmi, T., 228
O'Connor, Sally, 162
Ohara, Tomomi, 65
Okamoto, Ken-ichi, 314, 330
Oxford, C., 238
Park, H. C., 110
Pinnau, I., 1
Pintauro, Peter N., 162
Rhee, H. W., 110
Rieser, T., 189
Robertson, Gilles P., 137
Rota, M., 174
Scherer, G. G., 174
Simon, F., 189
Simpson, P. J., 205
Singh, S., 125
Suda, H., 295
Tak, T.-M., 228
Tam, Chung M., 137
Tanaka, Kazuhiro, 314, 330
Tang, Hao, 162
Teo, W. K., 96
Tsang, C., 125
Turturro, A., 252
Tyagi, R. K., 87
Uragami, T., 263, 280
Vigo, F., 252
Wang, D., 96
Wang, P., 228
Won, J., 110
Xu, X. L., 205
Yamasaki, A., 87
Yeom, Min-Oh, 42
Yoshikawa, Masakazu, 137
Yoshino, Makoto, 314

Subject Index

A

Acetone, solvent in immersion precipitation process, 14*t*
Acetylene-based monomers. *See* Poly(1-tri-methylsilyl-1-propyne) (PTMSP) membranes
Affinity membranes
purification of proteins and optical isomers, 228–229
scheme for preparing iminodiacetic acid (IDA), 158*f*
selective binding and fractionation of amino acids, 157
See also Facilitated transport metal affinity membranes (FTMAMs)
Alcohol derivatives, functionalized polymer, 144
Alcohols
polyetherimide hollow fiber membranes, 99, 103
polyethersulfone hollow fiber membranes, 104, 108
Alkene/alkane separation, performance of activated carbon molecular sieve membranes, 306–307
Amine derivatives, functionalized polymer, 144
Amino acids
optical resolution of racemic phenylalanine using immobilized BSA membranes, 232, 234–236
optical resolution procedure, 230
See also Facilitated transport metal affinity membranes (FTMAMs)
Anisotropic membranes. *See* Thermally-in-duced phase separation (TIPS) process
Annealing, membrane modification, 18, 19*t*
Aromatic polyimides
high gas selectivity for gas pairs, 79–80
See also Polyimide membranes, asymmetric
Asymmetric membranes
basic structures, 8, 12
immersion precipitation process, 12–15
integrally-skinned, 12–15

porous bulk structure and skin layer of polysulfone membrane by immersion precipitation, 10*f*, 11*f*
preparation, 80
schematic phase diagram of ternary polymer-solvent-non-solvent system, 13*f*
structure, 80
thin-film composite membranes, 15–18
uses of integral-asymmetric membranes with porous and dense skin layers, 8, 12
uses of thin-film composite membranes, 12
See also Morphology of asymmetric membranes; Polyimide gas separation membrane; Polyimide membranes, asymmetric; Polymer membranes; Polysulfone membranes for gas separation; Thermally-induced phase separation (TIPS) process
Asymmetric structure
capillary-type carbon molecular sieve (CMS) membranes, 307–310
SEM photograph of cross-sectional view of capillary CMS membrane, 309*f*
Atomic force microscopy (AFM)
image of surface and height profile of microporous membrane dried using supercritical CO_2, 117*f*
image of surface and height profile of microporous membrane dried without solvent exchange, 116*f*
images of surface and height profile of integrally-skinned asymmetric polysulfone membrane dried with and without solvent exchange, 122*f*, 123*f*
impact of ion irradiation on surface morphology of films, 218
lateral mean diameter from height profile of AFM image, 119*f*
monitoring evolution of surface morphology of irradiated polyimides, 215, 217
studying surface structure of polymeric membranes, 114, 118
See also Ion beam irradiation of polyimide thin films; Morphology of asymmetric membranes

Azide cycloadditions, functionalized polymer, 144, 145*f*
Azides
cycloaddition (2+3) reactions, 151
polysulfone, 150–151

B

Biochemical engineering, advances, 238–239
Block copolymers
hydrophilic and hydrophobic membranes, 283, 286
modification of poly(dimethylsiloxane) membranes, 281, 291, 293
results of polymerization of DEAA and NFHM by PDMS macro-azo-initiator, 282*t*
surface properties of block copolymer/ PDMS membranes, 289*t*
synthesis and characterization with monomers diethylacrylamide (DEAA) and nonafluorohexyl methacrylate (NFHM), 281–282
See also Poly(dimethylsiloxane) (PDMS) membranes
Bovine serum albumin (BSA)
immobilization procedure, 229–230
optical resolution of racemic phenylalanine using immobilized BSA membranes, 232, 234–236
See also Polyacrylonitrile hollow fibers
Bromination, lithiation of polysulfones, 141

C

Capillary-type membranes
carbon molecular sieve (CMS) membranes with asymmetric structure, 307–310
modifications to improve separation performance, 308, 310
permeances of H_2 and He gas, 310
SEM photograph of cross-sectional view, 309*f*
tailored microstructure, 310–311
See also Carbon molecular sieves (CMS) membranes
Carbinol derivatives, polysulfone, 150
Carbon dioxide (CO_2). *See* Polyimide gas separation membrane
Carbon molecular sieve (CMS) membranes
adsorption isotherms of polypyrrolone and its carbonized membranes, 318*f*
alkene/alkane separation performance of activated CMS membranes, 306–307
Arrhenius plots of permeability, diffusivity, and solubility, 305*f*
background, 296
... CMS membranes with asym-

chemical structures of precursor polyimide, polypyrrolone, and phenolic resin, 316*f*
comparisons of membrane performance toward several gas pairs, 322, 323*t*
dense CMS membrane fabrication method, 298
effects of other pyrolysis conditions on gas permeation properties, 301–302
fabrication methods and gas separation performance of CMS membranes, 297*t*
flat type CMS dense membranes from Kapton polyimide, 298, 301–303
flat-type CMS dense membranes from other precursors, 303, 306
gas adsorption and permeation properties, 315, 319–322
gas permeation experiments by vacuum time-lag method, 320
gas permeation properties of vacuum pyrolyzed CMS membranes, 301
gas separation properties of membrane from phenolic resin, 327–328
gelation, imidization, and pyrolization steps for preparation of CMS membranes, 299*f*
micropore volumes from Dubinin-Radushkevich equation of CO_2 adsorption, 319*t*
modification methods, 296
modification of capillary CMS to improve separation performance, 308, 310
order of gas permeabilities versus kinetic diameters of gas molecules, 302–303
performance dependence on pyrolysis conditions, 320
performance of propylene/propane separation of carbonized membranes, 328*t*
permeance and selectivity of membrane from phenolic resin, 327*t*
permeances of small gases like H_2 and He in capillary CMS membranes, 310
physical properties of carbonized membranes, 319*t*
plots of permeability ratio of CO_2/CH_4 versus CO_2 permeability for precursor polymers and carbonized membranes, 320, 321*f*
pore analysis of vacuum pyrolyzed CMS membrane, 301
preparation and characterization from phenolic resin, 326
preparation and characterization from polyimide and polypyrrolone, 315
propylene/propane separation, 322, 324
pyrolysis process of Kapton polyimide film, 298, 301
relationship between diffusivity and kinetic gas diameter, 304*f*
relationship between permeability and kinetic diameter of gases for polypyrrolone and its carbonized membrane, 320, 321*f*

relationship between permeability and kinetic gas diameter, 304f
relationship between solubility and Lennard-Jones potential, 304f
schematic of CMS membrane microstructure, 300f
SEM photograph of cross-sectional view of capillary CMS membrane, 309f
structure-gas permeation property relationships, 306
temperature dependence of permeability and selectivity of propylene/propane for carbonized polyimide membrane, 325f
thermograms of polypyrrolone and polyimide, 316f
time dependence of adsorption of propylene and propane in carbonized polyimide membranes, 324, 325f
tradeoff relationship between selectivity and permeability, 310–311, 312f
variation in atomic content of polypyrrolone and polyimide with heat treatment, 317f
Carboxylated polysulfone (CPSf), 143
Carboxylic acid derivative, polysulfone, 149
Casting solution method. See Polyimide gas separation membrane
Cellulose acetate
commercial membranes, 2t
multi-component casting or spinning solutions for immersion precipitation process, 14t
Cellulose nitrate, commercial membranes, 2t
Cellulose regenerated, commercial membranes, 2t
Cohesive energy density equation, 287
Composite membranes. See Thin-film composite membranes
Conductivity. See Proton conductivity
Crosslinking
chemical surface treatment, 20
effect on mechanical properties of membrane, 183
Cycloaddition reactions of azides, polysulfone, 151

D

Dense symmetric membranes, uses, 8
Diamine
one-pot preparation of polysulfone, 151
polysulfone, 151
Dibrominated polysulfone lithiation, 149
Diethylacrylamide (DEAA)
monomer in block copolymers, 281–282
See also Poly(dimethylsiloxane) (PDMS) membranes
Diethylene glycol dimethylether (DGDE) solvent in casting solution, 66

See also Polyimide gas separation membrane
Dimethylacetamide (DMAc), solvent in immersion precipitation, 14t
Dimethylformamide (DMF), solvent in immersion precipitation, 14t
Dry/wet phase inversion process
nature of coagulant, 96–97
See also Hollow fiber membranes; Phase inversion process; Polyimide membranes, asymmetric
Dual mode sorption model, solubility of gases in glassy polymers, 207

E

Electrophiles
functional groups on polymers, 143t
See also Polysulfones, functionalized
Ethanol, internal coagulant in polyetherimide, 99, 103

F

Facilitated transport metal affinity membranes (FTMAMs)
accomplishing desorption, 240–241
bioseparations method, 239
characteristics of metal affinity membranes, 250t
determination of mass transfer coefficients, 244, 246
diffusion apparatus for mass transfer experiments with histidine and phenylalanine, 245f
effect of pH gradient in future research, 250
experimental materials, 243
functionalization chemistry for attaching chelating ligand to poly(vinyl alcohol) (PVA) backbone via spacer arm, 245f
functionalizing PVA gel membranes using metal affinity chemistry, 246–247
histidine mass transfer rate, 250
histidine transport across control membrane of PVA-iminodiacetic acid (IDA), 247, 248f
histidine transport in PVA-IDA membranes with Cu^{2+}, Ni^{2+}, CO^{2+}, or Zn^{2+} charges, 247, 250
interaction mechanism of immobilized metal ions with imidazole ring of histidine, 242f
mechanism of transport of histidine in fixed site carrier FTMAMs, 242f
membrane formation, 243
membrane functionalization, 243–244

metal affinity chemistry, 239–241
metal and water content, 244
phenylalanine transport rate, 247
pH gradient altering facilitated transport of
  histidine, 241, 243
raw and normalized histidine permeate concentration versus time for control and
  copper charged membranes, 247, 249f
solutions for preparation of FTMAMs, 243t
transition metal ions as affinity ligands, 240
using metal affinity fixed site carriers, 241,
  243
Field-emission scanning electron microscopy
  (FE–SEM)
cross-section structure of membranes, 66,
  67f, 68f, 74f
FE–SEM images of microporous polysulfone membranes dried under different
  conditions, 114, 115f
See also Morphology of asymmetric membranes; Polyimide gas separation membrane
Flat-type membranes
carbon molecular sieve membranes from
  polyimides, 298, 301–303
CMS membranes from other precursors,
  303, 306
structure-gas permeation property relationships, 306
See also Carbon molecular sieves (CMS)
  membranes
Flory–Huggins equation, state of solvent in
  membrane, 289
Fluorination, chemical surface treatment,
  19–20
Fuel cells
generation of electricity and heat, 174
testing method, 178
tests of radiation-grafted membranes, 185
See also Radiation-grafted membranes
Functionalized polymers
cycloaddition (2+3) reactions of azides, 151
experimental methods for introducing,
  149–152
metal affinity chemistry, 243–244, 246–247
one-pot polymerization of polysulfone (ortho-sulfone) diamine, 151
polyphenylsulfone trimethylsilyl derivative,
  151–152
polysulfone carbinol derivatives, 150
polysulfone carboxylic acid derivative, 149
polysulfone (ortho-ether) azide, 150–151
polysulfone (ortho-ether) diamine, 151
polysulfone sulfinic acid derivatives, 150
surface reactions to introduce functional
  groups, 228
See also Facilitated transport metal affinity
  membranes (FTMAMs); Polysulfones,
  functionalized

G

Gas molecules, order of gas permeabilities
  versus kinetic diameters, 302–303
Gas pairs
comparisons of carbon molecular sieve
  (CMS) membranes toward several, 322,
  323t
gas transport properties in irradiated polyimides, 221–225
normalized O_2 permeance and O_2/N_2 selectivity as function of H^+ ion fluence, 223f,
  224
permeance and selectivity of CMS membranes from phenolic resin, 327t
permeation properties for He/N_2 gas pair in
  irradiated polyimide, 223f, 224–225
permeation properties of irradiated polyimide membranes prior to and following H^+
  ion irradiation, 221t, 224t
selectivity of aromatic polyimides, 79
See also Carbon molecular sieve (CMS)
  membranes; Ion beam irradiation of polyimide thin films
Gas separation
achieving high gas fluxes, 65
functionalized polysulfone (PSf) membranes, 154, 157
gas permeability and ideal O_2/N_2 selectivity
  for aminated polysulfones, 158f
highly permeable hollow fiber membranes,
  96
materials with high permeabilities and selectivities, 205–206
See also Carbon molecular sieve (CMS)
  membranes; Hollow fiber membranes;
  Ion beam irradiation of polyimide thin
  films; Polyimide gas separation membrane; Polyimide membranes, asymmetric; Polysulfone membranes for gas separation; Polysulfones, functionalized;
  Poly(1-trimethylsilyl-1-propyne)
  (PTMSP) membranes
Glass transition temperature
differential scanning calorimetry (DSC)
  method, 165
sulfonated polyphosphazenes, 168–170
variation in sulfonated polyphosphazene as
  function of ion-exchange capacity, 169f
Grafting
critical degree of, 184
forming negatively charged substrates, 199
plasma-induced, 195
radiation method, 175, 177
water permeation as function of grafting,
  198f
yields of irradiated polymers, 179–180
See also Polypropylene membranes; Radiation-grafted membranes

H

High performance membrane materials, development approaches, 205–206
Histidine
facilitated transport using metal affinity fixed site carriers, 241, 243
focus to metal affinity based recovery schemes, 239
increasing mass transfer rate, 250
interaction mechanism of immobilized metal ions with imidazole ring of histidine, 242f
mechanism of transport in fixed site carriers, 242f
metal affinity chemistry, 239–241
raw and normalized permeate concentration versus time, 247, 249f
transport across poly(vinyl alcohol)-iminodiacetic acid (PVA-IDA) membrane, 247, 248f
transport in charged PVA-IDA membranes, 247, 250
See also Facilitated transport metal affinity membranes (FTMAMs)
Hollow fiber membranes
characterization methods, 98–99
coagulation rate in relation to polymer solution state, 108
complicated fabrication process, 96–97
composition and properties of polymer spinning dopes, 98t
experimental materials, 97–98
internal coagulants in polyetherimide (PEI) membrane formation, 99, 103
internal coagulants in polyethersulfone (PESf) membrane formation, 104, 108
photomicrographs of PEI membranes cross-sections using internal coagulants, 102f
photomicrographs of PESf cross-sections using internal coagulants, 107f
polyetherimides (PEI), 99, 103
polyethersulfones (PESf), 103–104, 108
precipitation values of nonsolvents in PEI/NMP and PESf/NMP (N-methyl-2-pyrrolidone), 98t
pressure-normalized He flux of silicone-coated PEI membranes using internal coagulants, 100f
pressure-normalized He flux of silicone-coated PESf using internal coagulants, 105f
role of internal coagulants in membrane structure control and integrity, 104, 108
selectivity for He/N$_2$ of silicone-coated PEI using internal coagulants, 101f
selectivity for He/N$_2$ of silicone-coated PESf membranes using internal coagulants, 106f

spinning conditions for preparation, 99t
spinning method, 98
surface modification of polyacrylonitrile hollow fibers, 228–229
See also Polyacrylonitrile hollow fibers
Human serum albumin (HSA)
adsorption procedure, 192
filtration with modified polypropylene membranes, 199–200, 202
See also Polypropylene membranes
Hydrophilization, surface modification, 190

I

Iminodiacetic acid (IDA), chelating ligand for functionalizing poly(vinyl alcohol), 243–244, 245f
Immersion precipitation process
altering structure with multi-component quench media, 14–15
commercial membranes, 14
important parameters, 15
integrally-skinned asymmetric membranes, 12–15
multi-component casting or spinning solutions for membrane production, 14t
schematic phase diagram of ternary polymer-solvent-non-solvent system, 13f
structures of porous bulk and skin layer of polysulfone, 10f, 11f
time scale of membrane formation, 12, 14
See also Asymmetric membranes
Integrally-skinned asymmetric membranes
atomic force microscopy (AFM) image of surface and height profile of polysulfone dried with and without solvent exchange, 122f, 123f
effect of liquid surface tension on membrane surface roughness, 120t
effect of solvent exchange on gas transport properties, 120t
FE–SEM images of polysulfone dried under different conditions, 121f
immersion precipitation method, 12–15
membrane with thin gas-tight skin layer, 120
porous bulk structure and skin layer of polysulfone membrane by immersion precipitation process, 10f, 11f
See also Morphology of asymmetric membranes
Interaction parameters, water, ethanol, and surface-modified poly(1-trimethylsilyl-1-propyne) (PTMSP) membranes, 273
Interfacial polymerization method, thin-film composite membranes, 17
Internal coagulants
polyetherimide hollow fiber membranes, 99, 103

polyethersulfone hollow fiber membranes, 103–104, 108
  role in membrane structure and integrity control, 104, 108
  role in preparation of high-performance hollow fiber membranes, 108
  See also Hollow fiber membranes
Ion beam irradiation of polyimide thin films
  AFM (atomic force microscopy) for impact of irradiation on surface morphology, 218
  AFM monitoring surface morphology, 215, 217
  defect model of positron annihilation spectroscopy (PAS) spectra of 6FDA-pMDA samples, 220f
  depth profile of energy transfer mechanism of incident ions in polyimide 6FDA-6FpDA, 209f
  depth profile of energy transfer mechanism of incident ions in polyimide 6FDA-MDA, 210f
  depth profiles of energy loss for irradiating ions, 208, 211
  detect modeling clarifying microstructural evolution by ion irradiation, 218, 221
  dual mode sorption model, 207
  effective for synthesis and modification, 206
  evolution in relative permeance of He, O_2, N_2, CH_4 in 6FDA-6FpDA membranes following H^+ irradiation, 222f
  experimental fluorine-containing polyimides, 213–214, 216f
  factors determining polymer properties following irradiation, 206
  gas permeation measurements, 215
  gas permeation properties of polymer-ceramic membranes after H^+ ion irradiation, 224t
  gas permeation properties of polymer-ceramic membranes prior to H^+ ion irradiation, 221t
  general increase in permeance and selectivity of gas pairs, 225–226
  inducing changes in polymer backbone structure, 211
  ion beam modification of polymer structure and properties, 208, 211–213
  ion irradiation procedure, 215
  large variations in permeation properties with H^+ ion fluence, 225
  manipulating microstructure of materials, 208
  modification of gas transport properties, 221–225
  modification of polymer properties, 212
  modification of polymer structure and morphology, 217–221
  Monte Carlo simulation method (SRIM code), 208
  morphology determining transport of small molecules, 212–213
  normalized O_2 permeance and O_2/N_2 selectivity as function of H^+ ion fluence, 223f, 224
  PAS measurements, 217
  permeation measurements of H^+ irradiated 6FDA-6FpDA using polyimide-ceramic composites, 214–215
  permeation properties of He/H_2 gas pair in modified membranes, 223f, 224–225
  preparation of polymer films and composite membranes, 214–215
  S-parameters as function of positron energy of 6FDA-MDA samples, 219f
  scanning electron microscopy (SEM) image of 6FDA-6FpDA-ceramic following ion irradiation, 216f
  solubility of gases in glassy polymers, 207
  surface layer microporosity by PAS, 218
  transport in polymer matrix, 207–208
Ion containing membranes, method of sulfonating polysulfones, 157, 159
Ion-exchange capacity (IEC)
  determination method, 164
  radiation-grafted membranes, 183–184
  sulfonated polyphosphazenes, 167–168
Ion-exchange membranes. *See* Polyphosphazene-based cation-exchange membranes

Irradiation-etching process, symmetric membranes, 3, 6
Isotactic polypropylene (iPP)
  phase diagram of iPP/diphenyl ether (DPE) system, 25f, 35f
  See also Thermally-induced phase separation (TIPS) process

L

Lithiation
  bis-6F polysulfone (DS=2.0), 148
  bromination, 141, 142f
  dibrominated polysulfone (DS=2.0 and 3.0), 149
  direct, 138, 139f
  experimental methods, 146, 148–149
  polymer, 138–141
  polymers amenable to, 138, 141
  polyphenylsulfone (DS=2.0), 148
  polysulfone (DS=2.0), 146
  polysulfone (DS=3.0), 148
  See also Polysulfones, functionalized

M

Mechanism study. *See* Thermally-induced phase separation (TIPS) mechanism study
Melt extrusion/stretching process, symmetric membranes, 6

Membrane modification
  annealing, 18, 19t
  chemical surface treatments, 19–20
  methods, 18–20
  solvent-exchange methods, 18–19
  surface coating, 19
Membranes. *See* Polymer membranes
Metal affinity membranes
  chemistry, 239–241
  facilitated transport using fixed site carriers, 241, 243
  recovery methods, 239
  See also Facilitated transport metal affinity membranes (FTMAMs)
Metal ion chelation
  formation of oxine derivative, 160f
  functionalizing polysulfones, 159
Methane (CH_4). *See* Polyimide gas separation membrane
Methanol, internal coagulant mixtures in polyetherimide, 99, 103
Methanol/methyl *tert*-butyl ether (MTBE)
  comparing zeolite and silicalite membranes, 338f
  determining pervaporation performance, 340
  production of MTBE, 339
  properties of various pervaporation membranes for methanol/MTBE separation, 339t
  selectivity of NaX and NaY zeolite membranes, 337
  See also Zeolite membranes
N-Methyl-2-pyrrolidone (NMP)
  solvent in casting solution, 66
  solvent in immersion precipitation, 14t
  solvent in thin-film composite membranes, 126
  See also Polyimide gas separation membrane
Microporous membranes. *See* Morphology of asymmetric membranes; Polypropylene membranes
Molecular sieves
  background in separations, 296
  See also Carbon molecular sieve (CMS) membranes
Monte Carlo simulation method, depth profiles of energy loss for irradiating ions, 208, 209f, 210f
Morphology of asymmetric membranes
  AFM (atomic force microscopy) image of surface and height profile in integrally-skinned asymmetric polysulfone membrane (T2) dried after solvent exchange, 123f
  AFM image of surface and height profile in integrally-skinned asymmetric polysulfone membrane (T2) dried without solvent exchange, 122f
  AFM image of surface and height profile of microporous polysulfone membrane dried using supercritical CO_2, 117f
  AFM image of surface and height profile of microporous polysulfone membrane dried without solvent exchange, 116f
  AFM studying surface structure, 114, 118
  average roughness calculation, 114, 118
  changes due to capillary forces of coagulation medium, 110–111
  collapse of micropores in top skin layer, 111
  effect of liquid surface tension on membrane surface roughness of integrally-skinned asymmetric polysulfone (T2), 120t
  effect of solvent exchange on gas transport properties of integrally-skinned asymmetric polysulfone (T2), 120t
  effect of solvent exchange on gas transport properties of microporous polysulfone membranes (T1), 118t
  FE–SEM (field emission electron microscopy) images of microporous polysulfone membranes (T1 type) dried under different conditions, 114, 115f
  FE–SEM images of integrally-skinned asymmetric polysulfone membranes (T2) dried under different conditions, 121f
  investigation of membrane morphology and performance, 112
  lateral mean diameter from height profile of AFM image, 119f
  membrane preparation, 111–112
  membrane with porous skin layer (T1), 112, 114–120
  membrane with thin gas-tight skin layer (T2), 120
  model calculating capillary pressure of close-packed sphere structure, 111
  root-mean-square roughness, 118
  supercritical CO_2 drying, 112
  supercritical CO_2 drying path in pressure-temperature diagram, 113f
  surface roughness of microporous polysulfone membranes (T1), 118t

N

NaA, NaX, and NaY zeolites. *See* Zeolite membranes
Nitrogen (N_2). *See* Polyimide gas separation membrane
Nonafluorohexyl methacrylate (NFHM)
  monomer in block copolymers, 281–282
  See also Poly(dimethylsiloxane) (PDMS) membranes
Nucleation and growth, phase separation mechanisms, 6, 8, 9f

O

Oxygen (O₂). *See* Polyimide gas separation membrane

P

Permselectivity. *See* Polyimide membranes, asymmetric
Pervaporation
  applications, 330
  determining overall selectivity, 340
  experiments, 266, 283
  experiments for zeolite membranes, 331, 333
  flux and ethanol concentration in permeate through surface-modified PTMSP, 272f
  NaA zeolite membranes, 330–331
  performance of NaX, NaY, and NaA zeolite membranes, 337t
  process for separation of azeotropic and close-boiling point mixtures, 280–281
  separation of methanol/methyl *tert*-butyl ether (MTBE), 337, 339–340
  surface-modified PDMS membranes, 286–287, 288f
  surface-modified PTMSP membranes, 269, 271
  theoretical discussion for pervaporation characteristics of surface-modified PTMSP, 271, 273–274, 278
  zeolite membranes, 333, 337–340
  See also Poly(dimethylsiloxane) (PDMS) membranes; Poly(1-trimethylsilyl-1-propyne) (PTMSP) membranes; Zeolite membranes
Phase diagrams
  determination of polyimide gas separation membrane, 66, 69f, 70
  isotactic polypropylene-diphenyl ether (iPP/DPE) system, 25f, 35f
  ternary polymer-solvent-non-solvent system, 13f
Phase inversion process
  asymmetric structure of capillary-type carbon molecular sieve (CMS) membranes, 308, 309f
  changes in membrane morphology with drying step, 120, 124
  membranes for gas separation, 87
  polymer-coagulant interaction as variable in control of, 104, 108
  symmetric membranes, 6
  See also Dry/wet phase inversion process; Morphology of asymmetric membranes; Polyimide gas separation membrane; Polysulfone membranes for gas separation
Phenol formaldehyde, precursor to carbon molecular sieve membranes, 296, 297t

Phenolic resins
  gas separation properties of carbonized, 327–328
  permeance and selectivity of CMS membranes from, 327t
  preparation and characterization of CMS membranes, 326
  propylene/propane separation of carbonized, 328t
  See also Carbon molecular sieves (CMS) membranes
Phenylalanine
  optical resolution on bovine serum albumin (BSA) immobilized membranes, 232, 234–236
  optical resolution procedure, 230
  transport rate, 247
  See also Facilitated transport metal affinity membranes (FTMAMs)
Phosphazene polymers. *See* Polyphosphazene-based cation-exchange membranes
Plasma treatment, polypropylene membranes, 192–195
Polyacrylonitrile, commercial membranes, 2t
Polyacrylonitrile hollow fibers
  bovine serum albumin (BSA) immobilization procedure, 229–230
  chemical modification of polyacrylonitrile (PAN), 230, 232
  chemical modifications with Raney nickel and hydrazine, 229, 236–237
  comparing optical resolution of racemic phenylalanine by BSA solution and immobilized BSA membranes, 234
  concentration ratio of L-phenylalanine to BSA, 236
  dependence of rejection of solutes on their molecular weights, 233f
  dependence of substitution ratio of amine group on reaction temperature, 232, 233f
  experimental materials, 229
  immobilization of BSA onto surface of chemically modified, 231f
  mechanism for optical resolution of racemic amino acids, 234
  optical resolution of amino acid, 230
  optical resolution of phenylalanine by BSA solution and immobilized BSA membranes, 236t
  optical resolution of racemic phenylalanine using immobilized BSA membranes, 232, 234, 236
  reaction scheme of surface-modified, 231f
  reversible binding of amino acids to BSA, 236
  time dependence of separation factor of phenylalanine for BSA-2 membranes, 234, 235f
  time dependence of separation factor of

phenylalanine for unmodified and BSA-1 membranes, 232, 234, 235f
transport experiments, 232
transport measurements, 230
Polyamide, commercial membranes, 2t
Polycarbonate, commercial membranes, 2t
Poly(2,6-dimethyl-1,4-phenylene oxide), commercial membranes, 2t
Poly(dimethylsiloxane) (PDMS) membranes
cohesive energy density, 287
commercial membranes, 2t
contact angle measurements, 282
effect of block copolymer content on ethanol concentration in permeate and normalized flux through surface-modified PDMS, 287, 288f
effect of block copolymer content on theoretical ethanol concentration in copolymer/PDMS membranes, 289, 290f
effect of poly(diethylacrylamide) (PDEAA) content on ethanol concentration in permeate and normalized flux through PDMS-b-PDEAA, 284f
effect of poly(nonafluorohexyl methacrylate) (PNFHM) content on ethanol concentration in permeate and normalized flux through PDMS-b-PNFHM, 285f
experimental materials, 281
hydrophilic and hydrophobic block copolymer membranes, 283, 286
membrane preparation from PDMS-b-PDEAA and PDMS-b-PNFHM, 282
modification by addition of hydrophilic and hydrophobic block copolymers, 281, 291, 293
pervaporation experiments, 283
relationship between block copolymer content and contact angles of water on PDMS-b-PDEAA/PDMS and PDMS-b-PNFHM/PDMS, 286, 288f
relationship between membrane surface and selectivity, 287, 289–291
results of polymerization of DEAA and NFHM by PDMS macro-azo-initiator, 282t
schematic illustration of block copolymer/PDMS structures, 292f
selectivity relative to continuous or noncontinuous PDMS phase, 286
selectivity versus hydrophobic or hydrophilic modification, 291
separation of aqueous ethanol solutions by pervaporation, 280–281
solubility parameter from cohesive energy density, 287, 289
solubility parameter from surface free energy, 287
state of solvent in membrane by Flory-Huggins equation, 289
structures of monomers and initiator, 284f

surface-modified PDMS, 286–287
surface properties of block copolymer/PDMS membranes, 289t
synthesis and characterization of block copolymers with monomers DEAA and NFHM, 281–282
TEM of cross-sections of PDMS-b-PNFHM membranes, 285f
transmission electron micrographs (TEM), 283
Polyelectrolyte multilayers
build-up on polypropylene membranes, 195–196
See also Polypropylene membranes
Polyetherimide (PEI)
commercial membranes, 2t
features of cross-section hollow fiber structures, 103
hollow fiber membranes using internal coagulants, 99, 103
photomicrographs of membrane cross-sections, 102f
pressure-normalized He flux of silicone-coated PEI using internal coagulants, 100f
selectivity for He/N₂, 101f
See also Hollow fiber membranes
Polyethersulfone (PES)
commercial membranes, 2t
hollow fiber membranes using internal coagulants, 103–104, 108
multi-component casting or spinning solutions for immersion precipitation process, 14t
pressure-normalized He flux of silicone-coated PES using internal coagulants, 105f
role of internal coagulants in membrane structure and integrity control, 104, 108
selectivity for He/N₂, 106f
surface morphology of microporous, 127–130
See also Hollow fiber membranes
Poly(ethylene-alt-tetrafluoroethylene). See Radiation-grafted membranes
Polyfurfuryl alcohol, precursor to carbon molecular sieve membranes, 296, 297t
Polyimide
commercial membranes, 2t
flat type carbon molecular sieve membranes from Kapton type, 298, 301–303
gas adsorption and permeation properties of carbonized membranes, 315, 319–322
multi-component casting or spinning solutions for immersion precipitation process, 14t
physical properties of carbonized membranes, 319t
preparation and characterization of carbon molecular sieve (CMS) membranes, 315

propylene/propane separation on carbonized membranes, 322, 324

time dependence of adsorption of propylene and propane in carbonized membranes, 325*f*

See also Carbon molecular sieve (CMS) membranes; Ion beam irradiation of polyimide thin films

Polyimide gas separation membrane

2,2-bis(3,4-dicarboxyphenyl) hexafluoropropane dianhydride (6FDA) structure, 67*f*

casting of 6FDA-BAAF (2,2-bis(4-amino phenyl) hexafluoropropane) membranes on commercial-sized equipment, 72, 76

comparison of dielectric constant and/or dipole moment of solvents, 70*t*

cross-section structure of membranes using 6FDA-BAAF/DGDE (diethylene glycol dimethylether) dope by field-emission scanning electron microscopy (FE-SEM), 66, 67*f*

determination of phase diagram, 66, 69*f*, 70

experimental, 66–70

FE–SEM photographs of cross section of asymmetric 6FDA-BAAF membrane from *N*-methyl-2-pyrrolidone (NMP)-based dope, 68*f*

FE–SEM photographs of cross section of asymmetric 6FDA-BAAF membranes quenched in water at 38°C and 30°C, 74*f*

gas permeation properties of 6FDA-BAAF spiral-wound membrane element for various gases, 78*t*

gas permeation properties of asymmetric 6FDA-BAAF membrane made on continuous casting equipment, 76*t*

gas permeation properties of asymmetric 6FDA-BAAF membranes, 72

influence of quench medium temperature on skin layer, 72

influence of silicone coating on membrane properties, 76

influence of solvent properties on membrane structure, 70, 72

new casting solution method, 65–66

performance of polyimide spiral-wound membrane element, 76

permeation properties (CO_2 and CH_4) of machine-made asymmetric 6FDA-BAAF membrane at 30°C and 38°C, 75*f*, 77*f*

phase diagram of 6FDA-BAAF/DGDE/water system, 69*f*

photographs of mixing DGDE and NMP with water, 71*f*

preparation and analysis of asymmetric 6FDA-BAAF membranes, 66

surface photographs of 6FDA-BAAF/DGDE dope and 6FDA-BAAF/NMP dope in water, 73*f*

Polyimide membranes, asymmetric

apparent skin layer thickness calculation, 82

cross-section of asymmetric 6FDA-APPS by dry/wet phase inversion, 82, 83*f*

defect-free thin skin layers by dry/wet phase inversion process, 79–80

dependence of surface roughness on apparent skin layer thickness, 82, 83*f*

evaluating differences in morphology by surface roughness, 82, 84

gas permeances and selectivities of 6FDA-APPS membranes, 84*t*

influences of intermolecular and intramolecular charge transfer interactions, 84, 86

mechanism for surface skin layer formation by dry/wet phase inversion, 82

permeability measurements, 80, 82

preparation of 6FDA-APPS, 80

relationship between CO_2/CH_4 selectivity and apparent skin layer thickness, 85*f*

relationship between CO_2/CH_4 selectivity and CO_2 permeance, 85*f*

structure of 6FDA-APPS, 81*f*

structure of asymmetric membrane, 80

thermogravimetric analysis (TGA) for residual solvent, 82

Polymer electrolyte fuel cell (PEFC)

polarization properties, 185

proton-conducting polymer membrane, 174–175

See also Radiation-grafted membranes

Polymer matrix, transport of penetrant, 207–208

Polymer membranes

characteristics for industrial separation process, 1–2

classification scheme of synthetic membranes, 4*f*

common polymers for production of commercial, 2*t*

integrally-skinned asymmetric membranes, 12–15

membrane types and formation methods, 3

modification methods, 18–20

porous symmetric membranes, 3, 6–8

progress in tailor-made, 2–3

schematic of symmetric and asymmetric structures, 5*f*

symmetric membranes, 3, 6–8

thin-film composite membranes, 15–18

uses of porous, 2

Polymer structure and properties

ion beam modification, 208, 211–213

modified polyimides by irradiation, 217–221

See also Ion beam irradiation of polyimide thin films

Poly(methyl methacrylate) (PMMA), commercial membranes, 2*t*

Polyphenylene oxide, sulfonated. *See* Thin-film composite membranes

Polyphenylsulfone, lithiation, 148

Polyphosphazene-based cation-exchange membranes

Arrhenius plot of proton conductivity in water-equilibrated poly[bis(3-methyl phenoxy) phosphazene] membranes as function of reciprocal temperature, 171*f*

Arrhenius plot of water diffusion coefficient versus reciprocal temperature, 171*f*

chemical reaction sequence during sulfonation of poly[bis(3-methyl phenoxy) phosphazene] with SO_3, 166*f*

constants in conductivity expression for sulfonated polyphosphazene and Nafion 117 membranes, 170*t*

constants in diffusivity expression for sulfonated polyphosphazene and Nafion 117 membranes, 172*t*

determination of equilibrium water swelling, 164–165

differential scanning calorimetry (DSC) measurements, 165

diffusion coefficient of water by vapor-phase sorption/desorption experiments, 165

DSC of base polymer and sulfonated poly[bis(3-methyl phenoxy) phosphazene], 168–170

electrical conductivity of protons by AC impedance method, 165, 167

experimental, 164–167

ion-exchange capacity and swelling of sulfonated poly[bis(3-methyl phenoxy) phosphazene], 167*t*

IR and NMR spectra of sulfonated bis(3-methyl phenoxy) phosphazene polymer samples, 168

membrane ion-exchange capacity (IEC) and water swelling, 167–168

membrane water weight loss versus square-root of time during McBain balance vapor desorption experiment, 166*f*

method of determining ion-exchange capacity, 164

photocrosslinking phosphazene polymers, 163–164

potentially useful for ion-exchange, 162

proton conductivity in H^+ form, 170

sulfonating polyphosphazene procedure, 164

sulfonation of phosphazene polymers, 163

synthetic routes to water insoluble sulfonic acid membranes, 162–163

variation in glass transition temperature of sulfonated poly[bis(3-methyl phenoxy) phosphazene] as function of polymer IEC, 169*f*

water diffusion coefficient, 172

Polypropylene

commercial membranes, 2*t*

plasma treatment, 192–195

Polypropylene, isotactic (iPP)

phase diagram of iPP/diphenyl ether (DPE) system, 25*f*, 35*f*

See also Thermally-induced phase separation (TIPS) process

Polypropylene membranes

advantages of modified membranes, 190–191

characterization methods, 191–192

CO_2 plasma treatment for peroxide formation, 202

dependence of reduction on PAC graft yield, 200*t*

experimental materials, 191

filtration experiments, 196, 199–202

FTIR–ATR spectra proving incorporation of oxygen, 193*f*

functionalization scheme by plasma treatment and multilayer build-up, 193*f*

future design membranes for filtration, 202–203

graft yield and flux reduction as function of reaction time, 198*f*

human serum albumin adsorption procedure, 192

human serum albumin filtration, 199–200, 202

incorporation of oxygen functionalities in surface, 194

indirect proof of peroxide formation, 194

influence of consecutive polyelectrolyte layer deposition on zeta potentials of modified membranes, 198*f*

influence of plasma treatment and poly(acrylic acid) (PAC) graft yield on zeta potentials, 197*f*

influence of polyelectrolyte layer assemblies on membrane fouling, 201*f*

influence of polyelectrolyte layers on water flux, 201*f*

influence of reaction time on PAC graft yield, 195, 197*f*

irreversibly adsorbed protein, 200

modification methods, 191

modification of hydrophobic by polyelectrolytes, 190

negatively charged substrates from plasma treatment and acrylic acid grafting, 199

pH adjustment before layer-by-layer deposition, 196

plasma-induced graft polymerization, 195

plasma treatment, 192–195

polyelectrolyte multilayer build-up, 195–196

protein adsorption by FTIR-ATR studies, 202

relative water flux reduction equation, 196

results of XPS after CO_2 plasma treatment, 194t
retention equation, 200
studying filtration and fouling properties of microfiltration, 190
water flux enhancement, 199
water flux using dead-ended stirred filtration cells, 192
water permeation measurements, 196, 199
XPS measurements of plasma-activated membranes after derivatization with SO_2, 194t
Polypyrrolone
adsorption isotherms of, and its carbonized membrane, 318f
gas adsorption and permeation properties of carbonized membrane, 315, 319–322
initial pyrolysis process, 317f
physical properties of carbonized membranes, 319t
propylene/propane separation on carbonized membranes, 322, 324
relationship between permeability and kinetic diameter of gases for, and its carbonized membranes, 321f
See also Carbon molecular sieve (CMS) membranes
Polysulfone
commercial membranes, 2t
effect of liquid surface tension on membrane surface roughness, 120t
effect of solvent exchange on gas transport properties, 120t
effect of solvent exchange on gas transport properties of microporous membranes, 118t
membrane preparation, 111–112
membrane with porous skin layer (T1), 112, 114–120
membrane with thin gas-tight skin layer (T2), 120
multi-component casting or spinning solutions for immersion precipitation process, 14t
porous bulk structure and skin layer by immersion precipitation process, 10f, 11f
surface roughness of microporous membranes, 118t
See also Morphology of asymmetric membranes
Polysulfone membranes for gas separation
additives for controlling membrane formation process, 88
additive sodium dodecyl sulfate (SDS) in casting process, 88
casting process, 88
configurational change of polymer solution dependence on SDS content, 93, 95
effect of evaporation time on oxygen permeance of membranes, 89, 90f

effect of solvent evaporation time on O_2/N_2 selectivity, 89, 90f
effects of surfactant addition to casting solution, 93, 94f
gas permeation measurements, 89
growth of skin layer with solvent evaporation time, 89, 93
membrane preparation, 88–89
reduction in skin layer thickness with SDS, 95
scanning electron microscopy (SEM) images of cross-sections made without SDS, 89, 91f
schematic concentration profile in cast film, 94f
SEM images of cross-sections made with SDS, 89, 92f
X-ray photoelectron spectroscopy (XPS) for detecting SDS, 88
Polysulfones, functionalized
affinity membranes, 157
alcohol derivatives, 144
amine derivatives, 144
azide cycloadditions, 144, 145f
bis-6F polysulfone lithiation (DS=2.0), 148
bromination-lithiation, 141
carboxylated polysulfone, 143
chemical structures, 140f
cycloaddition (2+3) reactions of azides, 151
dibrominated polysulfone lithiation, 149
direct lithiation, 138, 139f
effect of polymer concentration on pore size in casting solutions with different solvents, 153f
electrophile selection, 141, 143
experimental methods for introducing functional groups, 149–152
experimental methods for lithiation, 146, 148–149
flux versus pore radius for carboxylated polysulfones (CPSf) and polysulfone (PSf) membranes, 155f
formation of metal ion chelating oxine derivative, 160f
gas permeability and ideal O_2/N_2 selectivity for aminated polysulfones, 158f
gas separation application, 154, 157
general experimental methods, 146
ion containing membranes, 157, 159
lithiation to desired degree of substitution (DS), 141
metal ion chelation, 159
nanofiltration (NF) rejection of solutes with flat-sheet thin-film composites from CPSf, 156t
one-pot preparation of polysulfone (ortho-sulfone) diamine, 151
ortho-ether lithiation by halogen-metal exchange, 142f

ortho-sulfone-directed lithiation of polysulfone, 139f

polymer functionalization, 141, 143–146

polymer lithiation, 138–141

polymers amenable to lithiation, 138, 141

polyphenylsulfone lithiation (DS=2.0), 148

polyphenylsulfone trimethylsilyl derivative, 151–152

polysulfone carbinol derivatives, 150

polysulfone carboxylic acid derivative, 149

polysulfone lithiation (DS=2.0), 146

polysulfone lithiation (DS=3.0), 148

polysulfone (ortho-ether) azide, 151

polysulfone (ortho-ether) diamine, 151

polysulfone sulfinic acid derivative, 150

reverse osmosis (RO) application, 154, 156t

scheme for preparing iminodiacetic acid (IDA) affinity membrane, 158f

silyl derivatives, 144

sulfinic acid derivative, 143–144

surface modified membranes, 154

technology applications, 137

triazole and triazoline derivatives by 2+3 cycloaddition reactions of polysulfone azides, 145f

tricyclic polymers, 146, 147f

ultrafiltration (UF) application, 152–154

variety of functional groups on polymers using simple electrophiles, 143t

water flux and salt rejection of asymmetric CPSf membranes, 156t

Polytetrafluoroethylene, commercial membranes, 2t

Poly(tetrafluoroethylene-co-hexafluoropropylene). See Radiation-grafted membranes

Poly(1-trimethylsilyl-1-propyne) (PTMSP) membranes

air permeability of membranes from PTMSP and TMSP/n-PrC(CSiMe$_3$ (1-trimethylsilyl-1-pentyne) as function of testing time, 261f

air permeability of membranes from PTMSP and TMSP/PhC(CSiMe$_3$ (1-phenyl-2-trimethylsilylacetylene) as function of testing time, 258f

characterization of surface-modified membranes, 266, 268–269

contact angle measurements, 266

copolymer membranes of TMSP/n-PrC(CSiMe$_3$, 260

copolymers of TMSP with n-PrC(CSiMe$_3$ or PhC(CSiMe$_3$, 254, 255t

effect of co-monomer unit with phenyl group on permeability, 257, 260

effect of PFA-g-PDMS and PMMA-g-PDMS on interaction parameters, 274, 276f

effect of PFA-g-PDMS content on contact

angle to water on surface-modified PTMSP, 267f

effect of PMMA-g-PDMS content on contact angle to water on surface-modified PTMSP, 268, 270f

efficient catalyst system, 254–255

ethanol/water selectivity, 263–264

experimental materials, 253, 264

flux and ethanol concentration in permeate through surface-modified PTMSP as function of additives content, 272f

high gas permeability, 252–253

homopolymer investigation of n-PrC(CSiMe$_3$ and PhC(CSiMe$_3$, 254

interaction parameters for water, ethanol, and surface-modified PTMSP, 273

membrane preparation, 254

modification of surface with additives, 264

oxygen/nitrogen selectivity of membranes from PTMSP and TMSP/n-PrC(CSiMe$_3$ as function of testing time, 261f

oxygen/nitrogen selectivity of membranes from PTMSP and TMSP/PhC(CSiMe$_3$ as function of testing time, 259f

performance in pervaporation applications, 278

permeability and selectivity measurements, 254

pervaporation experiments, 266

pervaporation properties of surface-modified PTMSP, 269, 271

physical aging, 253

polymer characterization, 253–254

polymer synthesis, 253

preparation of membranes with polymer additives, 265–266

properties of polymer additives for surface-modified PTMSP, 265t

Raman spectra of PTMSP homopolymer and TMSP/PhC(CSiMe$_3$ 70/30 copolymer, 256f

schematic model of interactions between water, ethanol, and polymer, 275f

solubility parameter of surface-modified PTMSP, 273

solubility tests and Raman spectroscopy characterization, 255, 257

structure of disubstituted acetylene-based monomers, 256f

structures of surface-modified PTMSP, 269, 270f

surface compositions of PTMSP and surface-modified PTMSP by XPS, 268t

surface free energy estimation by contact angles to water and formamide, 271, 273

synthesis of polymer additives PFA-g-PDMS (FA = heptadecafluorodecylacrylate and DMS = oligodimethylsiloxane)

and PMMA-*g*-PDMS (MMA = methyl methacrylate), 265, 267*f*
synthesis of PTMSP, 265
TGA traces of PTMSP, TMSP/PhC(CSiMe₃ 50/50, and TMSP/*n*-PrC(CSiMe₃ 50/50, 258*f*
theoretical discussion for pervaporation characteristics of surface-modified PTMSP, 271, 273–274, 278
theoretical ethanol concentration in surface-modified PTMSP, 274
theoretical ethanol concentrations in surface of air-side of surface-modified PTMSP, 274. 277*f*
thermal stability of copolymers, 257
transport properties of membranes, 257
X-ray photoelectron spectroscopy (XPS), 266
Poly(vinyl alcohol)
characteristics of metal affinity membranes, 250*t*
commercial membranes, 2*t*
functionalizing with metal affinity chemistry, 246–247
gel membrane formation, 243
membrane functionalization, 243–244, 245*f*
See also Facilitated transport metal affinity membranes (FTMAMs)
Poly(vinylidene chloride), precursor to carbon molecular sieve membranes, 296, 297*t*
Poly(vinylidene fluoride), commercial membranes, 2*t*
Polyvinylpyrrolidone (PVP), preparation of porous support membranes, 126
Positron annihilation spectroscopy (PAS)
method for determining depth profile vacancy-based structures of solids, 217
microporosity in surface layer of N⁺ ion irradiated films, 218, 219*f*
See also Ion beam irradiation of polyimide thin films
Post-synthesis method. *See* Ion beam irradiation of polyimide thin films
2-Propanol, internal coagulant in polyetherimide, 99
Propylene/propane separation
carbonized phenolic resin membranes, 328
CMS membranes from polyimide and polypyrrolone, 322, 324
performance of carbonized membranes, 324*t*
performance of CMS membranes, 328*t*
See also Carbon molecular sieve (CMS) membranes
Proton conductivity
AC impedance method, 165, 167
Arrhenius plot for polyphosphazenes as function of reciprocal temperature, 171*f*
sulfonated polyphosphazenes, 170

Pyrolysis
chemical surface treatment, 20
polyimide film as precursor to carbon molecular sieve membranes, 298, 301

R

Radiation-grafted membranes
advantages of ETFE (poly(ethylene-*alt*-tetrafluoroethylene)) base polymer over FEP (poly(tetrafluoroethylene-*co*-hexafluoropropylene)), 175, 177
changes in crystallinity upon irradiation, 179
comparing swollen Nafion 117 membrane, 183
critical degree of grafting, 184
cross-sectional view of polymer electrolyte fuel cell (PEFC), 176*f*
effect of crosslinking on mechanical properties, 183
e⁻/N₂ pre-irradiation dose as function of ETFE and FEP film thickness, 181*f*
experimental methods, 177–178
ex-situ area resistances as function of graft level of membranes, 186*f*
factors affecting membrane mechanical properties, 185
film and membrane characterization, 177–178
fuel cell testing method, 178
fuel cell tests, 185
grafted film and membrane synthesis, 177
grafting front mechanism, 179–180
grafting yields, 179–180
in-situ membrane resistances of ETFE- and FEP-based membranes, 186*f*
ion exchange capacity (IEC) of ETFE- and FEP-based membranes, 183–184
irradiated base polymer films, 178–179
lifetimes, 185
mechanical properties of crosslinked g/air radiation-grafted ETFE- and FEP-based grafted films and membranes, 181*f*
mechanical properties of ETFE and FEP, 180, 182–183
melt flow index (MFI) and DSC measurements on irradiated ETFE and FEP films, 179*t*
membrane properties, 183–184
polarization properties of PEFCs, 185
pre-irradiation grafting of styrene into ETFE film, 176*f*
radiation grafting method, 175, 177
Reverse osmosis
experiments using sulfonated polyphenylene oxide membranes, 133–135
functionalized polysulfone (PSf) membranes, 154, 156*t*

sulfonated poly(2,6-dimethyl-1,4-phenylene oxide) (SPPO) promising material, 125–126
See also Polysulfones, functionalized; Thin-film composite membranes

S

Selectivity
relationship between membrane surface and, 287, 289–291
See also Poly(dimethylsiloxane) (PDMS) membranes
Silicone coating, influence on membrane properties, 76
Silyl derivatives, functionalized polymer, 144
Skin layer
growth, 87–88
influence of quench medium temperature, 72
See also Polyimide gas separation membrane; Polyimide membranes, asymmetric
Sodium dodecyl sulfate (SDS)
additive in casting process, 88
See also Polysulfone membranes for gas separation
Solubility parameter
cohesive energy density, 287, 289
membrane surfaces, 287
surface-modified poly(1-trimethylsilyl-1-propyne) (PTMSP) membranes, 273
Solution coating method, thin-film composite membranes, 17–18
Solvent-exchange methods
membrane modification, 18–19
membrane with thin gas-tight skin layer, 120
See also Morphology of asymmetric membranes
Solvents, influence of solvent properties on membrane structure, 70, 72
Spinodal decomposition, phase separation mechanisms, 6, 8, 9f
Sulfinic acid derivatives
functionalized polymer, 143–144
polysulfone, 150
Sulfonated poly(2,6-dimethyl-1,4-phenylene oxide) (SPPO)
composite membrane preparation, 126–127
effect of ion-exchange from SPPOH hydrogen form to cation form on membrane performance, 133–135
intrinsic viscosities of SPPOH polymer, 131, 133t
promising for reverse osmosis applications, 125–126
surface morphology of composite SPPO membranes, 127–130
See also Thin-film composite membranes

Supercritical CO_2 drying
drying path in pressure-temperature diagram, 113f
method, 112
See also Morphology of asymmetric membranes
Surface coating, membrane modification, 19
Surface modified membranes
applications requiring functionalized or hydrophilic surfaces, 189
polysulfones, 154
tailor-made membranes, 189–190
See also Polyacrylonitrile hollow fibers; Poly(dimethylsiloxane) (PDMS) membranes; Polypropylene membranes; Polysulfones, functionalized; Poly(1-trimethylsilyl-1-propyne) (PTMSP) membranes;
Surface roughness
average roughness calculation, 114, 118
mean roughness calculation, 127
mean roughness parameter values from atomic force microscopy (AFM) images, 130
microporous polysulfone membranes dried under different conditions, 118t
root-mean-square roughness, 118
Surface tension, membrane with thin gas-tight skin layer, 120
Surfactants. *See* Polysulfone membranes for gas separation
Symmetric membranes
cross-section of porous polysulfone membrane from vapor-precipitation/evaporation process, 7f
dense, 8
irradiation-etching process, 3, 6
melt extrusion/stretching process using semi-crystalline polymers, 6
phase inversion process, 6
production methods, 3
schematic of binary polymer-solvent system with upper critical solution temperature (UCST), 7f
structures resulting from nucleation and growth of polymer-rich and polymer-poor phases and spinodal decomposition, 9f
thermally-induced phase inversion process, 6, 8
vapor-precipitation/evaporation process, 6
vitrification point, 8
See also Polymer membranes

T

Tailor-made polymers, progress, 2–3
Thermally-induced phase inversion process
phase separation mechanisms, 6, 8
structures resulting from nucleation and

growth of polymer-rich and polymer-poor phases and spinodal decomposition, 9f

symmetric membranes, 6, 7f, 8

thermodynamic equilibrium and glass transition temperature of binary polymer-solvent system as function of solution composition, 9f

vitrification point, 8

Thermally-induced phase separation (TIPS) mechanism study

applying Fowke's relationship for nonpolar compounds, 54, 56

bicontinuous phase after phase separation, 57

coalescence images at later stage of phase separation, 51f

coalescence or collision-combination mechanism, 56–57

comparison of droplet and cell sizes of 20/80 isotactic polypropylene/diphenyl ether (PP/DPE) sample, 62f

controlling structure requiring mechanistic understanding, 42–43

dependence of droplet growth on composition, 50

differential scanning calorimetry (DSC) for thermal analysis, 43, 45

droplet formation during liquid-liquid phase separation, 58

droplet growth and crystallization at 105°C for PP/DPE (20/80), 53f

droplet growth at different phase separation temperatures for 10/90 PP/DPE, 51f

droplet growth at different phase separation temperatures for 20/80 PP/DPE, 52f

droplet growth at different PP/DPE compositions at 115°C, 52f

droplet growth at different PP/DPE compositions at 125°C, 53f

droplet growth kinetics studies, 56

droplet size and growth rate with quench temperature, 57–58

effect of quench temperature on cell size, 58

estimating equilibrium droplet size using Laplace equation for spherical drop, 50, 54

experimental, 43, 45

experimental data for estimating interfacial tension of polymer-lean and polymer-rich phases for PP/DPE system at 125°C, 56t

experimental phase diagram of PP/DPE system, 45, 47f

geometry of pendant drop in selected plane, 55f

homogeneous melt solution of 10/90 PP/DPE at 210°C on 110°C hot-stage, 45, 50

images of droplets and cells, 58

images of phase separation for PP/DPE (10/90) at 110°C, 48f, 49f

interfacial tension between phases for estimating radius of droplet, 54

isotactic PP and DPE system, 43

liquid-liquid phase separation mechanism for TIPS membrane formation, 44f

phase diagram of PP/DPE at 10°C/min, 47f

PP crystallization interference with droplet growth at lower temperature, 50

PP/DPE system separation into polymer-lean and polymer-rich phases, 56

proportional constants and scaling exponents of Furukawa equation for PP/DPE systems, 57t

scanning electron microscopy (SEM) images of membranes at 110°C for different intervals, 59f

schematic of surface tension measurement system, 46f

schematic of thermo-optical microscope system, 46f

SEM images of membranes at different temperatures for 10 minutes, 60f

SEM images of membranes prepared from different PP/DPE compositions, 61f

size and growth rate of droplets with degree of supercooling for liquid-liquid phase separation, 58, 63

surface tension equation, 54

temperature dependence of growth rate, 50

theory by Furukawa with single length scale as function of time, 57–58

Thermally-induced phase separation (TIPS) process

analysis of evaporation process, 26–27

calculated relationship between polymer volume fraction across melt-blended solution and length of evaporation period, 29f

conservation equation of polymer in membrane solution, 27

cooling procedures, 28

detail structures at top and bottom surfaces of samples at 333.2 K quench, 36f

detail structures at top and bottom surfaces of samples at 373.2 K quench, 38f

detail structures at top and bottom surfaces of samples when only top immersed in ice-water, 40f

differential scanning calorimetry (DSC) for dynamic crystallization temperature, 28

dimensionless weight and thickness changes over time, 30f

diphenyl ether extraction method, 28

effects of thermal history on membrane structures, 33, 37–39

estimated cooling rates for isothermal quench conditions, 33t

experimental, 27–33
heat transfer process equation, 27
isotactic polypropylene (iPP)-diphenyl
    ether (DPE) system, 25f, 27–28
isothermal cooling in air, 33, 37–39
mass transfer equation, 26
membrane structure for samples cooled at
    room temperature, 31–33
micrographs of cross-section of 20 wt% sam-
    ple cooled in room temperature air, 32f
microporous materials, 23–24
non-isothermal cooling, 33
phase diagram considerations, 33, 35f
phase diagram for iPP/DPE system, 25f, 35f
principle of formation of anisotropic and
    asymmetric structures based on polymer
    concentration gradient, 24–26
production of anisotropic and asymmetric
    TIPS membranes, 24
schematic of glass container for preparation
    of TIPS membranes, 29f
simulation of evaporation process, 28, 31
structures at top and bottom surfaces for
    20 wt% sample cooled in room tempera-
    ture air, 34f
use of temperature and concentration gradi-
    ents, 39
Thermosetting polymers, precursor to carbon
    molecular sieve membranes, 296, 297t
Thin-film composite membranes
advantages over integrally-skinned asym-
    metric membranes, 15
basic types, 15, 16f
commercial production methods, 17
effect of counter-ions in hydrogen form
    SPPO (SPPOH)-alkali metal cation
    (Me+) composite membranes on separa-
    tion and flux in reverse osmosis[sulfo-
    nated poly(2,6-dimethyl-1,4-phenylene ox-
    ide) = SPPO], 135t
effect of counter-ions of SPPOH-alkali
    metal cation (Me+) membranes on sepa-
    ration of carbohydrates, 135t
effect of ion-exchange from SPPOH to cat-
    ion form on membrane performance,
    133–135
effect of solvent in coating solution on
    membrane performance, 131–133
effect of solvent system in coating solutions
    on flux of different electrolyte solutions,
    132f
effect of solvent system in coating solutions
    on separation of different electrolytes,
    132f
estimated nodule diameters of SPPOH coat-
    ing layer as function of chloroform con-
    tent in coating solution, 133t
interfacial polymerization method, 17

intrinsic viscosities of SPPOH polymer,
    131, 133t
ion-exchange equilibrium constant of sulfo-
    nated polystyrene, 134t
mean pore size and standard deviation
    from separation data and AFM images,
    129–130
mean roughness calculation, 127
mean roughness parameter values from
    AFM images, 130t
membrane characterization, 127
membrane material, 126
nodules by AFM topographical picture, 133
porous support membrane preparation
    from polyethersulfone (PES), 126
properties of optimum porous supports for,
    15, 17
proton-alkali metal cation exchange,
    133–134
pure water flux and solute separations data
    dependence on metal cation, 135
reverse osmosis and nanofiltration applica-
    tions, 126
reverse osmosis performance of SPPOH
    composite membranes for alkali chloride
    solutes, 134t
schematic of single-layer and multi-layer,
    16f
solute separation of polyethylene glycol
    (PEG) and polyethylene oxide (PEO) of
    PES and SPPO/PES membranes versus
    solute diameters, 128f
solution coating method, 17–18
SPPO composite membrane preparation,
    126–127
SPPO promising membrane for reverse os-
    mosis applications, 125–126
surface morphology of microporous poly-
    ethersulfone and composite SPPO mem-
    branes, 127–130
ultrafiltration of PEG and PEO, 127
uses, 12
See also Asymmetric membranes
Transition metal ions
affinity ligands, 240
See also Facilitated transport metal affinity
    membranes (FTMAMs)
Transport
polymer matrix, 207–208
See also Facilitated transport metal affinity
    membranes (FTMAMs)
Tricyclic polymers, functionalized polymer,
    146
Trimethylsilyl derivative, polyphenylsulfone,
    151–152
Two-step coagulation process, studying
    growth of skin layer during phase inversion
    process, 87–88

U

Ultrafiltration (UF)
  membrane applications, 152–154, 155f
  See also Polysulfones, functionalized
Ultrathin skin layer. *See* Polyimide gas separation membrane
Upper critical solution temperature (UCST), thermally-induced phase inversion process, 6, 7f

V

Vacuum pyrolysis
  effects of pyrolysis conditions on gas permeation properties, 301–302
  gas permeation properties of CMS membranes, 301
  pore analysis of carbon molecular sieve membranes, 301
Vapor-precipitation/evaporation process, symmetric membranes, 6, 7f
Vitrification point (VP), membrane formation, 8

W

Water
  internal coagulant in polyetherimide, 99
  internal coagulant mixtures in polyetherimide, 99, 103
  permeation measurements, 196, 199
Water diffusion coefficient
  Arrhenius plot of water diffusion coefficient versus reciprocal temperature, 171f
  sulfonated polyphosphazenes, 172
  vapor-phase sorption/desorption experiments for determining, 165
Water swelling
  determination method, 164–165
  sulfonated polyphosphazenes, 167–168

Y

Young-Laplace equation, model calculating capillary pressure of close-packed sphere structure, 111

Z

Zeolite membranes
  comparison of pervaporation for NaY and silicalite membranes, 338f
  high alcohol selectivity of NaX and NaY membranes, 337
  membrane characterization, 333
  methanol concentration in permeate as function of methanol concentration in feed, 338f
  methanol/methyl *tert*-butyl ether (MTBE) separation, 337, 339–340
  module for testing pervaporation membranes, 332f
  NaX and NaY membrane preparation, 331, 332f
  pervaporation experiments, 331, 333
  pervaporation performance of NaX, NaY, and NaA membranes, 337t
  pervaporation through zeolites, 333, 337–340
  properties of various pervaporation membranes for methanol/MTBE separation, 339t
  SEM micrographs of surface and cross-section of NaY membrane, 336f
  SEM micrographs of surface of porous support and NaY membrane, 334f, 335f
  temperature dependence of flux of pure methanol and 10/90 methanol/MTBE mixture, 340f
  X-ray diffraction (XRD) patterns of porous support, NaY membranes and NaY powder, 334f

Highlights from ACS Books

Bestsellers from ACS Books

The ACS Style Guide: A Manual for Authors and Editors (2nd Edition)
Edited by Janet S. Dodd
470 pp; clothbound ISBN 0–8412–3461–2; paperback ISBN 0–8412–3462–0

Writing the Laboratory Notebook
By Howard M. Kanare
145 pp; clothbound ISBN 0–8412–0906–5; paperback ISBN 0–8412–0933–2

Career Transitions for Chemists
By Dorothy P. Rodmann, Donald D. Bly, Frederick H. Owens, and Anne-Claire Anderson
240 pp; clothbound ISBN 0–8412–3052–8; paperback ISBN 0–8412–3038–2

Chemical Activities (student and teacher editions)
By Christie L. Borgford and Lee R. Summerlin
330 pp; spiralbound ISBN 0–8412–1417–4; teacher edition, ISBN 0–8412–1416–6

Chemical Demonstrations: A Sourcebook for Teachers, Volumes 1 and 2, Second Edition
Volume 1 by Lee R. Summerlin and James L. Ealy, Jr.
198 pp; spiralbound ISBN 0–8412–1481–6
Volume 2 by Lee R. Summerlin, Christie L. Borgford, and Julie B. Ealy
234 pp; spiralbound ISBN 0–8412–1535–9

The Internet: A Guide for Chemists
Edited by Steven M. Bachrach
360 pp; clothbound ISBN 0–8412–3223–7; paperback ISBN 0–8412–3224–5

Laboratory Waste Management: A Guidebook
ACS Task Force on Laboratory Waste Management
250 pp; clothbound ISBN 0–8412–2735–7; paperback ISBN 0–8412–2849–3

Reagent Chemicals, Eighth Edition
700 pp; clothbound ISBN 0–8412–2502–8

Good Laboratory Practice Standards: Applications for Field and Laboratory Studies
Edited by Willa Y. Garner, Maureen S. Barge, and James P. Ussary
571 pp; clothbound ISBN 0–8412–2192–8

For further information contact:
Order Department
Oxford University Press
2001 Evans Road
Cary, NC 27513
Phone: 1-800-445-9714 or 919-677-0977

MICHIGAN MOLECULAR INSTITUTE
1910 WEST ST. ANDREWS ROAD
MIDLAND, MICHIGAN 48640